Study Guide and Solutions Manual for Seager and Slabaugh's

Chemistry for Today
General, Organic, and Biochemistry
4th edition

Prepared by
Peter Krieger
Palm Beach Community College

 **Brooks/Cole**
Thomson Learning.

Australia • Canada • Mexico • Singapore • Spain • United Kingdom • United States

Assistant Editor: Melissa Henderson
Marketing Manager: Steve Catalano
Marketing Assistant: Christina DeVeto
Editorial Assistant: Dena Dowsett-Jones/Brandon Horn
Production Coordinator: Stephanie Andersen

Cover Design: Roy Neuhaus
Cover Photo: Art Wolfe/Tony Stone Images
Print Buyer: Micky Lawler
Typesetting: Erick and Mary Ann Reinstedt
Printing and Binding: Globus Printing Company

Contents

Preface

This Solutions Manual is written to accompany *Chemistry for Today: General, Organic and Biochemistry, Fourth Edition* and its two split volumes: *Introductory Chemistry for Today, Fourth Edition*, and *Organic and Biochemistry for Today, Fourth Edition*. Chapters 1–12 of *Introductory Chemistry for Today, Fourth Edition* correspond to the same chapter numbers in this Solutions Manual and in *Chemistry for Today: General, Organic and Biochemistry, Fourth Edition*. Chapters 1–15 of *Organic and Biochemistry for Today, Fourth Edition* correspond to chapters 11–25 in this Solutions Manual and in *Chemistry for Today: General, Organic and Biochemistry, Fourth Edition*.

Students frequently request some study aids for their chemistry courses: (1) completely worked-out solutions to end-of-chapter exercises; (2) an overview of the topics covered; and (3) sample examination questions. We have attempted to provide these aids in this manual. The appendix of the textbook gives a simple answer to the even-numbered, end-of-chapter exercises. This manual includes a complete, worked-out solution to those exercises, giving you an opportunity to check your problem-solving approach. A programmed review in the form of fill-in-the-blanks is provided to cover the major topics in each chapter of the text. The answers to those programmed reviews are given. Finally, there are sample examination questions provided with correct answers following the questions.

We hope these aids prove useful for you. Please provide us your suggestions, comments, and corrections.

The Author

Chapter 1

Matter, Measurements, and Calculations

CHAPTER OUTLINE

1.1 What is Matter
1.2 Properties and Changes
1.3 A Model of Matter
1.4 Classification of Matter
1.5 Measurement Units
1.6 The Metric System

1.7 Large and Small Numbers
1.8 Significant Figures
1.9 Using Units in Calculations
1.10 Percent Calculations
1.11 Density

LEARNING OBJECTIVES

When you have completed your study of this chapter, you should be able to:
1. Explain what matter is.
2. Explain differences between the terms physical and chemical as applied to:
   a. Properties of matter
   b. Changes in matter
3. Describe matter in terms of the accepted scientific model.
4. On the basis of observation or information given to you, classify matter into the correct category based on the following pairs:
   a. Heterogeneous–homogenous
   b. Solution–pure substance
   c. Element–compound
5. Recognize the units of the metric system.
6. Convert any measurement unit of the metric system into a related unit.
7. Express numbers using scientific notation.
8. Do calculations involving numbers expressed in scientific notation.
9. Express the results of measurements and calculations with the correct number of significant figures.
10. Use the factor–unit method to solve problems.
11. Do simple calculations involving percentage.
12. Do simple calculations involving density.

ANSWERS AND SOLUTIONS TO EVEN-NUMBERED PROBLEMS

What is Matter (Section 1.1)

1.2   How would moving the experiment of Exercise 1.1 to the moon influence the minimum strength of the wire that could be used to suspend the ball?

Solution:

The gravitational force will be less on the moon so a smaller wire with less strength could be used.

1.4   Which of the following do you think is likely to change the most when done on the earth and then on the moon? Carefully explain your reasoning.
a) the distance you can throw a bowling ball through the air
b) the distance you can roll a bowling ball on a flat, smooth surface

Solution:

a) The throwing force of the human body would be the same on the earth and on the moon. However, there is less gravity pulling the ball downward so the ball would travel farther before hitting the surface of the moon.
b) The rolling distance would be the same in both cases. Gravity has no significant effect on rolling motion.

1.6   The rotation of the earth causes it to bulge at the equator. How would the weights of people of equal mass differ when one was determined at the equator and one at the North Pole? (See Exercise 1.5)

Solution:

The gravitational force is greater the closer the object is to the center of the earth. Thus a person would weigh more at the poles (closer to the center) and less at the equator (the bulge makes it farther from the center).

Properties and Changes (Section 1.2)

1.8   Classify each of the following as a *physical* or *chemical* change and give at least one observation, fact, or reason to support your answer.
a) a stick is broken into two pieces
b) a candle burns
c) rock salt is crushed by a hammer
d) tree leaves change color in Autumn

Solution:

a) Physical. There is no change in the composition of the stick.
b) Chemical. Heat is produced; heat production is a sign of a chemical reaction.
c) Physical. There is no change in the color or chemical composition.
d) Chemical. A change in color is a sign that a chemical reaction has occurred.

1.10   Classify each of the following properties as *physical* or *chemical.* Explain your reasoning in each case.
a) mercury metal is a liquid at room temperature
b) sodium metal reacts vigorously with water
c) water freezes at 0°C
d) gold does not rust
e) chlorophyll molecules are green in color

Solution:
a) Physical. The physical state of a substance is not a chemical characteristic because the composition does not change with a change in state.
b) Chemical. There is heat released during the reaction, which is a sign of a chemical reaction. Further, the sodium disappears during the process which indicates a chemical reaction, especially with the generation of heat.
c) Physical. The state in which a substance exists is a physical characteristic as the composition does not change with state change.
d) Chemical. The statement implies that gold does not react chemically with other elements spontaneously at room temperature.
e) Physical. The color of chlorophyll is due to the way in which light is reflected, not due to a change in the chemistry of the chlorophyll.

A Model of Matter (Section 1.3)

1.12   Succinic acid, a white solid that melts at 182°C, is heated gently, and a gas is given off. After the gas evolution stops, a white solid remains that melts at a different temperature than 182°C.
a) Have the succinic acid molecules been changed by the process? Explain your answer.
b) Is the white solid that remains after the heating still succinic acid? Explain your answer.
c) In terms of the number of atoms contained, how do you think the size of the succinic acid molecules compares to the size of the molecules of white solid produced by this process? Explain your answer.
d) Classify molecules of succinic acid using the terms homoatomic or heteroatomic. Explain your reasoning.

Solution:
a) Yes. The melting point changes, so the molecules have changed into a different molecule and a gas is given off.
b) No. The change in melting point means that it must be different.
c) The succinic acid molecule must be larger. The difference must be the size of the gas molecule. The atoms that comprise the gas molecules must have come out of the succinic acid molecule.
d) Heteroatomic. Because the succinic acid changed into two different substances, molecules of succinic acid must contain at least two different kinds of atoms.

1.14   Hydrogen gas and solid sulfur are both made up of homoatomic molecules. Hydrogen and sulfur react to form a single substance, hydrogen sulfide. Use the terms *homoatomic* or *heteroatomic* to classify molecules of hydrogen sulfide. Explain your reasoning.

Solution:
Hydrogen gas (H_2) and sulfur (S_8) are homoatomic molecules because each is composed of atoms from one element. Hydrogen sulfide (H_2S) is composed of atoms from two different elements and is heteroatomic.

1.16   Water can be decomposed to hydrogen gas and oxygen gas by passing electricity through it. Use the terms *homoatomic* or *heteroatomic* to classify molecules of water. Explain your reasoning.

Solution:
Water (H_2O) is composed of atoms of two elements, and is heteroatomic. Water decomposes to hydrogen (H_2) and oxygen (O_2) gases whose molecules are each composed of atoms of the same type. Hydrogen and oxygen gases are homoatomic.

Classification of Matter (Section 1.4)

1.18   Classify each pure substance represented below by a capital letter as an *element* or a *compound*.
Indicate when such a classification cannot be made and explain why.
a) substance A is composed of heteroatomic molecules
b) substance D is composed of homoatomic molecules
c) substance E is changed into substances G and J when it is heated

Solution:
a) A is a compound. All substances having heteroatomic molecules are compounds.
b) D is an element. All homoatomic molecules are elements.
c) E is a compound. It was changed into two simpler substances. G and J cannot be classified. No tests were described to tell whether or not they could have been decomposed into simpler substances.

1.20   Consider the following experiments and answer the questions pertaining to classification:
a) A pure substance R is heated, cooled, put under pressure, and exposed to light but does not change into anything else. What can be said about classifying R as an element or a compound? Explain your reasoning.
b) Upon heating, solid pure substance T gives off a gas and leaves another solid behind. What can be said about classifying T as an element or compound? Explain your reasoning.
c) What can be said about classifying the solid left in part (b) as an element or compound? Explain your reasoning.

Solution:
a) R cannot be classified as either an element or a compound. The lack of change with a few tests does not prove that it will not change. The experiments did not include many methods of decomposing or combining substances.
b) T is a compound. It was changed into a simpler substance.
c) The remaining solid cannot be classified. No tests were made to see if it could be decomposed.

1.22   Classify each of the following as *homogenous* or *heterogenous*:
a) a gold chain   b) liquid eye drops   c) chunky peanut butter
d) a slice of watermelon   e) cooking oil   f) Italian salad dressing

Solution:
a) Homogeneous. Jewelry gold is a mixture of metals that is of consistent composition throughout.
b) Homogeneous. This is a solution in which water is the solvent in which various chemicals are dissolved.
c) Heterogeneous. The pieces of peanut are in the smooth peanut butter. There is no constant composition as one portion may have more peanut pieces than another.
d) Heterogeneous. The pits are placed consistently in specific locations in the meat of the watermelon toward the center. The meat of the watermelon has less and less pits as it is sampled moving out from the center.
e) Homogeneous. Many cooking oils are composed of only one, specific substance and, therefore, are of consistent composition. Other cooking oils are mixtures of oils that are completely miscible (mixable) and do not separate; these are homogeneous, also.
f) Heterogeneous. Notice that the label says "Shake Well Before Using" to indicate that the dressing separates. There are spice pieces in the bottle that are not distributed evenly throughout the dressing.

1.24   Classify as a *pure substance* or *solution* each of the materials of Exercise 1.22 that you classified as homogenous.

    Solution:
    a) solution
    b) solution
    e) pure substance or solution depending on the type

Measurement Units (Section 1.5)

1.26   Briefly discuss why a system of measurement units is an important part of our modern society.

    Solution:
    So many parts of our society involve trading, dispensing, and timing items. A consistent system of measurements is required for the social order to exist.

1.28   An old British unit used to express weight is a stone. It is equal to 14 lbs.
    a) What sort of weighings might be expressed in stones?
    b) Suggest some standard that might have been used to establish the unit.

    Solution:
    a) The unit of stone must be used for weighing large objects whose mass is between 1/10 of a stone and 10 stones.
    b) The standard was probably a field stone found in the area.

The Metric System (Section 1.6)

1.30   Which of the following quantities are expressed in metric units?
    a) freezing point of water: $0°C$
    b) amount of soft drink in a can: 12 fl oz
    c) amount of aspirin in a tablet: 325 g
    d) the length of a playing field: 85 yd
    e) the length of a piece of lumber: 8 ft
    f) volume of a cooking pot: 8 qt

    Solution:
    a) Degrees Celsius is metric.
    b) Fluid ounce is not a metric unit.
    c) The gram is a metric unit.
    d) The yard is not a metric unit.
    e) The foot is not a metric unit.
    f) A quart is not metric.

1.32   Refer to Table 1.2 and suggest an appropriate metric system unit for each nonmetric unit in Exercise 1.30.

    Solution:
    b) milliliters   e) meter
    d) meter   f) liter

1.34   Refer only to Table 1.2 and answer the following questions:
    a) Devices are available that allow liquid volumes as small as one microliter (μL) to be measured. How many microliters would be contained in 1.00 liters?

b) Electrical power is often measured in kilowatts. How many watts would equal 75 kilowatts?

c) Ultrasound is sound of such high frequency that it cannot be heard. The frequency is measured in hertz (vibrations per second). How many hertz corresponds to 15 megahertz?

d) A chlorine atom has a diameter of 200 picometers. How many meters is this diameter?

Solution:

a) Since micro (μ) is one-millionth, there would be a million (1,000,000) microliters in a liter.

b) 75 kilowatts is 75 thousand watts (75,000 watts)

c) 15 megahertz is 15 million hertz (15,000,000 hertz)

d) 200 picometers is 0.000 000 000 200 meters (200 $\times$ 10^{-12} meters)

1.36   Cookbooks are going metric. In such books, 1 cup is equal to 240 mL. Express 1.00 cup in terms of liters and cubic centimeters.

Solution:

$1.00 \, \cancel{\text{cup}} \times \frac{240 \, \cancel{\text{mL}}}{1 \, \cancel{\text{cup}}} \times \frac{1 \, \text{L}}{1000 \, \cancel{\text{mL}}} = 0.240 \, \text{L}$

$1.00 \, \cancel{\text{cup}} \times \frac{240 \, \cancel{\text{mL}}}{1 \, \cancel{\text{cup}}} \times \frac{1 \, \text{cm}^3}{1 \, \cancel{\text{mL}}} = 240 \, \text{cm}^3$

1.38   An $8\frac{1}{2} \times 11$ inch piece of writing paper measures 21.6×27.9 cm. Express these dimensions in meters.

Solution:

$21.6 \, \cancel{\text{cm}} \times \frac{1 \, \text{m}}{100 \, \cancel{\text{cm}}} = 0.216 \, \text{m}$ and $27.9 \, \cancel{\text{cm}} \times \frac{1 \, \text{m}}{100 \, \cancel{\text{cm}}} = 0.279 \, \text{m}$

1.40   Refer to Table 1.3 and answer the following questions:

a) Approximately how many inches longer is a meter stick than a yard stick?

b) A temperature increases by 65°C. How many kelvins would this increase be?

c) You have a 5 lb bag of sugar. Approximately how many kilograms of sugar do you have?

Solution:

Use the factors from Table 1.3.

a) A meter is about 1.1 yard or just over 3 inches longer than a yard.

b) 65 kelvins. A change of 1 degree Celsius is a 1 kelvin change.

c) $5 \, \cancel{\text{lb}} \times \frac{1 \, \text{kg}}{2.2 \, \cancel{\text{lb}}} = 2.27$ or about 2 kg

1.42   Do the following, using appropriate values from Table 1.3.

a) One kilogram of water has a volume of 1.0 dm^3. What is the mass of 1.0 cm^3 of water?

b) One quart is 32 fl oz. How many fluid ounces are contained in a 2.0 L bottle of soft drink?

c) Approximately how many milligrams of aspirin are contained in a 5 grain tablet?

Solution:

Use the factors from Table 1.3.

a) A dm^3 equals a liter or 1000 cm^3. A kilogram equals 1000 grams. 1.0 cm^3 is one thousandth of a liter, so it would weigh one thousandth of a kilogram, or 1.0 gram.

b) $2.0 \, \cancel{\text{L}} \times \frac{1000 \, \cancel{\text{mL}}}{1 \, \cancel{\text{L}}} \times \frac{0.0338 \, \text{fl oz}}{1 \, \cancel{\text{mL}}} = 67.6$ fl oz or about 68 fl oz

c) $5 \, \cancel{\text{grain}} \times \frac{1 \, \text{mg}}{0.015 \, \cancel{\text{grain}}} = 333$ mg or about 300 mg

1.44   In Chemistry Around Us 1.3, it was said that a normal body temperature might be as low 36.1°C in the morning and as high as 37.2°C at bedtime. What are these temperatures on the Fahrenheit scale?

Solution:

$°F = \frac{9}{5}°C + 32$

morning temp: $°F = \frac{9}{5}°(36.1) + 32 = 65.0 + 32 = 97.0°F$

bedtime temp: $°F = \frac{9}{5}°(37.2) + 32 = 67.0 + 32 = 99.0°F$

Large and Small Numbers (Section 1.7)

1.46   Which of the following numbers are written using scientific notation correctly? For those that are not, explain what is wrong.

a) 02.7×10^{-3} b) 4.1×10^2 c) 71.9×10^{-6}
d) 10^3 e) $.0405 \times 10^{-2}$ f) 0.00514

Solution:
a) Incorrect. The zero in front has no meaning. The number should be 2.7×10^{-3}.
b) Correct.
c) Incorrect. The format for scientific notation is a digit, a decimal point, then the rest of the number. The correct format is 7.19×10^{-5}.
d) Incorrect. The technically correct form is to include the one, as 1.0×10^3 or the one written with the appropriate number of zeros to indicate the uncertainty. A generally accepted informal way to write 1.0×10^3 is 10^3.
e) Incorrect. The decimal point should be moved to 4.05×10^{-4}.
f) Incorrect. This is not scientific notation. The correct form is 5.14×10^{-3}.

1.48   Write each of the following numbers using scientific notation:

a) 81 hundred   b) 17.99   c) 11 million
d) 0.0000773   e) 5280   f) 0.00514

Solution:
a) 8.1×10^3. This is 8.1 thousands, which is the same as 81 hundreds.
b) 1.799×10^1. The value of 10^1 is 10, and 1.799×10 is 17.99.
c) 1.1×10^7. One million is 10^6, but the decimal is moved one more place to the left to give an exponent of 10^7.
d) 7.73×10^{-5}. One way to look at this is to write 7.73, then multiply by the number to return to the original as $7.73 \times 0.00001 = 7.73 \times 10^{-5}$.
e) 5.280×10^3. This is 5.280×1000, which is 5,280.
f) 5.14×10^{-3}. This value is determined in the same manner as explained in parts d and e: $5.14 \times 0.001 = 5.14 \times 10^{-3}$.

1.50   The speed of light is about 186 thousand mi/s or 1100 million km/h. Write both numbers using scientific notation.

Solution:
186,000 mi/s $= 1.86 \times 10^5$ mi/s
1,100,000,000 km/h $= 1.100 \times 10^9$ km/h

1.52   A single water molecule has a mass of 2.99×10^{-23} g. Write this number in a decimal form without using scientific notation.

Solution:
0.0000000000000000000000299. There are 22 zeros in front of 299.

1.54   Do the following multiplications and express each answer using scientific notation:
   a) $(2.1 \times 10^2)(1.6 \times 10^3)$
   b) $(8.5 \times 10^{-2})(2.9 \times 10^3)$
   c) $(4.2 \times 10^{-4})(5.9 \times 10^{-3})$
   d) $(2.9 \times 10^4)(3.7 \times 10^{-1})$
   e) $(8.7 \times 10^8)(1.9 \times 10^3)$

Solution:
   a) $2.1 \times 1.6 \times 10^{(2+3)} = 3.36 \times 10^5 = 3.4 \times 10^5$
   b) $8.5 \times 2.9 \times 10^{(-2+3)} = 24.65 \times 10^1 = 2.5 \times 10^2$
   c) $4.2 \times 5.9 \times 10^{(-4-3)} = 24.78 \times 10^{-7} = 2.5 \times 10^{-6}$
   d) $2.9 \times 3.7 \times 10^{(4-1)} = 10.73 \times 10^3 = 1.1 \times 10^4$
   e) $8.7 \times 1.9 \times 10^{(8+3)} = 16.53 \times 10^{11} = 1.7 \times 10^{12}$

1.56   Express each of the following numbers using scientific notation, then carry out the multiplication. Express the answer using scientific notation.
   a) $(144)(0.0876)$ b) $(751)(106)$ c) $(0.0422)(0.00119)$ d) $(128{,}000)(0.0000316)$

Solution:
   a) $(1.44 \times 10^2)(8.76 \times 10^{-2}) = 1.44 \times 8.76 \times 10^{(2-2)} = 12.6 \times 10^0 = 1.26 \times 10^1$
   b) $(7.51 \times 10^2)(1.06 \times 10^2) = 7.51 \times 1.06 \times 10^{(2+2)} = 7.9606 \times 10^4 = 7.96 \times 10^4$
   c) $(4.22 \times 10^{-2})(1.19 \times 10^{-3}) = 4.22 \times 1.19 \times 10^{(-2-3)} = 5.0218 \times 10^{-5} = 5.02 \times 10^{-5}$
   d) $(1.28 \times 10^5)(3.16 \times 10^{-5}) = 1.28 \times 3.16 \times 10^{(5-5)} = 4.0448 \times 10^0 = 4.04$

1.58   Do the following divisions and express each answer using scientific notation:
   a) $\frac{3.1 \times 10^{-3}}{1.2 \times 10^2}$ b) $\frac{7.9 \times 10^4}{3.6 \times 10^2}$ c) $\frac{4.7 \times 10^{-1}}{7.4 \times 10^2}$ d) $\frac{0.00229}{3.16}$
   e) $\frac{119}{3.8 \times 10^3}$

Solution:
   a) $\frac{3.1}{1.2} \times 10^{(-3-2)} = 2.58333 \times 10^{-5} = 2.6 \times 10^{-5}$
   b) $\frac{7.9}{3.6} \times 10^{(4-2)} = 2.19444 \times 10^2 = 2.2 \times 10^2$
   c) $\frac{4.7}{7.4} \times 10^{(-1-2)} = 0.635135 \times 10^{-3} = 6.4 \times 10^{-4}$
   d) $\frac{2.29}{3.16} \times 10^{-3} = 0.72468 \times 10^{-3} = 7.25 \times 10^{-4}$
   e) $\frac{119}{3.8 \times 10^3} = 31.31579 \times 10^{-3} = 3.1 \times 10^{-2}$

1.60   Do the following divisions and express each answer using scientific notation:
   a) $\frac{(4.4 \times 10^3)(1.1 \times 10^2)}{7.9 \times 10^2}$ b) $\frac{5.6 \times 10^7}{(2.1 \times 10^{-2})(8.2 \times 10^{-3})}$ c) $\frac{(3.9 \times 10^3)(7.3 \times 10^{-6})}{(8.8 \times 10^{-4})(1.4 \times 10^{-3})}$
   d) $\frac{(0.19)(2.1)}{(7.2)(3.4)}$ e) $\frac{2.8}{(0.019)(7.1)}$

Solution:
   a) $\frac{(4.4 \times 10^3)(1.1 \times 10^2)}{7.9 \times 10^2} = \frac{4.4 \times 1.1 \times 10^{(3+2)}}{7.9 \times 10^2} = 0.613 \times 10^{(3+2-2)} = 0.613 \times 10^3 = 6.1 \times 10^2$
   b) $\frac{5.6 \times 10^7}{(2.1 \times 10^{-2})(8.2 \times 10^{-3})} = \frac{5.6 \times 10^7}{2.1 \times 8.2 \times 10^{(-2-3)}} = 0.32520 \times 10^{(7+5)} = 3.3 \times 10^{11}$
   c) $\frac{(3.9 \times 10^3)(7.3 \times 10^{-6})}{(8.8 \times 10^{-4})(1.4 \times 10^{-3})} = \frac{(3.9 \times 7.3 \times 10^{(3-6)})}{(8.8 \times 1.4 \times 10^{(-4-3)})} = \frac{28.47 \times 10^{-3}}{12.32 \times 10^{-7}} = 2.3109 \times 10^4 = 2.3 \times 10^4$
   d) $\frac{(0.19)(2.1)}{(7.2)(3.4)} = \frac{0.399}{24.48} = 0.01629902 = 1.6 \times 10^{-2}$
   e) $\frac{2.8}{(0.019)(7.1)} = \frac{2.8}{0.1349} = 20.7561 = 2.1 \times 10^1$

Significant Figures (Section 1.8)

1.62   Indicate to what decimal position readings should be estimated and recorded (nearest 0.1, 0.01, etc.) for measurements made with the following devices.
a) a ruler with smallest scale marking of 0.1 cm
b) a measuring telescope with smallest scale marking of 0.1 mm
c) a protractor with smallest scale marking of 1°
d) a tire pressure gauge with smallest scale marking of 1 lb/in.^{2}

Solution:
The reading should include 1 decimal position to the right of the smallest marking.
a) The marking is 0.1 cm. The reading is made to the nearest 0.01 cm.
b) The marking is 0.1 mm. The reading is made to the nearest 0.01 mm.
c) The marking is 1°. The reading is made to the nearest 0.1°.
d) The marking is 1 lb/in.^{2}. The reading is made to the nearest 0.1 lb/in^2.

1.64   Write the following measured quantities as you would record them, using the correct number of significant figures based on the device used to make the measurements.
a) a temperature that appears to be halfway between 29 and 30 on a thermometer with 1°C scale divisions
b) exactly 4 mL of water measured with a graduated cylinder that has smallest markings of 1 mL
c) fifteen and one-half degrees measured with a protractor that has 1° scale markings
d) a tire pressure of exactly 29 lb/in.^{2} measured with a tire pressure gauge that has smallest markings of 1 lb/in.^{2}

Solution:
The reading should include 1 decimal position to the right of the smallest marking.
a) 29.5°C   b) 4.0 mL   c) 15.5°   d) 29.0 lb/in.^{2}

1.66   In each of the following, identify the measured numbers and exact numbers. Do the indicated calculation and write your answer using the correct number of significant figures.
a) A bag of potatoes is found to weigh 5.06 lb. The bag contains 16 potatoes. Calculate the weight of an average potato.
b) A women's basketball team has a starting team of players with the following heights: 6 ft, 6 in; 5 ft, 6 in; 5 ft, 3 in; 5 ft, 8 in; and 5 ft, 1 in. What is the average height per member of the starting five players?

Solution:
The average is the total divided by the number of objects.
a) Measured number is 5.06 lb; exact number is 16. $\frac{5.06 \text{ lb}}{16 \text{ potatoes}} = \frac{0.31625 \text{ lb each}}{\text{potato}} = 3.16 \times 10^{-1}$ lb/potato
b) Measured numbers are players, heights; exact number is 5.
$\frac{6'2'' + 5'6'' + 5'3'' + 5'8'' + 5'1''}{5} = \frac{26'20''}{5} = \frac{27'8''}{5} = \frac{27.75'}{5} = 5.55$ feet average height. This is a little more than 5'6" average height.

1.68   Determine the number of significant figures in each of the following:
a) 210   b) 0.011   c) 3.72×10^{-2}
d) 5.152   e) 0.0170   f) 7.09

Solution:
a) Three. The trailing zero is significant.
b) Two. The leading zero is not significant.

c) Three. The 10^{-2} indicates the number of places to move the decimal.
d) Four. There is no question as there are no zeros to analyze as in part e.
e) Three. The leading zero is not significant, but the trailing zero is significant.
f) Three. Buried zeros are significant.

1.70   Do the following calculations and use the correct number of significant figures in your answer. Assume all numbers are the results of measurements.

a) $(3.71)(1.4)$

b) $(0.0851)(1.2262)$

c) $\frac{(3.9\times10^3)(7.3\times10^{-6})}{(8.8\times10^{-4})(1.4\times10^{-3})}$

d) $(3.3 \times 10^4)(3.09 \times 10^{-3})$

e) $\frac{(760)(2.00)}{6.02\times10^{20}}$

Solution:

a) $(3.71)(1.4) = 3.71 \times 1.4 = 5.194 = 5.2$

b) $(0.0851)(1.2262) = 0.0851 \times 1.2262 = 0.10434962 = 0.104$

c) $\frac{(3.9\times10^3)(7.3\times10^{-6})}{(8.8\times10^{-4})(1.4\times10^{-3})} = \frac{3.9\times7.3\times10^{(3-6)}}{8.8\times1.4\times10^{(-4-3)}} = \frac{28.47\times10^{-3}}{12.32\times10^{-7}} = 2.31088 \times 10^4 = 2.3 \times 10^4$

d) $(3.3 \times 10^4)(3.09 \times 10^{-3}) = 3.3 \times 3.09 \times 10^{(4-3)} = 10.197 \times 10^1 = 1.0 \times 10^2$

e) $\frac{(760)(2.00)}{6.02\times10^{20}} = \frac{760\times2.00}{6.02\times10^{20}} = \frac{1520\times10^{-20}}{6.02} = 252.4917 \times 10^{-20} = 2.52 \times 10^{-18}$

1.72   Do the following calculations and use the correct number of significant figures in your answers. Assume all numbers are the results of measurements.

a) $3.7 + 6.104 + 33.12$
b) $1.11 + 0.927 + 0.00082$
c) $11.327 - 10.648$
d) $(4.2 \times 10^{-3}) + (7.8 \times 10^{-1})$ (Hint: Write in decimal form first, then add.)
e) $146.05 - 36.182$
f) $27.09 - 0.0146$

Solution:

a) $3.7 + 6.104 + 33.12 = 42.924 = 42.9$
b) $1.11 + 0.927 + 0.00082 = 203782 = 2.04$
c) $11.327 - 10.648 = 0.679$
d) $(4.2 \times 10^{-3}) + (7.8 \times 10^{-1}) = 0.0042 + 0.78 = 0.7842 = 0.78$
e) $146.05 - 36.182 = 109.87$
f) $27.09 - 0.0146 = 27.11$

1.74   Do the following calculations and use the correct number of significant figures in your answers. Assume all numbers are the result of measurements. In calculations involving both addition/subtraction and multiplication/division, the procedure is to do additions and subtractions first.

Solution:

a) $\frac{(0.0267+0.00119)(4.626)}{28.7794}$

b) $\frac{212.6-21.88}{86.37}$

c) $\frac{27.99-18.07}{4.63-0.88}$

d) $\frac{18.87}{2.46} - \frac{18.07}{0.88}$

e) $\frac{(8.46-2.09)(0.51+0.22)}{(3.74+0.07)(0.16+0.2)}$

f) $\frac{12.06-11.84}{0.271}$

Solution:

a) $\frac{(0.0267+0.00119)(4.626)}{28.7794} = \frac{(0.02789)(4.626)}{28.7794} = \frac{(0.0279)(4.626)}{28.7794} = \frac{0.129065}{28.7794} = \frac{0.129}{28.7794} = 00.00448 = 4.48 \times 10^{-3}$

b) $\frac{212.6-21.88}{86.37} = \frac{190.72}{86.37} = \frac{190.7}{86.37} = 2.207943 = 2.208$

c) $\frac{27.99-18.07}{4.63-0.88} = \frac{9.92}{3.75} = 2.6453 = 2.65$

d) $\frac{18.87}{2.46} - \frac{18.07}{0.88} = 7.67073 - 20.53409 = 7.67 - 21 = -13$

e) $\frac{(8.46-2.09)(0.51+0.22)}{(3.74+0.07)(0.16+0.2)} = \frac{(6.37)(0.73)}{(3.81)(0.4)} = \frac{4.6501}{1.524} = \frac{4.7}{2} = 2.35 = 2$

f) $\frac{12.06-11.84}{0.271} = \frac{0.22}{0.271} = 0.811808 = 0.81$

1.76   The following measurements were obtained for the length and width of a series of rectangles. Each measurement was made using a ruler with a smallest scale marking of 0.1 cm.

  Black rectangle: l = 12.00 cm, w = 10.40 cm
  Red rectangle: l = 20.20 cm, w = 2.42 cm
  Green rectangle: l = 3.18 cm, w = 2.55 cm
  Orange rectangle: l = 13.22 cm, w = 0.68 cm

a) Calculate the area (length × width) and perimeter (sum of all four sides) for each rectangle and express your results in square centimeters and centimeters, respectively, and give the correct number of significant figures in the result.

b) Change all measured values to meters and then calculate the area and perimeter of each rectangle. Express your answers in square meters and meters, respectively, and give the correct number of significant figures.

c) Does changing the units used change the number of significant figures in the answer?

Solution:

a) Area = l × w
  Area of Black = 12.00 cm × 10.40 cm = 124.8 cm^2 (keep 4 sig figs)
  Area of Red = 20.20 cm × 2.42 cm = 48.884 cm^2 (round to 3 sig figs) = 48.9 cm^2
  Area of Green = 3.18 cm × 2.55 cm = 8.109 cm^2 (round to 3 sig figs) = 8.11 cm^2
  Area of Orange = 13.22 cm × 0.68 cm = 8.9896 cm^2 (round to 2 sig figs) = 9.0 cm^2
  Perimeter = 2(l + w) where the 2 is an exact number.
  Perimeter of Black = 2(12.00 cm + 10.40 cm) = 44.80 cm (keep 2 decimal places)
  Perimeter of Red = 2(20.20 cm + 2.42 cm) = 45.24 cm (keep 2 decimal places)
  Perimeter of Green = 2(3.18 cm + 2.55 cm) = 11.46 cm (keep 2 decimal places)
  Perimeter of Orange = 2(13.22 cm + 0.68 cm) = 27.80 cm (keep 2 decimal places)

b) To change cm to m, divide by 100. (Conversion factor is 1 m/100 cm
  Area = l × w
  Area of Black = 0.1200 m × 0.1040 m = 0.01248 m^2 (keep 4 sig figs)
  Area of Red = 0.2020 m × 0.242 m = 0.0048884 (round to 3 sig figs) = 0.00489 m^2
  Area of Green = 0.0318 m × 0.0255 m = .0008109 (round to 3 sig figs) = 0.000811 m^2
  Area of Orange = 0.1322 m × 0.0068 m = 0.00089896 (round to 2 sig figs) = 0.00090 m^2
  Perimeter = 2(l + w) where the 2 is an exact number.
  Perimeter of Black = 2(0.1200 m + 0.1040 m) = 0.4480 m (keep 4 decimal places)
  Perimeter of Red = 2(0.2020 m + 0.0242 m) = 0.4524 m (keep 4 decimal places)
  Perimeter of Green = 2(0.0318 m + 0.0255 m) = 0.1146 m (keep 4 decimal places)
  Perimeter of Orange = 2(0.1322 m + 0.0068 m) = 0.2780 m (keep 4 decimal places)

c) No. The number of significant figures remains the same.

Using Units In Calculations (Section 1.9)

1.78   Determine a single factor derived from Table 1.3 that could be used as a multiplier to make each of the following conversions:

a) 12 fl oz to cubic centimeters
b) 35 BTU to joules
c) 110 mm to meters
d) 8 lb to kilograms

Solution:

a) $12 \text{ fl oz} \times \frac{1 \text{ mL}}{0.0338 \text{ fl oz}}$

b) $35 \text{ BTU} \times \frac{1 \text{ joule}}{0.000949 \text{ BTU}}$

c) $110 \text{ mm} \times \frac{1 \text{ m}}{1,000 \text{ mm}}$

d) $8 \text{ lb} \times \frac{1 \text{ kg}}{2.20 \text{ lb}}$

1.80   A marathon race is about 26 miles. Obtain a factor from Table 1.3 and use the factor-unit method to calculate the distance of a marathon in kilometers.

Solution:

$\frac{1 \text{ km}}{0.621 \text{ mi}}$ is the factor.

$26 \text{ mi} \times \frac{1 \text{ km}}{0.621 \text{ mi}} = 41.86795$ (round to 2 sig figs) $= 42 \text{ km}$

1.82   A metric cookbook calls for 250 mL of milk. Your measuring cup is in English units. About how many cups of milk should you use:
(NOTE: you will need two factors, one from Table 1.3 and one from the fact that 1 cup = 8 fl oz.)

Solution:

$250 \text{ mL} \times \frac{0.0338 \text{ fl oz}}{1 \text{ mL}} \times \frac{1 \text{ cup}}{8 \text{ fl oz}} = 1.05625 \text{ cups} = 1 \text{ cup}$

1.84   You have a 40 lb baggage limit for a transatlantic flight. When your baggage is put on the scale, you think you are within the limits because it reads 18.0. But then you realize that weight is in kilograms. Do a calculation to determine whether your baggage is overweight.

Solution:

$18.0 \text{ kg} \times \frac{2.20 \text{ lb}}{1 \text{ kg}} = 39.6 \text{ lbs}$

The baggage is less than 40 1bs, so it is not overweight.

1.86   During a glucose tolerance test, the serum glucose concentration of a patient was found to be 139 mg/dL. Convert the concentration to grams per liter.

Solution:

$\frac{139 \text{ mg}}{\text{dL}} \times \frac{1 \text{ g}}{1000 \text{ mg}} \times \frac{10 \text{ dL}}{1 \text{ L}} = 1.39 \text{ g/L}$

Percent Calculations (Section 1.10)

1.88   A salesperson made a sale of $212.00 and received a commission of $10.60. What percent commision was paid?

Solution:

The Commission amount is the Sale price multiplied by the Percent commission, then the

Percent commission $= \frac{\text{Commission amount}}{\text{Sale price}} \times 100 = \frac{\$10.60}{\$212.60} \times 100 = 0.049859 \times 100 = 4.986\%$ commission

1.90   The recommended daily intake of thiamine is 1.4 mg for a male adult. Suppose a person takes in only 1.2 mg/day. What percentage of the recommended intake is he receiving?

Solution:

$\% = \frac{\text{parts}}{\text{total}} \times 100 = \frac{1.2 \text{ mg/day}}{1.4 \text{ mg/day}} \times 100 = 85.71429$ (round to 2 sig figs) $= 86\%$

1.92   Immunoglobulin antibodies occur in five forms. A sample of serum is analyzed with the following results. Calculate the percentage of total immunoglobulin represented by each type.

Type:	IgG	IgA	IgM	IgD	IgE
Amount (mg):	987.1	213.3	99.7	14.4	0.1

Solution:

$\% = \frac{parts}{total} \times 100$

total$= 987.1 + 213.3 + 99.7 + 14.4 + 0.1 = 1314.6$ mg

$\%IgG = \frac{987.1 \text{ mg}}{1314.6 \text{ mg}} \times 100 = 75.08748$ (round to 4 sig figs) $= 75.09\%$

$\%IgA = \frac{213.3 \text{ mg}}{1314.6 \text{ mg}} \times 100 = 16.22547$ (round to 4 sig figs) $= 16.23\%$

$\%IgM = \frac{99.7 \text{ mg}}{1314.6 \text{ mg}} \times 100 = 7.58406$ (round to 3 sig figs) $= 7.58\%$

$\%IgD = \frac{14.4 \text{ mg}}{1314.6 \text{ mg}} \times 100 = 1.09539$ (round to 3 sig figs) $= 1.10\%$

$\%IgE = \frac{0.1 \text{ mg}}{1314.6 \text{ mg}} \times 100 = 0.007607$ (round to 1 sig figs) $= 0.008\%$

Density (Section 1.11)

1.94   Calculate the density of the following materials for which the mass and volume of samples have been measured. Express the density of liquids in g/mL, the density of solids in g/cm^3, and the density of gases in g/L.

a) A 50.0 mL sample of liquid acetone has a mass of 39.6 g.

b) A 1.00 cup (236 mL) sample of homogenized milk has a mass of 243 g.

c) 20.0 L of dry carbon dioxide gas (CO_2) has a mass of 39.54 g.

d) A 25.0-cm^3 block of nickel metal (Ni) has a mass of 222.5 g.

Solution:

a) Density $= \frac{g}{mL} = \frac{39.6 \text{ g}}{50 \text{ mL}} = 0.792$ g/mL

b) Density $= \frac{g}{mL} = \frac{243 \text{ g}}{236 \text{ mL}} = 1.03$ g/mL

c) Density $= \frac{g}{L} = \frac{39.54}{20.0 \text{ L}} = 1.98$ g/L

d) Density $= \frac{g}{cm^3} = \frac{222.5 \text{ g}}{25.0 \text{ cm}^3} = 8.90$ g/cm^3

1.96   Calculate the volume and density of a cube of lead metal (Pb) that has a mass of 373.6 g and has edges that measure 3.20 cm.

Solution:

Density $= \frac{mass}{volume}$ and Volume $= L \times W \times H$; then Density $= \frac{mass}{L \times W \times H}$. Since a cube has sides that are the same size, $L \times W \times H = (3.20 \text{ cm})^3$. We substitute into the formula as Density $= \frac{373.6 \text{ g}}{(3.20 \text{ cm})^3}$ which is $\frac{373.6 \text{ g}}{32.768 \text{ cm}^3} = 11.4$ g/cm^3.

1.98   The density of ether is 0.736 g/mL. What is the volume of 500 g of ether?

Solution:

The density can be thought of as a conversion factor and can be written in either of these forms:

$\frac{0.736 \text{ g}}{1 \text{ mL}}$ or $\frac{1 \text{ mL}}{0.736 \text{ g}}$

$500 \text{ g} \times \frac{1 \text{ mL}}{0.736 \text{ g}} = 679.3478$ (round to 3 sig figs) $= 679$ mL

PROGRAMMED REVIEW

Section 1.1 What is Matter

Matter is anything that (a) _has_ _mass_ and (b) _occupies space_ . (c) _mass_ is a measure of the amount of matter present. The gravitational force pulling an object toward the earth is the object's (d) _weight_.

Section 1.2 Properties and Changes

(a) _physical_ properties can be observed or measured without changing or attempting to change the (b) _chemical_ of the matter in question. (c) _composition_ properties are demonstrated when attempts are made to change the composition of matter. Changes can be classified as (d) _chemical_ changes or (e) _physical_ changes.

Section 1.3 A Model of Matter

Explanations for observed behavior are called scientific (a) _models_ . A (b) _molecule_ is the smallest unit of a pure substance. Molecules that are made up of one kind of atom are called (c) _homoatomic_ molecules The term polyatomic is used to describe molecules containing (d) _two_ _or more_ atoms. An atom is the limit of (e) _chemical_ subdivision.

Section 1.4 Classification of Matter

Matter that has a constant composition and fixed properties is called a (a) _pure substance_. A (b) _mixture_ can be physically separated into two or more components. (c) _homogeneous_ matter has the same appearance and properties throughout. Homogenous mixtures of two or more substances are called (d) _solutions_. Most matter found in nature is classified as (e) _mixtures_ . A pure substance is either a (f) _compound_ or an (g) _element_.

Section 1.5 Measurement Units

Measurements are based on (a) _units_ that have been agreed upon. A measurement is expressed as some (b) _multiple_ of a specific (c) _unit_ . The earliest measurements were based on dimensions of the (d) _human body_ .

Section 1.6 The Metric System

A specific unit from which other units are obtained by multiplication or division is called a (a) _basic unit_ of measurement. Units obtained from basic units are called (b) _derived units_ . The basic units of length, volume, and mass, respectively, in the metric system are the (c) _meter_ , (d) _cubic decimeter_ and (e) _kilogram_. Units of the metric system that are larger or smaller than a basic unit are indicated by (f) _prefixes_ attached to the basic unit. The basic unit of temperature in the metric system is the (g) _kelvins_

Section 1.7 Large and Small Numbers

In scientific notation, a number is represented as a product of a (a) _coefficient_ and (b) _10_ raised to a whole number exponent. The standard position for a decimal in scientific notation is to the (c) _right_ of the first (d) _non-zero_ digit in the coefficient. In scientific notation, an exponent of −3 indicates the original decimal position is (e) _three_ places to the (f) _left_ of the standard position.

Section 1.8 Significant Figures

Every measurement contains an (a) _uncertainty_ that depends on the measuring device. When a measurement is represented using significant figures correctly, the last number is an (b) _estimate_. The number 0.0219 contains (c) _3_ significant figures. When the number 8.42149 is properly rounded to four significant figures, the last number will be (d) _1_ . When (e) _exact numbers_ are used in calculations, they do not influence the number of significant figures in the calculated results.

Section 1.9 Using Units in Calculations

The (a) __factor-unit__ method is a systematic approach to solving numerical problems. The (b) __factors__ used in the factor-unit method are (c) __fractions__ which are equal to 1, and are derived from fixed relationships between quantities. In a factor-unit calculation, the units of the factor cancel out the units of the (d) __known__ quantity, and generate the units of the (e) __unknown__ quantity.

Section 1.10 Percent Calculations

The word percent means per (a) __one__ __hundred__ A percent is the (b) __number__ of specific items found in a "total" group of (c) __100__ such items. A basket of fruit contains 3 oranges, 2 apples, and 4 pears. In a calculation of the percent of pears in the basket, the "total" in the calculation would be (d) __9__.

Section 1.11 Density

Density is the ratio obtained by dividing the (a) __mass__ of the object by its (b) __volume__. The common scientific units for the density of a solid is (c) g/cm^3, for a liquid is (d) __g/ml__, and for a gas is (e) __g/L__.

SELF-TEST QUESTIONS

Multiple Choice

1. Which of the following involves a chemical change?
    a) stretching a rubber band
    c) lighting a candle
    b) breaking a stick
    d) melting an ice cube

2. Which of the following terms could not be properly used in the description of a compound?
    a) solution
    c) pure substance
    b) polyatomic
    d) heteroatomic

3. A solid substance is subjected to a number of tests and observations. Which of the following would be classified as a chemical property of the substance?
    a) it is gray in color
    b) it has a density of 2.04 grams per milliliter
    c) it dissolves in acid and a gas is liberated
    d) it is not attracted to either pole of a magnet

4. Which of the following is an example of heterogenous matter?
    a) water containing sand
    c) a sample of salt water
    b) a sample of pure table salt
    d) a pure sample of iron

5. When a substance undergoes a physical change which of the following is always true?
    a) it melts
    b) a new substance is produced
    c) heat is given off
    d) the molecules of the substance remain unchanged

6. Which of the following is *not* a chemical change?
    a) burning of magnesium
    c) pulverizing of some sulfur
    b) exploding of some nitroglycerine
    d) rusting of iron

7. Which of the following is the basic unit of length in the metric system?
    a) centimeter   b) meter   c) millimeter   d) kilometer

8. Which of the following is a derived unit?
   a) calorie   b) cubic decimeter   c) joule   d) kilogram

9. In the number 3.91×10^{-3}, the original decimal position is located
   a) 3 places to the right from 3.91   b) 2 places to the right from 3.91
   c) 3 places to the left from 3.91   d) 2 places to the left from 3.91

10. How many significant figures are included in the number 0.02102?
    a) two   b) three   c) four   d) five

11. Twenty-one (21) students in a class of 116 got a B grade on an exam. What percent of the students in the class got B's?
    a) 21.0   b) 22.1   c) 15.3   d) 18.1

12. What single factor derived from Table 1.3 would allow you to calculate the number of quarts in a 2.0 L bottle of soft drink?
    a) 1.057 quart/1 L   b) 1 L/1.057 quart
    c) 0.0338 fl oz/ 1 mL   d) 1 mL/0.0338 fl oz

13. On a hot day, a Fahrenheit thermometer reads 97.3°F. What would this reading be on a Celsius thermometer?
    a) 118°C   b) 22.1°C   c) 36.3°C   d) 143°C

14. The density of a 1 mL sample of a patient's blood is 1.08 g/mL. The density of a pint of blood taken at the same time from the same patient would be _____ 1.08 g/mL.
    a) greater than   b) less than
    c) equal to   d) more than one possible answer

15. A 120 mL urine specimen weighs 130.6 g. The density of the specimen is:
    a) 1.09   b) 0.917   c) 250.6   d) 10.6

Matching

Match the type of measurement given on the right as responses to the measurement units given on the left.

16. __d__ kelvin   a) mass
17. __b__ milliliter   b) volume
18. __a__ gram   c) length
19. __c__ millimeter   d) temperature
20. __b__ cubic decimeter   e) density
21. __c__ kilometer
22. __e__ pounds per cubic foot

True-False

23. The mass of an object is the same as its weight.

24. A physical property can be observed without attempting any composition changes.

25. The cooking of food involves chemical changes.

26. The smallest piece of water that has the properties of water is called an atom.

27. Carbon monoxide molecules are diatomic and heteroatomic.

28. The prefix *milli-* means one thousand times.

29. A pure substance containing sulfur and oxygen atoms must be classified as a compound.

30. One meter is shorter than one yard.

31. The calorie and joule are both units of energy.

32. In scientific notation, the exponent on the 10 cannot be larger than 15.

33. The correctly rounded sum resulting from adding 13.0, 1.094, and 0.132 will contain five significant figures.

34. If an object floats in water, it must have a higher density than water.

35. Most gases are less dense than liquids.

SOLUTIONS

A. Answers to Programmed Review

1.1	a) has mass	b) occupies space	c) mass	d) weight
1.2	a) physical e) physical	b) composition	c) chemical	d) chemical
1.3	a) models e) chemical	b) molecule	c) homoatomic	d) two or more
1.4	a) pure substance e) mixtures	b) mixture f) compound	c) homogenous g) element	d) solutions
1.5	a) units	b) multiple	c) unit	d) human body
1.6	a) basic unit e) kilogram	b) derived units f) prefixes	c) meter g) kelvins	d) cubic decimeter
1.7	a) coefficient e) three	b) 10 f) left	c) right	d) non-zero
1.8	a) uncertainty e) exact numbers or counted numbers	b) estimate	c) three	d) 1
1.9	a) factor-unit e) unknown	b) factors	c) fractions	d) known
1.10	a) one hundred	b) number	c) 100	d) 9
1.11	a) mass e) g/L	b) volume	c) g/cm^3	d) g/mL

B. Answers to Self-Test Questions

1. c	13. c	25. T
2. a	14. c	26. F
3. c	15. a	27. T
4. a	16. d	28. F
5. d	17. b	29. T
6. c	18. a	30. F
7. b	19. c	31. T
8. a	20. b	32. F
9. c	21. c	33. F
10. c	22. e	34. F
11. d	23. F	35. T
12. a	24. T	

Chapter 2

Atoms and Molecules

CHAPTER OUTLINE

2.1   Symbols and Formulas
2.2   Inside the Atom
2.3   Isotopes
2.4   Relative Masses of Atoms and Molecules

2.5   Isotopes and Atomic Weights
2.6   Avogadro's Number: The Mole
2.7   The Mole and Chemical Formulas

LEARNING OBJECTIVES

When you have completed your study of this chapter, you should be able to:
1. Use chemical element symbols to write formulas for compounds.
2. Use atomic weights of the elements to calculate molecular weights of compounds.
3. Know the characteristics of protons, neutrons, and electrons.
4. Use the concept of atomic number and mass number to determine the number of subatomic particles in isotopes and to write symbols for isotopes.
5. Use isotope percent abundances and masses to calculate atomic weights of elements.
6. Use the mole concept to obtain relationships between number of moles, number of grams, and number of atoms for elements and use those relationships to obtain factors for use in factor unit calculations.
7. Use the mole concept and molecular formulas to obtain relationships between number of moles, number of grams, and number of atoms or molecules for compounds and use those relationships to obtain factors for use in factor unit calculations.

ANSWERS AND SOLUTIONS TO EVEN-NUMBERED PROBLEMS

Symbols and Formulas (Section 2.1)

2.2   Draw a "formula" for each of the following molecules using circular symbols of your choice to represent atoms.
a) a triatomic molecule of a compound
b) a molecule of a compound containing 2 atoms of one element and 2 atoms of a second element

c) a molecule of a compound containing two atoms of one element, one atom of a second element, and four atoms of a third element
d) a molecule containing two atoms of one element, six atoms of a second element, and one atom of a third element

Solution:
Note: Formulas other than the ones shown are possible. The arrangements shown are based on the answers to Exercise 2.4.

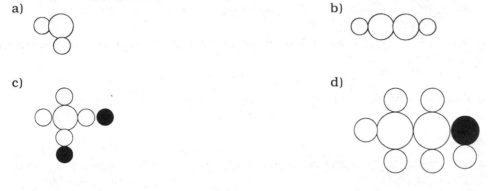

a)

b)

c)

d)

2.4   Write formulas for the following molecules using elemental symbols from Table 2.1 and subscripts. Compare these formulas to those of Exercise 2.2.
a) a molecule of water (two hydrogen atoms and one oxygen atom)
b) a molecule of hydrogen peroxide (two hydrogen atoms and two oxygen atoms)
c) a molecule of sulfuric acid (two hydrogen atoms, one sulfur atom, and four oxygen atoms)
d) a molecule of ethyl alcohol (two carbon atoms, six hydrogen atoms, and one oxygen atom)

Solution:
a) H_2O b) H_2O_2 c) H_2SO_4 d) C_2H_6O

2.6   Determine the number of each type of atom in molecules represented by the following formulas:
a) sulfur dioxide (SO_2)   b) butane (C_4H_{10})
c) chlorous acid ($HClO_2$)   d) boron triflouride (BF_3)

Solution:
a) 1 sulfur, 2 oxygen   b) 4 carbon, 10 hydrogen
c) 1 hydrogen, 1 chlorine, 2 oxygen   d) 1 boron, 3 fluorine

2.8   Tell what is wrong with each of the following formulas and write the correct formula.
a) HSH (hydrogen sulfide)
b) $HCLO_2$ (chlorous acid)
c) $2HN_2$ (hydrazine–two hydrogen and four nitrogen)
d) C2H6 (ethane)

Solution:
a) Combine hydrogens with subscript (H_2S).
b) The symbol for chlorine is Cl ($HClO_2$).
c) Use subscripts, not coefficients, to indicate how many atoms (N_2H_4).
d) The numerals should be subscripts (C_2H_6).

Inside the Atom (Section 2.2)

2.10   Determine the charge and mass (in u) of nuclei made up of the following particles:
  a) 5 protons and 4 neutrons
  b) 11 protons and 10 neutrons
  c) 36 protons and 50 neutrons
  d) 50 protons and 68 neutrons

  Solution:
  Each proton has a positive charge and each proton and neutron has a mass of about 1 u.
  a) charge $= 5+$; mass $=$ about 9 u
  b) charge $= 11+$; mass $=$ about 21 u
  c) charge $= 36+$; mass $=$ about 86 u
  d) charge $= 50+$; mass $= 118$ u

2.12   Determine the number of electrons that would have to be associated with each nucleus described in Exercise 2.10 to produce a neutral atom.

  Solution:
  To produce a neutral atom, the number of electrons must equal the number of protons.
  a) 5   b) 11   c) 36   d) 50

Isotopes (Section 2.3)

2.14   Determine the number of electrons and protons contained in an atom of the following elements:
  a) silicon   b) Sn   c) Element number 74

  Solution:
  The number of protons is the atomic number; the number of electrons must equal the number of protons in an atom.
  a) 14 electrons, 14 protons
  b) 50 electrons, 50 protons
  c) 74 electrons, 74 protons

2.16   Determine the number of protons, number of neutrons, and number of electrons in atoms of the following isotopes:
  a) $^{25}_{12}$Mg   b) $^{58}_{28}$Ni   c) $^{119}_{48}$Cd

  Solution:
  The number of protons is the atomic number; the number of neutrons is the mass number minus the atomic number; and the number of electrons equals the number of protons in an atom.
  a) 12 protons, 13 neutrons, 12 electrons
  b) 28 protons, 30 neutrons, 28 electrons
  c) 48 protons, 71 neutrons, 48 electrons

2.18   Write symbols like those given in Exercises 2.15 and 2.16 for the following isotopes:
  a) silicon-28   b) argon-40   c) strontium-88

  Solution:
  a) $^{28}_{14}$Si   b) $^{40}_{18}$Ar   c) $^{88}_{38}$Sr

2.20   Determine the mass number and atomic number for atoms containing the nuclei described in Exercise 2.10. Write symbols for each atom like those given in Exercises 2.15 and 2.16.

  Solution:
  a) atomic number $= 5$, mass number $= 9$; $^{9}_{5}$B
  b) atomic number $= 11$, mass number $= 21$; $^{21}_{11}$Na
  c) atomic number $= 36$, mass number $= 86$; $^{86}_{36}$Kr
  d) atomic number $= 50$, mass number $= 118$; $^{118}_{50}$Sn

2.22   Write isotope symbols for atoms with the following characteristics:
   a) contains 17 electrons and 20 neutrons
   b) a copper atom with a mass number of 65
   c) a zinc atom that contains 36 neutrons

   Solution:
   a) $^{37}_{17}Cl$ b) $^{65}_{29}Cu$ c) $^{66}_{30}Zn$

Relative Masses of Atoms and Molecules (Section 2.4)

2.24   How many average helium atoms would be needed to balance the mass of a single average carbon atom?

   Solution:
   Three. Since the average helium atom has a relative mass of about 4.0 u and the average carbon atom is about 12.0 u, three helium atoms would have a mass about the same as a single carbon atom.

2.26   What is the symbol and name of an element whose average atoms have a mass that is 77.1% of the mass of an average chromium atom?

   Solution:
   77.1% of the mass of the chromium atom would be 0.771×52.00 u $= 40.092$
   Calcium is close with 40.08 u

2.28   What is the symbol and name of the element whose average atoms have a mass very nearly one-half the mass of an average silicon atom?

   Solution:
   An average silicon atom has a mass of 28.09 u. One-half of that is 14.04 u. Nitrogen is close with u $= 14.01$

2.30   Determine the molecular weights of the following in u:
   a) nitrogen dioxide (NO_2)   b) ammonia (NH_3)   c) glucose ($C_6H_{12}O_6$)
   d) ozone (O_3)   e) ethylene glycol ($C_2H_6O_2$)

   Solution:
   a) $(1 \times 14.01$ u$) + (2 \times 16.00$ u$) = 46.01$ u
   b) $(1 \times 14.01$ u$) + (3 \times 1.008$ u$) = 17.034$ u (round to 2 decimal places) $= 17.03$ u
   c) $(6 \times 12.01$ u$) + (12 \times 1.008$ u$) + (6 \times 16.00$ u$) = 180.156$ u (round to 2 decimal places) $= 180.16$ u
   d) 3×16.00 u $= 48.00$ u
   e) $(2 \times 12.01$ u$) + (6 \times 1.008$ u$) + (2 \times 16.00$ u$) = 62.068$ u (round to 2 decimal places) $= 62.07$ u

2.32   A flammable gas is known to contain only carbon and hydrogen. Its molecular weight is determined and found to be 28.05 u. Which of the following is a likely identity of the gas? acetylene (C_2H_2), ethylene (C_2H_4), ethane (C_2H_6)

   Solution:
   The molecular weights for the three gases are:
   $C_2H_2 = (2 \times 12.01$ u$) + (2 \times 1.008$ u$) = 26.036$ u (round to 2 decimal places) $= 26.04$ u
   $C_2H_4 = (2 \times 12.01$ u$) + (4 \times 1.008$ u$) = 28.052$ u (round to 2 decimal places) $= 28.05$ u
   $C_2H_6 = (2 \times 12.01$ u$) + (6 \times 1.008$ u$) = 30.068$ u (round to 2 decimal places) $= 30.07$ u
   The unknown gas has a molecular weight equal to C_2H_4.

2.34   Serine, an amino acid found in proteins, has a molecular weight of 105.10 u and is represented by the formula $C_yH_7NO_3$. What number does y stand for in the formula?

Solution:
Add the relative masses of 7 hydrogen, 1 nitrogen, and 3 oxygen. Subtracting that sum from the molecular weight of serine will give the mass of the carbon atoms.
$(7 \times 1.008 \text{ u}) + (1 \times 14.01 \text{ u}) + (3 \times 16.00 \text{ u}) = 69.066 \text{ u}$ (round to 2 decimal places) $= 69.07 \text{ u}$
105.10 u − 69.07 u = 36.03 u must come from the carbon. That is the mass of 3 carbon atoms.
y = 3

Isotopes and Atomic Weights (Section 2.5)
2.36   Naturally occurring sodium has only a single isotope. Determine the following for the naturally occurring atoms of sodium:
a) The number of neutrons in the nucleus.
b) The mass (in u) of the nucleus (to three significant figures)

Solution:
a) Twelve. Since sodium has only one naturally occurring isotope, the mass number of that isotope must be the nearest integer to the atomic weight, 23. The atom of sodium has 11 protons so it must have (23 − 11 = 12) 12 neutrons.
b) 23.0 u. Look up the atomic mass value in the periodic table.

2.38   Calculate the atomic weight of gallium on the basis of the following percentage composition and atomic weights of the naturally occurring isotopes. Compare the calculated value with the atomic weight listed for gallium in the periodic table.
gallium-69 = 60.40% (68.9257 u), gallium-71 = 39.60% (70.9249 u)

Solution:
Atomic weight = 60.40% of 68.9257 u + 39.60% of 70.9249 u
= $(0.6040 \times 68.9257 \text{ u}) + (0.3960 \times 70.9249 \text{ u}) = 41.63 \text{ u} + 29.09 = 69.72 \text{ u}$
Actual weight from periodic table = 69.72 u

2.40   Calculate the atomic weight of copper on the basis of the following percentage composition and atomic weights of the naturally occurring isotopes. Compare the calculated value with the atomic weight listed for copper in the periodic table.
copper-63 = 69.09% (62.9298 u) copper-65 = 30.91% (64.9278 u)

Solution:
Atomic weight = 69.09% of 62.9298 u + 30.91% of 64.9278 u
= $(0.6909 \times 62.9298 \text{ u}) + (0.3091 \times 64.9278 \text{ u}) = 43.48 + 20.07 = 63.55 \text{ u}$
Actual weight from periodic table = 63.55 u

Avogadro's Number: The Mole (Section 2.6)
2.42   Refer to the periodic table and determine how many grams of fluorine contain the same number of atoms as 1.60 g of oxygen.

Solution:
If the ratio of the grams of F to the grams of O is the same as the ratio of the atomic weights, the mass of F atoms equals the mass of the O times the ratio of the atomic weights.
$F = 1.60 \text{ g O} \times \frac{19.00 \text{ u}}{16.00 \text{ u}} = 1.90 \text{ g F}$

2.44 Write three relationships (equalities) based on the mole concept for each of the following elements:

a) phosphorus   b) aluminum   c) krypton

Solution:

a) 1 mol P atoms $= 6.021 \times 10^{23}$ P atoms
   6.02×10^{23} P atoms $= 30.97$ g P
   1 mol P atoms $= 30.97$ g P

b) 1 mol Al atoms $= 6.021 \times 10^{23}$ Al atoms
   6.02×10^{23} Al atoms $= 26.98$ g Al
   1 mol Al atoms $= 26.98$ g Al

c) 1 mol Kr atoms $= 6.021 \times 10^{23}$ Kr atoms
   6.02×10^{23} Kr atoms $= 83.80$ g Kr
   1 mol Kr atoms $= 83.80$ g Kr

2.46 Use a factor derived from the relationships in Exercise 2.44 and the factor-unit method to determine the following:

a) the mass in grams of one phosphorus atom

b) the number of grams of aluminum in 1.65 mol of aluminum

c) the total mass in grams of one-half Avogadro's number of krypton atoms

Solution:

a) 1 P $\cancel{\text{atom}}$ $\times \dfrac{30.97 \text{ g P}}{6.02 \times 10^{23} \text{ P } \cancel{\text{atoms}}} = 5.145 \times 10^{-23}$ (round to 3 sig fig) $= 5.15 \times 10^{-23}$ g/atom

b) 1.65 $\cancel{\text{mol Al}}$ $\times \dfrac{26.98 \text{ g Al}}{1 \cancel{\text{ mol Al}}} = 44.517$ (round to 3 sig fig) $= 44.5$ g Al

c) $\frac{1}{4}(6.02 \times 10^{23}$ $\cancel{\text{atoms Kr}}) \times \dfrac{83.80 \text{ g Kr}}{6.02 \times 10^{23} \cancel{\text{ atoms Kr}}} = 20.95$ (round to 3 sig fig) $= 21.0$ g Kr

The Mole and Chemical Formulas (Section 2.7)

2.48 For each formula given below, write statements equivalent to Statements 1–6 (see Section 2.7):

a) ethyl ether ($C_4H_{10}O$)   b) fluoroacetic acid ($C_2H_3O_2F$)   c) aniline (C_6H_7N)

Solution:

a) 2 $C_4H_{10}O$ molecules contain 8 C atoms, 20 H atoms, and 2 O atoms
   10 $C_4H_{10}O$ molecules contain 40 C atoms, 100 H atoms, and 10 O atoms
   100 $C_4H_{10}O$ molecules contain 400 C atoms, 1000 H atoms, and 100 O atoms
   6.02×10^{23} $C_4H_{10}O$ molecules contain 24.08×10^{23} C atoms, 60.2×10^{23} H atoms, and 6.02×10^{23} O atoms
   1 mol $C_4H_{10}O$ molecules contains 4 mol C atoms, 10 mol H atoms, and 1 mol O atoms
   74.12 g $C_4H_{10}O$ molecules contain 48.04 g C, 10.08 g H, and 16.00 g O

b) 2 $C_2H_3O_2F$ molecules contain 4 C atoms, 6 H atoms, 4 O atoms, and 2 F atoms
   10 $C_2H_3O_2F$ molecules contain 20 C atoms, 30 H atoms, 20 O atoms, and 10 F atoms
   100 $C_2H_3O_2F$ molecules contain 200 C atoms, 300 H atoms, 200 O atoms, 100 F atoms
   6.02×10^{23} $C_2H_3O_2F$ molecules contain 12.04×10^{23} C atoms, 18.06×10^{23} H atoms, 12.04×10^{23} O atoms, and 6.02×10^{23} F atoms
   1 mol $C_2H_3O_2F$ molecules contains 2 mol C atoms, 3 mol H atoms, 2 mol O atoms, and 1 mol F atoms
   78.04 g $C_2H_3O_2F$ molecules contain 24.02 g C, 3.02 g H, 32.00 g O, and 19.00 g F

c) 2 C_6H_7N molecules contain 12 C atoms, 14 H atoms, and 2 N atoms
   10 C_6H_7N molecules contain 60 C atoms, 70 H atoms, and 10 N atoms
   100 C_6H_7N molecules contain 600 C atoms, 700 H atoms, and 100 N atoms

6.02×10^{23} C_6H_7N molecules contain 36.12×10^{23} C atoms, 42.14×10^{23} H atoms, and 6.02×10^{23} N atoms

1 mol C_6H_7N molecules contains 6 mol C atoms, 7 mol H atoms, and 1 mol N atoms

93.13 g C_6H_7N molecules contain 72.06 g C, 7.06 g H, and 14.01 g N

2.50   Answer the following questions based on information contained in the statements you wrote for Exercise 2.48.
a) How many moles of hydrogen atoms are contained in 0.50 mol of ethyl ether?
b) How many carbon atoms are contained in 0.25 mol of $C_2H_3O_2F$?
c) How many grams of hydrogen are contained in 2.00 mol of C_6H_7N?

Solution:

a) $0.50 \text{ mol } C_4H_{10}O \times \frac{10 \text{ mol H}}{1 \text{ mol } C_4H_{10}O} = 5.0 \text{ mol H}$

b) $0.20 \text{ mol } C_2H_3O_2 F \times \frac{2 \text{ mol C}}{1 \text{ mol } C_2H_3O_2 F} \times \frac{6.02 \times 10^{23} \text{ C atoms}}{1 \text{ mol C}} = 3.01 \times 10^{23} \text{ C atoms}$

c) $2.00 \text{ mol } C_6H_7N \times \frac{7 \text{ mol H}}{1 \text{ mol } C_6H_7N} \times \frac{1.008 \text{ g H}}{1 \text{ mol H}} = 14.1 \text{ g H}$

2.52   How many grams of C_2H_6O contain the same number of oxygen atoms as 0.75 mol of H_2O?

Solution:

One mole of C_2H_6O contains 1 mol oxygen atoms as does H_2O on the basis of the chemical formulas. Therefore, 0.75 mol C_2H_6O contains the same amount of oxygen as 0.75 mol H_2O.

$0.75 \text{ mol } C_2H_6O \times \frac{46.07 \text{ g } C_2H_6O}{1 \text{mol } C_2H_6O} = 34.5525 \text{ (round to 2 sig fig)} = 35 \text{ g } C_2H_6O$

2.54   Determine the mass percentage of oxygen in CO and CO_2.

Solution:

CO: 28.0 g CO contain 12.0 g C and 16.0 g O

$\% \text{ O} = \frac{16.0 \text{ g O}}{28.0 \text{ g CO}} \times 100 = 57.1\% \text{ O}$

CO_2:   44.0 g CO_2 contain 12.0 g C and 32.0 g O.

$\% \text{ O} = \frac{32.0 \text{ g O}}{44.0 \text{ g CO}_2} \times 100 = 72.7\% \text{ O}$

2.56   Any of the statements based on a mole of substance (Statements 4–6) can be used to obtain factors for problem solving by the factor-unit method. Write statements equivalent to 4, 5, and 6 for fructose ($C_6H_{12}O_6$). Use a single factor obtained from the statements to solve each of the following. A different factor will be needed in each case.
a) How many grams of oxygen are contained in 43.5 g of $C_6H_{12}O_6$?
b) How many moles of hydrogen atoms are contained in 1.50 mol of $C_6H_{12}O_6$?
c) How many atoms of carbon are contained in 7.50×10^{22} molecules of $C_6H_{12}O_6$?

Solution:

Statements:

180.16 g $C_6H_{12}O_6$ contain 72.06 g C, 12.096 g H, and 96.00 g O

6.02×10^{23} $C_6H_{12}O_6$ molecules contain 3.61×10^{24} C atoms, 7.22×10^{24} H atoms, and 3.61×10^{23} O atoms

1 mole $C_6H_{12}O_6$ molecules contains 6 mol C atoms, 12 mol H atoms, and 6 mol O atoms

a) $43.5 \text{ g } C_6H_{12}O_6 \times \frac{96.00 \text{ g O}}{180.16 \text{ g } C_6H_{12}O_6} = 23.1794 \text{ g O (round to 3 sig fig)} = 23.2 \text{ g O}$

b) $1.50 \text{ mol } C_6H_{12}O_6 \times \frac{12 \text{ mol H}}{1 \text{ mol } C_6H_{12}O_6} = 18.0 \text{ mol H}$

c) 7.50×10^{22} ~~$C_6H_{12}O_6$ molecules~~ $\times \frac{3.61 \times 20^{24} \text{ C atoms}}{6.02 \times 10^{23} \text{ } \overline{C_6H_{12}O_6} \text{ molecules}} = 4.4975 \times 10^{23}$ (round to 3 sig

fig) $= 4.50 \times 10^{23}$ C atoms

2.58   Two iron ores that have been used as sources of iron are magnetite (Fe_3O_4) and hematite (Fe_2O_3). Which one contains the higher mass percentage of iron?

Solution:

Magnetite: 231.55 g Fe_3O_4 contains 167.55 g Fe

% Fe $= \frac{167.55 \text{ g Fe}}{231.55 \text{ g Fe}_3O_4} \times 100 = 72.36$ %

Hematite: 159.7 g Fe_3O_4 contains 111.7 g Fe

% Fe $= \frac{111.7 \text{ g Fe}}{159.7 \text{ g Fe2O3}} \times 100 = 69.94$ % Fe

The magnetite (Fe_3O_4) has a higher % Fe by mass.

PROGRAMMED REVIEW

Section 2.1 Symbols and Formulas

Each element has been assigned a unique (a) _name_ and (b) _Symbol_. The (c) _symbols_ assigned to elements are based on the element names, and consist of a (d) _capital letter_ or a (e) _capital letter_ followed by a (f) _small letter_. In the (g) _formula_ for a compound, each atom in the compound molecule is represented by a (h) _symbol_ with a (i) _subscript_ to tell how many atoms of that kind are in the (j) _molecule_.

Section 2.2 Inside the Atom

Most of the mass of an atom is found in the (a) _nucleus_ which is made up of particles called (b) _protons_ and (c) _neutrons_. Of the three fundamental particles, (d) _neutrons_ have no charge and a mass of about (e) _one_ u. (f) _protons_ have a +1 charge and a mass of about (g) _1_ u, and (h) _electrons_ have a −1 charge and a mass of (i) _$\frac{1}{1836}$_ u.

Section 2.3 Isotopes

The number of (a) _protons_ in the nucleus of an atom is given by the atomic number. The (b) _mass number_ of an atom is the sum of the number of protons and neutrons in the nucleus. Atoms having the same atomic number but different mass numbers are called (c) _isotopes_.

Section 2.4 Relative Masses of Atoms and Molecules

The atomic weight of an element consisting of a single isotope is about equal to the (a) _mass number_ _average_ of the isotope. The atomic weight of an element consisting of a mixture of isotopes is the (b) _____ weight of the isotope.

Section 2.5 Isotopes and Atomic Weights

A sample of 20.18 g of neon (Ne) and a sample of (a) _79.9_ g of bromine contain the same number of atoms.

Section 2.6 Avogadro's Number: The Mole

The number of atoms in a sample of an element that weighs the same in grams as the atomic weight of the element is called a (a) _mole_, and is equal to (b) _6.02×10^{23}_. One mol of krypton atoms (Kr) weighs (c) _83-8_ grams.

Section 2.7 The Mole and Chemical Formulas

The number of moles of nitrogen atoms (N) in 1.5 mol of N_2O is (a) _three_. One mol of N_2O weighs (b) _44.02_ grams and contains (c) _28.02_ grams of nitrogen and (d) _16.00_ grams of oxygen.

SELF-TEST QUESTIONS

Multiple Choice

1. Which of the following is an incorrect symbol for an element?
   a) Ce   b) Au   c) K   d) CR

2. Which of the following is an incorrect formula for a compound?
   a) CO_2 b) CO_1 c) N_2O d) NO_2

3. Two objects have masses of 3.2 g and 0.80 g. What is the relative mass of the 3.2 g object compared to the other?
   a) 4.0 to 1   b) 2.0 to 1   c) 0.50 to 1   d) 0.25 to 1

4. Suppose the atomic weights of the elements were assigned in such a way that the atomic weight of helium, He, was 1.00 u. What would be the atomic weight of oxygen, O, in u on this scale?
   a) 16.0   b) 8.00   c) 4.00   d) 0.250

5. What is the molecular weight of phosphoric acid, H_3PO_4, in u?
   a) 48.0   b) 50.0   c) 96.0   d) 98.0

6. How many neutrons are there in the nucleus of a potassium-39 atom?
   a) 1   b) 19   c) 20   d) 39

7. What is the mass in grams of 1.00 mole of chlorine molecules, Cl_2?
   a) 6.02×10^{23} b) 70.9   c) 35.5   d) 1.18×10^{-22}

8. Calculate the weight percent of sulfur, S, in SO_2.
   a) 50.1   b) 33.3   c) 66.7   d) 25.0

Matching

Match the molecules represented on the left with the terms on the right to the correct classification given.

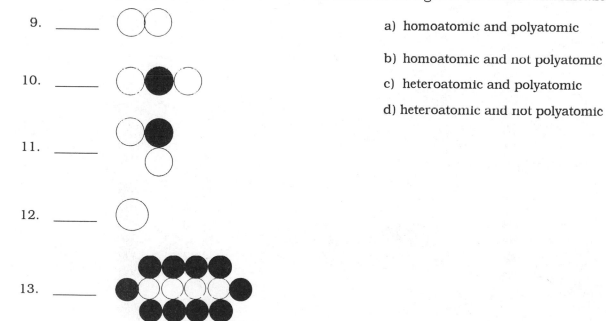

9. _____

10. _____

11. _____

12. _____

13. _____

a) homoatomic and polyatomic

b) homoatomic and not polyatomic

c) heteroatomic and polyatomic

d) heteroatomic and not polyatomic

Match the numbers given as responses to the following:

14. The number of moles of oxygen atoms in 2 moles of NO_2.   a) 1

15. The number of moles of NH_3 that contain 3 moles of nitrogen atoms.   b) 2

16. The number of moles of nitrogen atoms in one-half mole of N_2O_5.   c) 3

17. The number of moles of electrons in one mole of helium atoms.   d) 4

18. The number of moles of neutrons in one mole of $^{3}_{1}H$.

True-False

19. In some instances, two different elements are represented by the same symbol.

20. The mass of a single atom of silicon, Si, is 28.1 g.

21. All isotopes of a specific element have the same atomic number.

22. One mole of water molecules, H_2O, contain two moles of hydrogen atoms, H.

23. 1.00 mol of sulfur, S, contains the same number of atoms as 14.0 grams of nitrogen, N.

SOLUTIONS

A. Answers to Programmed Review

2.1	a) name	b) symbol	c) symbols	d) capital letter

2.1 a) name   b) symbol   c) symbols   d) capital letter
     e) capital letter f) small letter   g) formula   h) symbol
     i) subscript   j) molecule

2.2 a) nucleus   b) protons   c) neutrons   d) neutrons
     e) one   f) protons   g) one   h) electrons
     i) 1/1836

2.3 a) protons   b) mass number   c) isotopes

2.4 a) mass number   b) average

2.5 a) 79.90

2.6 a) mole   b) 6.02×10^{23} c) 83.80

2.7 a) three   b) 44.02   c) 28.02   d) 16.00

B. Answers to Self-Test Questions

1. d	9. a	17. d
2. b	10. c	18. b
3. a	11. c	19. F
4. c	12. b	20. F
5. d	13. c	21. T
6. c	14. d	22. T
7. b	15. c	23. T
8. a	16. a	

Chapter 3

Electronic Structure and the Periodic Law

CHAPTER OUTLINE

3.1 The Periodic Law and Table
3.2 Electronic Arrangements in Atoms
3.3 The Shell Model and Chemical Properties

3.4 Electronic Configurations
3.5 Another Look at the Periodic Table
3.6 Property Trends Within the Periodic Table

LEARNING OBJECTIVES

When you have completed your study of this chapter, you should be able to:
1. Locate elements in the periodic table on the basis of group and periodic designations.
2. Determine numerical relationships such as the number of electrons in designated shells and the number of subshells in designated shells.
3. Determine the number of electrons in the valence shell of elements.
4. Relate electronic structure to the location of elements in the periodic table.
5. Write electronic configurations for elements.
6. Determine the number of unpaired electrons for elements.
7. Identify elements from their electronic configurations.
8. Determine the shell and subshell location of the distinguishing electron in elements.
9. Classify elements into correct categories from the following: representative element, transition element, inner-transition element, noble gas, metal, nonmetal, and metalloid.
10. Recognize property trends within the periodic table.
11. Use trends to predict selected properties of elements.

ANSWERS AND SOLUTIONS TO EVEN-NUMBERED PROBLEMS

The Periodic Law and Table (Section 3.1)
3.2 Identify the group and period to which each of the following elements belong:
   a) element 27   b) Pb   c) arsenic   d) Ba

Solution:
a) group VIIB (9), period 4   b) group IVA (14), period 6
c) group VA (15), period 4   d) group IIA (2), period 6

3.4   Write the symbol and name for the elements located in the periodic table as follows:
a) The noble gas belonging to period 4
b) The fourth element (reading down) in group IVA (14)
c) Belongs to group VIB and period 5
d) The sixth element (reading left to right) in period 6

Solution:
a) Kr, krypton   b) Sn, tin   c) Mo, molybdenum   d) W, tungsten

3.6   Answer the following:
a) How many elements are located in group VIIA (7) of the periodic table?
b) How many total elements are found in periods 2 and 3 of the periodic table?
c) How many elements are found in period 6 of the periodic table?

Solution:
a) 5 elements   b) $8 + 8 = 16$ c) $18 + 14 = 32$

3.8   The following statements either define or are closely related to the terms *periodic law, period,* or *group.* Match the terms to the appropriate statements.
a) this is a horizontal arrangement of elements in the periodic table
b) element 11 begins this arrangement in the periodic table
c) the element nitrogen is the first member of this arrangement
d) elements 9, 17, 35, and 53 belong to this arrangement.

Solution:
a) period   b) period   c) group   d) group

Electronic Arrangements in Atoms (Section 3.2)
3.10   What particles in the nucleus cause the nucleus to have a positive charge?

Solution:
The protons in the nucleus have a positive charge.

3.12   What is the maximum number of electrons that can be contained in each of the following?
a) a $2p$ orbital   b) a $2p$ subshell   c) the second shell

Solution:
a) 2 Each orbital can contain 2 electrons.
b) 6 The $2p$ subshell has 3 orbitals. Each orbital can contain 2 electrons.
c) 8 The second shell has two subshells. The $2s$ subshell has 1 orbital, the $2p$ subshell has 3 orbitals. Each orbital can contain 2 electrons.

3.14   How many orbitals are found in the fourth shell? Write the designations for the orbitals.

Solution:
The maximum number of orbitals in the fourth shell is 16; the designations are one $4s$, three $4p$, five $4d$, and seven $4f$.

3.16   How many orbitals are found in a $4f$ subshell? What is the maximum number of electrons that could be located in this subshell?

Solution:
There are seven $4f$ orbitals. Each orbital can hold up to 2 electrons for a maximum of 14 electrons in the $4f$ subshell.

The Shell Model and Chemical Properties (Section 3.3)

3.18   Look at the periodic table and tell how many electrons are in the valence shell of the following elements:
a) element number 54
b) the first element (reading down) in group VA (15)
c) Sn
d) the fourth element (reading left to right) in period 3

Solution:
The number of electrons in the A groups is the same as the group number.
a) 8   b) 5   c) 4   d) 4

3.20   What period 6 element has chemical properties most like sodium? How many valence shell electrons does this element have? How many valence shell electrons does sodium have?

Solution:
Cesium is the period 6 element with 1 valence shell electron. Sodium also has 1. Cesium will have chemical properties similar to sodium.

3.22   If you discovered an ore deposit containing copper, what other two elements might you also expect to find in the ore? Explain your reasoning completely.

Solution:
The other two elements in group IB, silver and gold, would have similar chemical properties, and should be found with copper.

Electronic Configurations (Section 3.4)

3.24   Write the electronic configuration for each of the following elements, using the form $1s^2 2s^2 2p^6$, and so on. Indicate how many electrons are unpaired in each case.
a) element number 37   b) Si   c) titanium   d) Ar

Solution:
a) $1s^2 2s^2 2p^6 3s^2 3p^6 4s^2 3d^{10} 4p^6 5s^1$ with one unpaired electron
b) $1s^2 2s^2 2p^6 3s^2 3p^2$ with 2 unpaired electrons
c) $1s^2 2s^2 2p^6 3s^2 3p^6 4s^2 3d^2$ with 2 unpaired electrons
d) $1s^2 2s^2 2p^6 3s^2 3p^6$ with no unpaired electrons

3.26   Write electronic configurations and answer the following.
a) How many total s electrons are found in magnesium?
b) How many unpaired electrons are in nitrogen?
c) How many subshells are completely filled in Al?

Solution:
a) $1s^2 2s^2 2p^6 3s^2$ There are a total of six electrons in s orbitals; two in each of the first, second, and third shells.
b) $1s^2 2s^2 2p^3$ There are three unpaired electrons, the last three.
c) $1s^2 2s^2 2p^6 3s^2 3p^1$ There are four subshells completely filled and one partially filled.

3.28 Write the symbol and name for each of the elements described. More than one element will fit some descriptions.
a) contains only two $2p$ electrons
b) contains an unpaired $3s$ electrons
c) contains two unpaired $3p$ electrons
d) contains three $4d$ electrons
e) contains three unpaired $3d$ electrons

Solution:
a) C, carbon
b) Na, sodium
c) Si, silicon or S, sulfur
d) Nb, niobium
e) V, vanadium or Co, cobalt

3.30 Write the abbreviated electron configurations for the following:
a) selenium
b) element number 23
c) Ca
d) carbon

Solution:
a) [Ar] $4s^2 3d^{10} 4p^4$
b) [Ar] $4s^2 3d^3$
c) [Ar] $4s^2$
d) [He] $2s^2 2p^2$

3.32 Refer to the periodic table and write abbreviated, electronic configurations for all elements in which the noble gas symbol used will be [Ne].

Solution:
Na: [Ne] $3s^1$ Mg: [Ne] $3s^2$ Al: [Ne] $3s^2 3p^1$ Si: [Ne] $3s^2 3p^2$
P: [Ne] $3s^2 3p^3$ S: [Ne] $3s^2 3p^4$ Cl: [Ne] $3s^2 3p^5$ Ar: [Ne] $3s^2 3p^6$

Another Look at the Periodic Table (Section 3.5)
3.34 Classify each of the following elements into the s, p, d, or f area of the periodic table on the basis of the distinguishing electron.
a) nickel
b) Rb
c) element 51
d) Cm

Solution:
a) d
b) s
c) p
d) f

3.36 Classify the following elements as *representative, transition, inner-transition,* or *noble gases.*
a) iron
b) element 15
c) U
d) xenon
e) tin

Solution:
a) transition
b) representative
c) inner transition
d) noble gas
e) representative

3.38 Classify the following as *metals, nonmetals, metalloids,* or *noble gases.*
a) element 51
b) iodine
c) Al
d) radon
e) Pt

Solution:
a) metalloid   b) nonmetal   c) metal   d) nonmetal
e) metal

Property Trends Within the Periodic Table (Section 3.6)

3.40   Use trends within the periodic table to predict which member of each of the following pairs is more metallic.
a) Na or Mg   b) Pb or Ge   c) Mg or Ba   d) Cs or Li

Solution:
Elements to the left in a row or down in a column are more metallic.
a) Na   b) Pb   c) Ba   d) Cs

3.42   Use trends within the periodic table and indicate which member of each of the following pairs has the larger atomic radius.
a) Ga or Se   b) N or Sb   c) O or C   d) Te or S

Solution:
a) Ga   b) Sb   c) C   d) Te

3.44   Use trends within the periodic table to indicate which member of each of the following pairs gives up one electron more easily.
a) Li or K   b) C or Sn   c) Mg or S   d) Li or N

Solution:
For representative elements, the ionization energy decreases from top to bottom in a column and decreases from right to left in a row. The lower ionization energy is easier to give up an electron.
a) K   b) Sn   c) Mg   d) Li

PROGRAMMED REVIEW

Section 3.1 The Periodic Law and Table

According to the (a) _periodic law_ , the properties of elements arranged by increasing atomic numbers repeat at regular intervals. In a modern (b) _periodic table_ , elements with similar properties are found in vertical columns called (c) _groups_ or (d) _families_ The (e) _horizontal_ rows in the periodic table are called periods.

Section 3.2 Electronic Arrangements in Atoms

Neils Bohr improved our understanding of atomic structure by suggesting a modification to the (a) _solar system_ model proposed by Ernest Rutherford. Bohr suggested that (b) _electrons_ could occupy only (c) _orbits_ located (d) _specific distance_ from the nucleus. He also theorized that electrons changed (e) _orbits_ only by absorbing or releasing energy. However, additional research suggested that electrons did not follow specific paths around the nucleus, but instead moved in specific (f) _volumes_ of space called (g) _atomic orbitals_ Atomic orbitals occur in groups called (h) _subshells_ which are designated by the same number and letter used to designate the (i) _orbitals_ within the subshell. Subshells occur in groups called (j) _shells_ .

Section 3.3 The Shell Model and Elemental Properties

Similar elemental properties result when elements have identical numbers of electrons in the (a) _valence shells_ of their atoms. Elements in groups IIA(2), VA(15) and VIIA(17) have respectively the following numbers of electrons in their valence shells: (b) __2__ , __5__ , and __7__ . With the exception of helium, all noble gases have (c) __8__ electrons in their valence shells.

Section 3.4 Electronic Configurations

The detailed arrangements of electrons in atoms are called (a) _electronic configerations_. As electrons are added to atoms, they will occupy the subshell of (b) _lowest energy_ that is available. According to (c) _Hund's_ rule, electrons will not pair up in orbitals as long as empty orbitals of the same energy are available. Two electrons that occupy the same orbital must be spinning in opposite (d) _directions_, in compliance with the (e) _Pauli exclusion principal_ An electronic configuration that ends with a completely filled p subshell is called a (f) _nobel gas configerati_ Noble gas configurations can be used to write shortened electronic configurations by letting the noble gas symbol enclosed in brackets represent the (g) _electrons_ found in the noble gas configuration. In an (h) _electron dot formula_ or Lewis structure for an atom or ion, the valence electrons are represented by dots arranged around the elemental symbol.

Section 3.5 Another Look at the Periodic Table

The last or highest energy electron in an atom is called the (a) _distinguishing electron_ Elements found in the s and p areas of the periodic table are called (b) _representative electrons_ The (c) _d_ area of the periodic table contains (d) _transition elements._ Elements classified as (e) _metals_ are found in the left two-thirds of the periodic table and in the f area. Elements classified as (f) _metalloids_ form a diagonal separation zone between metals and nonmetals.

Section 3.6 Property Trends in the Periodic Table

For the representative elements, the size (radius) of the atom increases from (a) _top_ to (b) _bottom_ in any group, and decreases from (c) _left_ to (d) _right_ in any period. For the representative elements, ionization energy of the atom increases from (e) _bottom_ to (f) _top_ in any group, and decreases from (g) _right_ to (h) _left._ in any period.

SELF-TEST QUESTIONS

Multiple Choice

1. Elements in the same group
    a) have similar chemical properties
    c) have consecutive atomic numbers
    b) are called isotopes
    d) constitute a period of elements

2. Which element with the following atomic numbers should have properties similar to those of oxygen (element number 8)?
    a) 15   b) 4   c) 2   d) 34

3. Which of the following elements is found in period 3 of the periodic table?
    a) Al
    c) Ga
    b) B
    d) more than one response is correct

4. The electronic configuration for an element containing 15 protons would be
    a) $1s^2 2s^2 2p^6 3s^2 3p^6$ b) $1s^2 2s^2 2p^6 3s^2 4p^3$ c) $1s^2 2s^2 2p^6 3s^2 3p^6 4s^2$ d) $1s^2 2s^2 2p^6 3s^2 3p^3$

5. Which of the following is a true statement for an electronic configuration of $1s^2 2s^2 2p^6 3s^2 3p^6$?
   a) there are 6 electrons in the $3p$ orbital
   b) there are 6 electrons in the $3p$ subshell
   c) there are 6 electrons in the $3p$ shell
   d) more than one response is correct

6. The maximum number of electrons which may occupy a $4d$ orbital is
   a) 10   b) 4   c) 2   d) 8

7. How many unpaired electrons are found in titanium (element number 22)?
   a) 1   b) 2   c) 3   d) 4

8. Which element has the electronic configuration $1s^2 2s^2 2p^6 3s^2 3p^6$?
   a) Ne   b) Ar   c) K   d) Kr

9. The element with unpaired electrons in a d subshell is element number
   a) 33   b) 27   c) 20   d) 53

10. Which of the following elements has an electronic configuration ending in $4d^7$?
    a) Co
    b) Rh
    c) Ir
    d) more than one response is correct

11. How many electrons are found in the valence shell of oxygen (element number 8)?
    a) 4   b) 3   c) 5   d) 6

12. The distinguishing electronic configuration of np^4 is characteristic of which group in the periodic table?
    a) IIA (2)   b) IVA (14)   c) VIA (16)   d) noble gases

13. The element with the electronic configuration $1s^2 2s^2 2p^3$ will be found in
    a) period 1, group VIA (16)
    b) period 2, group VA (15)
    c) period VA (15), group 2
    d) period IIIA (13), group VA (15)

14. How many electrons would be contained in the valence shell of an atom in group VIA (16)?
    a) 6   b) 2   c) 4   d) 16

15. Which of the following atoms has the largest radius?
    a) Sr   b) Ge   c) Br   d) Rb

16. Which of the following atoms has the largest radius?
    a) Si   b) C   c) Sn   d) Ge

17. Which of the following atoms would have the lowest ionization potential?
    a) Ca   b) Na   c) K   d) Rb

True-False

18. Elements 20 and 21 are in the same period of the periodic table.

19. Hund's rule states that electrons within a subshell remain unpaired if possible.

20. A $2p$ and a $3p$ subshell contain the same number of orbitals.

21. The maximum number of electrons an orbital may contain does not vary with the type of orbital.

22. The distinguishing electron in Br is found in a p orbital.

Matching

From the list on the right, choose a response that is consistent with the description on the left as far as electronic configurations are concerned. You may use a response more than once.

23. _____ a noble gas

24. _____ Mg

25. _____ a transition element

26. _____ an element just completing the filling of the shell where n = 2

a) $1s^2 2s^2 2p^6 3s^2$

b) $1s^2 2s^2 2p^6$

c) $1s^2 2s^2 2p^6 3s^2 3p^6 4s^2 3d^3$

d) $1s^2 2s^2 2p^4$

Match each of the categories on the right with an element on the left.

27. _____ neon (Ne)

28. _____ phosphorus (P)

29. _____ calcium (Ca)

30. _____ element number 47

31. _____ element number 82

32. _____ top element of group VA (15)

a) a representative metal

b) a noble gas nonmetal

c) a nonmetal but not a noble gas

d) a transition

SOLUTIONS

A. Answers to Programmed Review

3.1 a) periodic law   b) periodic table   c) groups   d) families
     e) horizontal

3.2 a) solar system   b) electrons   c) orbits   d) specific distances
     e) orbits   f) volumes   g) atomic orbitals   h) subshells
     i) orbitals   j) shells

3.3 a) valence shells   b) two, five and seven   c) eight

3.4 a) electronic configurations   b) lowest energy
     c) Hund's   d) directions
     e) Pauli exclusion principle   f) noble gas configuration
     g) electrons   h) electron dot formula

3.5 a) distinguishing electron
     b) representative elements
     c) d
     d) transition elements
     e) metals
     f) metalloids

3.6 a) top   b) bottom   c) left   d) right
     e) bottom   f) top   g) right   h) left

B. Answers to Self-Test Questions

1. a	12. c	23. b
2. d	13. b	24. a
3. a	14. a	25. c
4. d	15. d	26. b
5. b	16. c	27. b
6. c	17. d	28. c
7. b	18. T	29. a
8. b	19. T	30. d
9. b	20. T	31. a
10. b	21. T	32. c
11. d	22. T	

Chapter 4

Forces Between Particles

CHAPTER OUTLINE

LEARNING OBJECTIVES

When you have completed your study of this chapter, you should be able to:

1. Use electronic configuration to predict the gain or loss of electrons by atoms as they achieve noble gas configurations.
2. Use the octet rule to predict the ions formed during the formation of ionic compounds.
3. Write formulas for the following types of ionic compounds:
   a. Binary, containing representative elements.
   b. Those containing representative metals and polyatomic ions.
4. Name binary ionic and covalent compounds and compounds containing representative metals and polyatomic ions.
5. Determine formula weights for ionic compounds.
6. Draw Lewis structures for molecules and polyatomic ions and apply VSEPR theory to predict molecular and ionic shapes.
7. Use electronegativities to determine the type of bonding that is likely to occur between pairs of representative elements.
8. Represent simple covalent molecules and polyatomic ions by Lewis structures and related formulas.
9. Determine whether a given covalent molecule is polar.

10. Relate melting points and boiling points of pure substances to the strengths and types of interparticle forces present.

ANSWERS AND SOLUTIONS TO EVEN-NUMBERED PROBLEMS

Noble Gas Configurations (Section 4.1)

4.2   Refer to the group numbers of the periodic table and draw Lewis structures for atoms of the following:
a) arsenic   b) silicon   c) lead   d) barium

Solution:
a) Arsenic is in group VA (15). The Lewis structure is:

$$\bullet \overset{\bullet\bullet}{\underset{\bullet}{\text{As}}} \bullet$$

b) Silicon is in group IVA (14). The Lewis structure is:

$$\bullet \overset{\bullet}{\underset{\bullet}{\text{Si}}} \bullet$$

c) Lead is in group IVA (14). Two possible Lewis structures are:

$$\overset{\bullet\bullet}{\underset{\bullet}{\text{Pb}}} \bullet \quad \text{and} \quad \bullet \overset{\bullet}{\underset{\bullet}{\text{Pb}}} \bullet$$

d) Barium is in group IIA (2). The Lewis structure is:

$$\overset{\bullet}{\text{Ba}} \bullet$$

4.4   Write abbreviated electronic configurations for the following:
a) element number 50   b) Se   c) cesium   d) iodine

Solution:
a) [Kr] $4d^{10}5s^25p^2$ b) [Ar] $3d^{10}4s^24p^4$ c) [Xe] $6s^1$ d) [Kr] $4d^{10}5s^25p^5$

4.6   Draw Lewis structures for the elements given in Exercise 4.4.

Solution:
a) $\overset{\bullet\bullet}{\underset{\bullet}{\text{Sn}}}\bullet$ b) $\bullet\,\overset{\bullet\bullet}{\underset{\bullet}{\text{Se}}}\,\colon$ c) Cs$\bullet$ d) $\bullet\,\overset{\bullet\bullet}{\underset{\bullet}{\text{I}}}\,\colon$

4.8   Use the symbol E to represent an element in a general way and draw Lewis structures for atoms of the following:
a) any group IIIA (13) element   b) any group VIA (16) element

Solution:
a) $\overset{\bullet\bullet}{\text{E}} \bullet$ The hybridized structure is: $\overset{\bullet}{\underset{\bullet}{\text{E}}} \bullet$

b) $\bullet\,\overset{\bullet\bullet}{\underset{\bullet}{\text{E}}}\,\colon$

Ionic Bonding (Section 4.2)

4.10   Indicate both the minimum number of electrons that would have to be added and the minimum number that would have to be removed to change the electronic configuration of each element listed in Exercise 4.4 to a noble gas configuration.

Solution:
a) add 14 or lose 14   b) add 2 or lose 16   c) add 31 or lose 1   d) add 1 or lose 17

4.12   Use the periodic table and predict the number of electrons that will be lost or gained by the following elements as they change into simple ions. Write an equation using elemental symbols, ionic symbols, and electrons to represent each change.
a) Cs   b) oxygen   c) element number 7   d) iodine

Solution:
a) lose 1; $Cs \rightarrow Cs^+ + e^-$ b) gain 2; $O + 2e^- \rightarrow O^{2-}$
c) gain 3; $N + 3e^- \rightarrow N^{3-}$ d) gain 1; $I + e^- \rightarrow I^-$

4.14   Write a symbol for each of the following ions:
a) A selenium atom that has gained two electrons
b) A rubidium atom that has lost one electron
c) An aluminum atom that has lost three electrons

Solution:
A gain of electrons produces a negative ion. Conversely, a loss of electrons produces a positive ion.
a) Se^{2-} b) Rb^+ c) Al^{3+}

4.16   Identify the element in period 3 that would form each of the following ions. E is used as a general symbol for an element.
a) E^{2-} b) E^{3+} c) E^+ d) E^-

Solution:
a) sulfur, making S^{2-} b) aluminum, making Al^{3+}
c) sodium, making Na^+ d) chlorine, making Cl^-

4.18   Identify the noble gas that is isoelectronic with each of the following ions:
a) Li^+ b) I^- c) S^{2-} d) Sr^{2+}

Solution:
a) An electron has been lost from the atom giving the configuration of helium, He.
b) An electron has been gained by the atom giving the configuration of xenon, Xe.
c) Two electrons have been gained by the atom giving the configuration of argon, Ar.
d) The atom has lost two electrons giving the configuration of krypton, Kv.

Ionic Compounds (Section 4.3)

4.20   Write equations to represent positive and negative ion formation for the following pairs of elements. Then write a formula for the ionic compound that results when the ions combine.
a) Ba and F   b) potassium and bromine   c) elements number 13 and 35

Solution:

a) Barium, Group IIA, will lose $2e^-$ $Ba \rightarrow Ba^{2+} + 2e^-$
    Fluorine, Group VIIA, will gain $1e^-$ $F + e^- \rightarrow F^-$
    BaF_2 is the formula.

b) Potassium, Group IA, will lose $1e^-$ $K \rightarrow K^+ + e^-$
    Bromine, Group VIIA will gain $1e^-$ $Br + e^- \rightarrow Br^-$
    KBr is the formula.

c) Element 13, aluminum, Group IIIA, will lose $3e^-$ $Al \rightarrow Al^{3+} + 3e^-$
    Element 35, bromine, Group VIIA will gain $1e^-$ $Br + e^- \rightarrow Br^-$
    $AlBr_3$ is the formula.

4.22   Write the formula for the ionic compound formed from Ba and each of the following ions:

a) Te^{2-} b) N^{3-} c) F^- d) P^{3-}

Solution:

The charge of the barium ion in each compound is 2+.

a) $BaTe$ b) Ba_3N_2 c) BaF_2 d) Ba_3P_2

4.24   Classify each of the following as a binary compound or not a binary compound:

a) PbO_2 b) $CuCl_2$ c) KNO_3 d) Be_3N_2
e) $CaCO_3$

Solution:

Binary compounds are composed of only two elements; the ratio of the elements is not used in this classification.

a) binary   b) binary   c) not binary   d) binary
e) not binary

Naming Binary Ionic Compounds (Section 4.4)

4.26   Name the following metal ions:

a) Li^+ b) Mg^{2+} c) Ba^{2+} d) Cs^+

Solution:

a) lithium ion   b) magnesium ion   c) barium ion   d) cesium ion

4.28   Name the following nonmetal ions:

a) Br^- b) O^{2-} c) P^{3-} d) Te^{2-}

Solution:

a) bromide ion   b) oxide ion   c) phosphide ion   d) telluride ion

4.30   Name the following binary ionic compounds:

a) MgO b) CaS c) ZnO d) $AlCl_3$
e) Na_3N

Solution:

a) magnesium oxide   b) calcium sulfide   c) zinc oxide   d) aluminum chloride
e) sodium nitride

4.32   Name the following binary ionic compounds using a roman numeral to indicate the charge on the metal ion:

    a) SnS and SnS_2 b) $FeCl_2$ and $FeCl_3$ c) Cu_2O and CuO d) $AuCl$ and $AuCl_3$

Solution:

    a) tin II sulfide and tin IV sulfide   b) iron II chloride and iron III chloride

    c) copper I oxide and copper II oxide   d) gold I chloride and gold III chloride

4.34   Name the binary compounds of Exercise 4.32 by adding the endings *-ous* and *-ic* to indicate the lower and higher ionic charges of the metal ion in each pair of compounds. The non-English root for gold (Au) is *aur-*, and that of tin (Sn) is *stann-*.

Solution:

    a) stannous sulfide and stannic sulfide   b) ferrous chloride and ferric chloride

    c) cuprous oxide and cupric oxide   d) aurous chloride and auric chloride

4.36   Write formulas for the following binary ionic compounds:

    a) mercury(I) oxide   b) lead(II) oxide   c) platinum(IV) iodide   d) copper(I) nitride

    e) cobalt(II) sulfide

Solution:

    a) Hg_2O b) PbO c) PtI_4 d) Cu_3N

    e) CoS

The Smallest Unit of Ionic Compounds (Section 4.5)

4.38   Determine the formula weight in atomic mass units for each of the following binary ionic compounds:

    a) KF b) Be_3N_2 c) Li_3P d) Cu_2O

Solution:

    a) $FW = 39.10 + 19.00 = 58.10\,u$

    b) $FW = 3(9.012) + 2(14.01) = 55.056 = 55.06\,u$

    c) $FW = 3(6.941) + 30.97 = 51.793 = 51.79\,u$

    d) $FW = 2(63.55) + 16.00 = 143.10\,u$

4.40   Identify the ions that would occupy lattice sites in a solid sample of each compound given in Exercise 4.38.

Solution:

    a) one K^+ ion and one F^- ion

    b) three Be^{2+} ions and two N^{3-} ions

    c) three Li^+ ions and one P^{3-} ion

    d) two Cu^+ ions and one O^{2-} ion

4.42   Calculate the mass in grams of positive ions and negative ions contained in 1 mol of each compound given in Exercise 4.38.

Solution:

    a) One mol KF contains 1 mol K^+ ions and 1 mol F^- ions. The mass of the positive or negative ions is the same as the mass of the original atoms.

$$1 \text{ mol } K^+ \times \frac{39.10 \text{ g } K^+}{1 \text{ mol } K^+} = 39.10 \text{ g } K^+$$

$$1 \text{ mol } F^- \times \frac{19.00 \text{ g } F^-}{1 \text{ mol } F^-} = 19.00 \text{ g } F^-$$

b) One mol Be_3N_2 contains 3 mol Be^{2+} ions and 2 mol N^{3-} ions.

$$3 \text{ mol } Be^{2+} \times \tfrac{9.012 \text{ g } Be^{2+}}{1 \text{ mol } Be^{2+}} = 27.04 \text{ g } Be^{2+}$$

$$2 \text{ mol } N^{3-} \times \tfrac{14.01 \text{ g } N^{3-}}{1 \text{ mol } N^{3-}} = 28.02 \text{ g } N^{3-}$$

c) One mol Li_3P contains 3 mol Li^+ ions and 1 mol P^{3-} ions.

$$3 \text{ mol } Li^+ \times \tfrac{6.941 \text{ g } Li^+}{1 \text{ mol } Li^+} = 20.82 \text{ g } Li^+$$

$$1 \text{ mol } P^{3-} \times \tfrac{30.97 \text{ g } P^{3-}}{1 \text{ mol } P^{3-}} = 30.97 \text{ g } P^{3-}$$

d) One mol Cu_2O contains 2 mol Cu^+ ions and 1 mol O^{2-} ions.

$$2 \text{ mol } Cu^+ \times \tfrac{63.55 \text{ g } Cu^+}{1 \text{ mol } Cu^+} = 127.1 \text{ g } Cu^+$$

$$1 \text{ mol } O^{2-} \times \tfrac{16.00 \text{ g } O^{2-}}{1 \text{ mol } O^{2-}} = 16.00 \text{ g } O^{2-}$$

4.44   Calculate the number of positive ions and negative ions contained in 1.00 mol of each compound given in Exercise 4.38.

Solution:
Using the results of Exercise 4.40 and knowing that 1 mol $= 6.02 \times 10^{23}$ objects,
a) 6.02×10^{23} K^+ ions and 6.02×10^{23} F^- ions
b) 18.1×10^{23} Be^{2+} ions and 12.0×10^{23} N^{3-} ions
c) 18.1×10^{23} Li^+ ions and 6.02×10^{23} P^{3-} ions
d) 12.0×10^{23} Cu^+ ions and 6.02×10^{23} O^{2-} ions
Note: Some of the numbers are not shown using correct scientific notation in order to see the relationships more clearly.

Covalent Bonding (Section 4.6)

4.46   Represent the following reaction using Lewis structures:
$8S \rightarrow S_8$

Solution:
Let each S atom share one electron with each adjacent S atom in the ring. The non-bonding electrons are not shown on the S atoms in the ring.

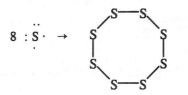

4.48   Represent the following molecules by Lewis structures:
a) H_2S (each H atom is bonded to the S atom)
b) ClF
c) HBr
d) HClO (the H and Cl are each bonded to the O)

Solution:
a) H : S̈ : H or H—S—H

b) : C̈l : F̈ : or Cl—F

c) H : $\overset{\cdot\cdot}{\underset{\cdot}{Br}}$: or H—Br

d) : $\overset{\cdot\cdot}{\underset{\cdot\cdot}{Cl}}$: $\overset{\cdot\cdot}{\underset{\cdot\cdot}{O}}$: H or Cl—O—H

Polyatomic Ions (Section 4.7)

4.50   Draw Lewis structures for the following polyatomic ions:
a) PH_4^+ (each H atom is bonded to the P atom)
b) HPO_4^{2-} (each O atom is bonded to the P atom and the H atom is bonded to an O atom)
c) HSO_4^- (each O atom is bonded to the S atom and the H atom is bonded to an O atom)

Solution:

a) 8 electrons $\left[H : \overset{H}{\underset{H}{P}} : H \right]^+$

b) 32 electrons: $\left[: \overset{\cdot\cdot}{\underset{\cdot\cdot}{O}} : \overset{:\overset{\cdot\cdot}{O}:}{\underset{:\overset{\cdot\cdot}{O}:}{P}} : \overset{\cdot\cdot}{\underset{\cdot\cdot}{O}} : H \right]^{2-}$

c) 32 electrons: $\left[: \overset{\cdot\cdot}{\underset{\cdot\cdot}{O}} : \overset{:\overset{\cdot\cdot}{O}:}{\underset{:\overset{\cdot\cdot}{O}:}{S}} : \overset{\cdot\cdot}{\underset{\cdot\cdot}{O}} : H \right]^{-}$

Shapes of Molecules and Polyatomic Ions (Section 4.8)

4.52   Predict the shape of each of the following molecules by first drawing a Lewis structure, then applying the VSEPR theory:
a) H_2S (each H atom is bonded to the S atom)
b) PCl_3 (each Cl atom is bonded to the P atom)
c) OF_2 (each F atom is bonded to the O atom)
d) SnF_4 (each F atom is bonded to the Sn atom)

Solution:

a) H — $\overset{\cdot\cdot}{S}$ — H
There are four pairs of electrons around the central S atom. They will be arranged tetrahedrally, with the H–S–H atoms forming an angular molecule.

b) Cl—$\overset{\cdot}{P}$—Cl
      |
     Cl
There are four pairs of electrons tetrahedrally arranged around the central P atom. The atoms form a triangular pyramid shaped molecule.

c) : $\overset{\cdot\cdot}{F}$—$\overset{\cdot\cdot}{O}$—$\overset{\cdot\cdot}{F}$:
There are four pairs of electrons tetrahedrally arranged around the central O atom. The F—O—F atoms form an angular molecule.

d) : $\overset{\cdot\cdot}{F}$ — $\underset{\overset{|}{:\overset{\cdot\cdot}{F}:}}{\overset{:\overset{\cdot\cdot}{F}:}{\overset{|}{Sn}}}$ — $\overset{\cdot\cdot}{F}$:
There are four pairs of electrons tetrahedrally arranged around the Sn. The molecule is tetrahedral.

4.54   Predict the shape of each of the following polyatomic ions by first drawing a Lewis structure, then applying the VSEPR theory:

a) NO_2^- (each O atom is bonded to the N atom)

b) ClO_3^{3-} (each O atom is bonded to the Cl atom)

c) CO_3^{2-} (each O atom is bonded to the C atom)

d) H_3O^+ (each H atom is bonded to the O atom) (Note the positive charge; compare to NH_4^+)

Solution:

a) $[:\overset{..}{\underset{..}{O}}:\overset{.}{N}::\overset{..}{O}:]^-$

There are three groups: one lone pair and two O atoms, trigonally arranged around the N atom. The shape of the ion is angular.

b) $\left[:\overset{..}{\underset{..}{O}}:\underset{:\overset{..}{\underset{..}{O}}:}{\overset{.}{\underset{.}{Cl}}}:\overset{..}{\underset{..}{O}}:\right]^-$

There are four groups, one lone pair and three O atoms, tetrahedrally arranged around the Cl atom. The shape of the ion is a triangular pyramid, with the Cl atom at the top.

c) $\left[:\overset{..}{\underset{..}{O}}:\underset{:\overset{..}{\underset{..}{O}}:}{C}::\overset{..}{O}:\right]^{2-}$

There are three O atoms trigonally arranged around the C atom. The shape of the ion is triangular or trigonal planar, with the C atom in the center.

d) $\left[H:\underset{\overset{|}{H}}{\overset{..}{O}}:H\right]^+$

There are four groups, one lone pair and three H atoms, tetrahedrally arranged around the O atom. The shape of the ion is a triangular pyramid, with the O atom at the top.

Polarity of Covalent Molecules (Section 4.9)

4.56   Use the periodic table and Table 4.3 to determine which of the following bonds will be polarized. Show the resulting charge distribution in those molecules that contain polarized bonds.

a) H—I

b)
$$\underset{O}{\overset{/}{S = O}}$$

c)
$$\underset{O}{\overset{\backslash\backslash}{O - O}}$$

Solution:

a) Polar. The electronegativity of I is greater than H. $^{\delta+}H—I^{\delta-}$

b) Both bonds are polar. The electronegativity of O is greater than S.

$$\underset{^{\delta-}O}{\overset{/}{^{\delta+}S = O\ ^{\delta-}}}$$

c) Neither bond is polar. There is no difference in electronegativity.

4.58   Use Table 4.3 and classify the bonds in the following compounds as *nonpolar covalent, polar covalent,* or *ionic.*

a) KI   b) NH_3 c) CO   d) CaO

e) NO

Solution:

a) The electronegativity difference is 1.7. Bond is polar covalent.

b) The electronegativity difference is 0.9. Bond is polar covalent.
c) The electronegativity difference is 1.0. Bond is polar covalent.
d) The electronegativity difference is 2.5. Bond is ionic.
e) The electronegativity difference is 0.5. Bond is polar covalent.

4.60   On the basis of the charge distributions you drew for the molecules of Exercise 4.56, classify each of the molecules as *polar* or *nonpolar*.

Solution:
a) polar; the polar bonds are not symmetrical in HI.
b) polar; the polar bonds are not symmetrical in SO_2.
c) nonpolar; There are no polar bonds in O_3.

4.62   Use Table 4.4 and predict the type of bond you would expect to find in compounds formed from the following elements:
a) nitrogen and oxygen   b) magnesium and oxygen   c) N and H

Solution:
The difference of electronegativities, ΔEN, is an indicator of the nature of the bond:
ΔEN = 0, the bond tends to be covalent and nonpolar.
ΔEN is above 0 and below 2.1, the bond tends to be polar covalent.
ΔEN is 2.1 or above, the bond tends to be ionic.
a) ΔEN = 3.5 − 3.0 = 0.5   The bond tends to be polar covalent.
b) ΔEN = 3.5 − 1.2 = 2.3   The bond tends to be ionic.
c) ΔEN = 3.0 − 2.1 = 0.9   The bond tends to be polar covalent.

4.64   Show the charge distribution in the following molecules and predict which are polar molecules.
a) C═O   b) H—Te   c)
                                      | I
                                      H |
                                                                    Al
                                                                   / \
                                                                  I   I

Solution:
a) polar; $^{\delta+}C = O^{\delta-}$
b) polar; $^{\delta\mid} H—Te^{\delta-}—H^{\delta+}$ (angular)
c) nonpolar; The polar bonds are symmetrical about the center.

More about Naming Compounds (Section 4.10)
4.66   Name the following binary covalent compounds:
a) NCl_2 b) P_2O_6 c) BrCl   d) SF_4
e) ClO_2

Solution:
a) nitrogen trichloride
b) phosphorous hexoxide
c) bromine monochloride
d) sulfur tetrafluoride
e) chlorine dioxide

4.68   Write formulas for the following binary covalent compounds:
a) sulfur tetrafluoride
b) oxygen difluoride
c) dinitrogen monoxide
d) phosphorus trichloride

Solution:
a) SF_4 b) OF_2 c) N_2O d) PCl_3

4.70   Write the formulas and names for compounds composed of ions of the following metals and the indicated polyatomic ions.
a) calcium and the hypochlorite ion
b) cesium and the nitrite ion
c) Mg and SO_4^{2-}
d) K and $Cr_2O_7^{2-}$

Solution:
a) $Ca(ClO)_2$ b) $CsNO_2$ c) $MgSO_4$ d) $K_2Cr_2O_7$

4.72   Write the formulas for the following compounds:
a) magnesium hydroxide
b) calcium sulfite
c) ammonium phosphate
d) sodium hydrogen carbonate
e) barium sulfate

Solution:
a) $Mg(OH)_2$
b) $CaSO_3$
c) $(NH_4)_3PO_4$
d) $NaHCO_3$, also named sodium bicarbonate
e) $BaSO_4$

4.74   Write a formula for the following compounds, using M with appropriate charges to represent the metal ion:
a) any group IA(1) element and SO_3^{2-}
b) any group IA(1) element and $C_2H_3O_2^-$
c) any metal that forms M^{2+} ions and $Cr_2O_7^{2-}$
d) any metal that forms M^{3+} ions and PO_4^{3-}
e) any metal that forms M^{3+} ions and NO_3^-

Solution:
a) M_2SO_3 b) $MC_2H_3O_2$ c) MCr_2O_7 d) MPO_4
e) $M(NO_3)_3$

Other Interparticle Forces (Section 4.11)

4.76   The covalent compounds, ethyl alcohol and dimethyl ether, both have the formula C_2H_6O. However, the alcohol melts at $-117.3°C$ and boils at $78.5°C$, whereas the ether melts and boils at $-138.5°C$ and $-23.7°C$, respectively. How could differences in forces between molecules be used to explain these observations?

Solution:
The stronger the attractive forces between molecules, the higher the melting point and boiling point will be. Thus, it can be concluded that forces attracting dimethyl ether molecules to each other are weaker

than the forces attracting ethyl alcohol molecules to each other. Because both compounds are covalently bonded, it is likely that the attractive forces are polar forces and we might conclude that the ethyl alcohol is more polar than is the dimethyl ether.

4.78   Describe the predominant forces that exist between molecules of the noble gases. Arrange the noble gases in a predicted order of increasing boiling point (lowest first) and explain the reason for the order.

Solution:
The predominant forces that exist between molecules of noble gases are dispersion forces because noble gas molecules are nonpolar. Dispersion forces increase with increasing mass of the particles involved, so the boiling points of the noble gases would increase in the order: He, Ne, Ar, Kr, Xe, and Rn.

4.80   Table sugar, sucrose, melts at about 185°C. Which interparticle forces do you think are unlikely to be the predominant ones in the lattice of solid sucrose.

Solution:
The relatively high melting point indicates that strong interparticle forces are present. It is unlikely that the weak dispersion forces are the predominant forces present. Dipolar forces are also quite weak, and so are also unlikely to be the dominant forces present. Hydrogen bonding must be the dominant interparticle forces.

PROGRAMMED REVIEW

Section 4.1 Noble-gas Configurations

The electronic structure of noble gases represents a (a) _stable_ configuration. A noble gas configuration is characterized by (b) _two_ electrons in the valence shell of helium and (c) _8_ valence-shell electrons for other members of the group.

Section 4.2 Ionic Bonding

According to the (a) _octet rule_, atoms tend to interact by achieving noble gas electronic configurations. Some atoms achieve noble gas configurations by transferring electrons and becoming charged atoms called (b) _simple ions_. The attraction between oppositely charged ions is called an (c) _ionic bond_. During ionic bond formation, metals generally (d) _lose_ electrons and nonmetals generally (e) _gain_ them. Atoms and simple ions that have the same electronic configurations are said to be (f) _isoelectronic_ with each other.

Section 4.3 Ionic Compounds

Ionic compounds containing ions of only two elements are called a) _binary_ compounds. Formulas for ionic compounds do not represent the formulas of molecules, but only the simplest (b) _combining ratio_ of the ions in the compounds. The stable form of ionic compounds is a (c) _crystal_ in which ions of opposite charge occupy (d) _lattice_ sites in a crystal lattice.

Section 4.4 Naming Binary Ionic Compounds

The name of binary ionic compounds consists of the name of the (a) _metal_, then the (b) _stem_ of the name of the non-metal with an (c) _-ide_ suffix. The name of those metals that form more than one type of charged ion must include a (d) _roman numeral_ in parentheses. The number in the parentheses of the name tells the positive (e) _charge_ on the metal ion in that compound. The correct name of Fe_2S_3 is (f) _iron (III) sulphide_. Using older methods of naming, you could also name it (g) _ferric sulphide_.

Section 4.5 The Smallest Unit of Ionic Compounds

The formulas of ionic compounds represent only the simplest (a) _combining ratio_ of ions in the compound. In solid ionic compounds, the ions occupy (b) _lattice sites_ in the crystal lattice. The sum of the atomic weights of the atoms in the ionic formula gives the (c) _formula_ weight of the ionic compound.

Section 4.6 Covalent Bonding

Some elements combine by (a) _sharring_ electrons rather than giving them up or accepting them. The net attractive force that results between atoms that share electrons is called a (b) _covalent bnd_. Covalent bonds in which bonding electrons are shared equally are called (c) _non polar_ covalent bonds. Differences in the (d) _electronegativ_ of covalently bonded atoms cause bonding electrons to be shared unequally; such bonds are called (e) _polar_ covalent bonds. (f) _polar molecules_ result when polarized bonds create an unsymmetrical charge distribution in molecules. The covalent bonds resulting from the sharing of two pairs and three pairs of electrons are referred to as (g) _double_ bonds and (h) _triple_ bonds, respectively.

Section 4.7 Polyatomic Ions

Covalently bonded groups of atoms that carry a net charge are called (a) _polyatomic_ ions. With the exception of the ammonium ion, the common polyatomic ions have a (b) _negative_ charge.

Section 4.8 Shapes of Molecules and Polyatomic Ions

The acronym used for the model used to predict the shape is (a) _VSPER_. The atom to which two or more other atoms are covalently bonded is called the (b) _central_ atom. To determine the shape, count all (c) _valence_ shell electron (d) _pairs_ around the central atom, both (e) _lone_ and shared (f) _pairs_. Multiple bonding pairs are counted as (g) _one_ pair. When a central atom has 4 bonding pairs and no lone pairs surrounding it, the shape of the molecules is (h) _tetrahedral_

Section 4.9 Polarity of Covalent Molecules

When two atoms do not share the electron pair evenly, a (a) _polar_ covalent bond results. If a molecule contains (b) _polar_ bonds, and the charge distribution is (c) _non symmetric_ the molecule is a (d) _polar_ molecule.

Section 4.10 More About Naming Compounds

The major difference in naming binary covalent compounds compared to binary ionic compounds is that covalent compounds have (a) _prefixes_ in their name to tell how many atoms are in the molecule. The prefix "di-" in the name of carbon dioxide tells how many (b) _oxygen_ atoms are in the molecule. The name of SO_3 is (c) _sulphur trioxide_. To name ionic compounds involving polyatomic ions, name the (d) _positive_ ion and then name the (e) _negative_ ion. The name of Na_2CO_3 is (f) _sodium carbonate_ $Ca(NO_3)_2$ is named (g) _calcium nitrate_ and NH_4Br is named (h) _ammonium bromide_

Section 4.11 Other Interparticle Forces

A solid in which lattice sites are occupied by atoms that are covalently bonded to each other is called a (a) _network_ solid. The (b) _metallic_ bond is the name given to the forces resulting from the attraction of positive (c) _kernals_ to mobile electrons in a crystal lattice. The attractive forces between positive and negative ends of polar molecules are called (d) _dipolar_ forces. (e) _hydrogen bonding_ results from the attractions of polar molecules in which hydrogen is covalently bonded to very electronegative elements. The weakest interparticle forces are called (f) _dispersion_ forces.

SELF-TEST QUESTIONS
Multiple Choice

1. Which of the following elements has the lowest electronegativity?
   a) As   b) P   c) Br   d) Cl

2. In describing the strength of interparticle forces we discover that the weakest forces or bonds are
   a) covalent   b) ionic   c) dipolar   d) dispersion

3. The formula for the compound formed between the elements Ba and O would be
   a) BaO   b) Ba_2O c) BaO_2 d) Ba_2O_3

4. The formula for the ionic compound containing Al^{3+} and SO_4^{2-} ions would be
   a) $Al(SO_4)_2$ b) $AlSO_4$ c) $Al_3(SO_4)_2$ d) $Al_2(SO_4)_3$

5. The expected formula of the molecule formed when nonmetals C and H combine in compliance with the octet rule is
   a) CH_4 b) CH_2 c) C_4H d) CH_3

6. The name of the covalent compound PCl_3 is
   a) trichlorophosphide   b) phosphorus trichloride
   c) phosphorus trichlorine   d) phosphorus chloride

7. The compound $MgSO_4$ is correctly named
   a) magnesium sulfur tetroxide   b) magnesium sulfoxide
   c) magnesium sulfide   d) magnesium sulfate

8. This bond is found in molecules such as HCl and H_2O.
   a) nonpolar covalent bond   b) polar covalent bond
   c) ionic bond   d) metallic bond

9. If the electronegativity difference between two elements A and B is 1.0, what type of bond is A—B?
   a) nonpolar covalent   b) polar covalent
   c) ionic   d) metallic

10. Which of the following is a correct electron dot formula for sulfur (element 16)?
    a) $: \overset{\cdot}{\underset{\cdot}{S}} :$ b) $\overset{\cdot}{S}$ c) $\cdot \overset{\cdot}{\underset{\cdot}{S}} \cdot$ d) $\overset{\cdot}{\underset{\cdot}{S}} :$

11. In the structures below, each bonding electron pair is denoted by a dash. A correct structure of SO_3^{2-} is

    a) $\left[: \overset{O}{\underset{O}{S}} = O \right]^{2-}$ b) $\left[: \overset{O}{\underset{O}{S}} - O \right]^{2-}$ c) $\left[\overset{O}{\underset{: S = O}{O}} \right]^{2-}$ d) $\left[: \overset{O}{\underset{O}{S}} - O \right]^{2-}$

12. A covalent molecule forms between elements A and B. B is more electronegative. Which of the following molecules would be polar?

    a) B—A—B   b) A—B   c) 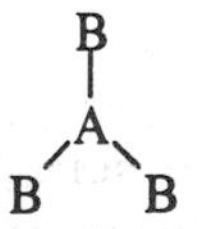 $\overset{B}{\underset{B \quad B}{A}}$ d) 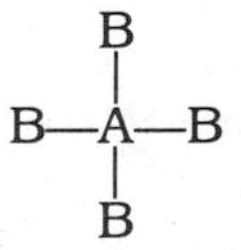 $B - \overset{B}{\underset{B}{A}} - B$

13. Which of the following contains polar bonds, but is a nonpolar molecule?
    a) 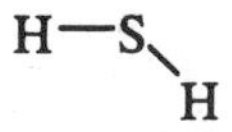 $H - \overset{S}{\diagdown}_H$ b) O=C=O   c) H—Cl   d) F—F

$\overset{..}{O} : \overset{..}{\underset{..}{S}} : \overset{..}{O}$

$\overset{}{\underset{O}{}}$

16
—
3

48
32
—
8 6

True-False

14. No more than one pair of electrons can be shared to form covalent bonds between atoms.

15. Dispersion forces between particles are correctly classified as very strong.

16. A compound between elements with atomic numbers 7 and 8 will contain covalent bonds.

17. All covalent bonds are polar.

18. The interparticle forces in a solid noble gas would have to be polar in nature.

19. Neon (Ne) has a higher boiling point than krypton (Kr).

Matching

An ionic compound is formed from each of the pairs given on the left. For each pair, choose the correct formula for the resulting compound from the responses on the right.

20. _____ X^+ and Y^{2-} a) X_3Y_2

21. _____ X^+ and Y^- b) XY

22. _____ X^{3+} and Y^{3-} c) XY_2

23. _____ X is a group IIA (2) ion and Y is a group   d) X_2Y
           VIA (16) ion

For each molecule given on the left, predict the molecular geometry based on VSEPR theory.

24. _____ NH_3 a) linear

25. _____ BrCl   b) planar triangle

26. _____ H_2S c) triangular pyramid

27. _____ CO_2 d) tetrahedral

28. _____ CH_4 e) bent

For each of the molecules on the left, choose the statement from the right that correctly gives the polarity of the bonds in the molecule and the polarity of the molecule as a whole.

29. _____ H_2S, 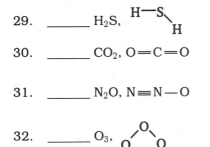   a) a polar molecule containing all polar bonds

30. _____ CO_2, O=C=O   b) a nonpolar molecule containing all nonpolar bonds

31. _____ N_2O, N≡N—O   c) a nonpolar molecule containing all polar bonds

32. _____ O_3,   d) a polar molecule containing polar bonds and nonpolar bonds

SOLUTIONS

A. Answers to Programmed Review

4.1 a) stable   b) two   c) eight

4.2 a) octet rule   b) simple ions   c) ionic bond   d) lose
     e) gain   f) isoelectronic

4.3 a) binary   b) combining ratio   c) crystal   d) lattice

4.4 a) metal   b) stem   c) -ide   d) roman numeral
     e) charge   f) iron (III) sulfide g) ferric sulfide

4.5 a) combining ratio   b) lattice sites   c) formula

4.6 a) sharing   b) covalent bond   c) nonpolar   d) electronegativity
     e) polar   f) polar molecules   g) double   h) triple

4.7 a) polyatomic   b) negative

4.8 a) VSEPR   b) central   c) valence   d) pairs
     e) lone   f) pairs   g) one   h) tetrahedral

4.9 a) polar   b) polar   c) nonsymmetrical   d) polar

4.10 a) prefixes   b) oxygen   c) sulfur trioxide   d) positive
     e) negative   f) sodium carbonate   g) calcium nitrate   h) ammonium bromide

4.11 a) network   b) metallic   c) kernels   d) dipolar
     e) hydrogen bonding f) dispersion

B. Answers to Self-Test Questions

1. a	9. b	17. F	25. a
2. d	10. d	18. F	26. e
3. a	11. d	19. F	27. a
4. d	12. b	20. d	28. d
5. a	13. b	21. b	29. a
6. b	14. F	22. b	30. c
7. d	15. F	23. b	31. d
8. b	16. T	24. c	32. b

Blank page

Chapter 5

Chemical Reactions

CHAPTER OUTLINE

5.1   Chemical Equations
5.2   Types of Reactions
5.3   Redox Reactions
5.4   Decomposition Reactions
5.5   Combination Reactions
5.6   Replacement Reactions

5.7   Ionic Equations
5.8   Energy and Reactions
5.9   The Mole and Chemical Equations
5.10   The Limiting Reactant
5.11   Reaction Yields

LEARNING OBJECTIVES

When you have completed your study of this chapter, you should be able to:
1. Identify the reactants and products in written reaction equations.
2. Balance simple reaction equations by inspection.
3. Assign oxidation numbers to elements in chemical formulas.
4. Classify reactions as redox or nonredox types.
5. Identify the oxidizing and reducing agents in redox reactions.
6. Classify reactions as decomposition, combination, single-replacement or double-replacement types.
7. Write molecular equations in total ionic and net ionic forms.
8. Classify reactions as exothermic or endothermic.
9. Use the mole concept to do calculations based on chemical equations.
10. Use the mole concept to do calculations based on the limiting-reactant concept.
11. Use the mole concept to do percentage yield calculations.

ANSWERS AND SOLUTIONS TO EVEN-NUMBERED PROBLEMS

Chemical Equations (Section 5.1)

5.2   Identify the reactants and products in each of the following reactions:

a) $H_{2(g)} + Cl_{2(g)} \rightarrow 2HCl_{(g)}$

b) $2KClO_{3(s)} \rightarrow 2KCl_{(s)} + 3O_{2(g)}$
c) magnesium oxide + carbon → magnesium + carbon monoxide
d) ethane + oxygen → carbon dioxide + water

Solution:
a) reactants = $H_{2(g)}$ and $Cl_{2(g)}$ product = $HCl_{(g)}$
b) reactant = $KClO_{3(s)}$ products = $KCl_{(s)}$ and $O_{2(g)}$
c) reactants = magnesium oxide and carbon   products = magnesium and carbon monoxide
d) reactants = ethane and oxygen   products = carbon dioxide and water

5.4   Identify which of the following are consistent with the law of conservation of matter. For those that are not, explain why they are not.
a) $ZnS_{(s)} + O_{2(g)} \rightarrow ZnO_{(s)} + SO_{2(g)}$
b) $Cl_{2(aq)} + 2I^-_{(aq)} \rightarrow I_{2(aq)} + 2Cl^-_{(aq)}$
c) 1.60 g oxygen + 1.00 g carbon → 2.80 g carbon monoxide
d) $2C_2H_{6(g)} + 7\,O_{2(g)} \rightarrow 4CO_{2(g)} + 6H_2O_{(g)}$

Solution:
a) Not consistent. There are more atoms of oxygen in the products than in reactants.
b) Consistent
c) Not consistent. The product weighs more than the sum of the reactants.
d) Consistent

5.6   Determine the number of atoms of each element on each side of the following equations and decide which equations are balanced:
a) $2AgO_{(s)} \rightarrow 2Ag_{(s)} + O_{2(g)}$
b) $Al(s) + O_2(g) \rightarrow Al_2O_{3(s)}$
c) $2AgNO_{3(aq)} + K_2SO_{4(aq)} \rightarrow Ag_2SO_{4(s)} + 2KNO_{3(aq)}$
d) $SO_{2(g)} + O_{2(g)} \rightarrow SO_{3(g)}$

Solution:
a) On the L side: 2 silver atoms, 2 oxygen atoms
   On the R side: 2 silver atoms, 2 oxygen atoms
       The equation is balanced.
b) On the L side: 1 aluminum atom, 2 oxygen atoms
   On the R side: 2 aluminum atoms, 3 oxygen atoms
       The equation is not balanced.
c) On the L side: 2 silver atoms, 2 nitrogen atoms, 10 oxygen atoms, 2 potassium atoms, 1 sulfur atom
   On the R side: 2 silver atoms, 2 nitrogen atoms, 10 oxygen atoms, 2 potassium atoms, 1 sulfur atom
       The equation is balanced.
d) On the L side: 1 sulfur atom, 4 oxygen atoms
   On the R side: 1 sulfur atom, 3 oxygen atoms
       The equation is not balanced.

5.8   Balance the following equations:
a) $C_2H_{6(g)} + O_{2(g)} \rightarrow CO_{2(g)} + H_2O_{(l)}$
b) hydrogen + chlorine → hydrogen chloride
c) $H_2S_{(g)} + O_{2(g)} \rightarrow SO_{2(g)} + H_2O_{(l)}$

d) sulfur + oxygen → sulfur dioxide

e) $Na_2CO_{3(aq)} + Ca(NO_3)_{2(aq)} \rightarrow NaNO_{3(aq)} + CaCO_{3(s)}$

f) $NaBr_{(aq)} + Cl_{2(aq)} \rightarrow NaCl_{(aq)} + Br_{2(aq)}$

g) $Ag_2CO_{3(s)} \rightarrow Ag_{(s)} + CO_{2(g)} + O_{2(g)}$

h) $H_2O_{2(aq)} + H_2S_{(aq)} \rightarrow H_2O_{(l)} + S_{(s)}$

Solution:

a) $2C_2H_{6(g)} + 7O_{2(g)} \rightarrow 4CO_{2(g)} + 6H_2O_{(l)}$

b) $H_{2(g)} + Cl_{2(g)} \rightarrow 2HCl(g)$

c) $2H_2S_{(g)} + 3O_{2(g)} \rightarrow 2SO_{2(g)} + 2H_2O_{(l)}$

d) $S_{(s)} + O_{2(g)} \rightarrow SO_{2(g)}$

e) $Na_2CO_{3(aq)} + Ca(NO_3)_{2(aq)} \rightarrow 2NaNO_{3(aq)} + CaCO_{3(s)}$

f) $2NaBr_{(aq)} + Cl_{2(aq)} \rightarrow 2NaCl_{(aq)} + Br_{2(aq)}$

g) $2Ag_2CO_{3(s)} \rightarrow 4Ag_{(s)} + 2CO_{2(g)} + O_{2(g)}$

h) $H_2O_{2(aq)} + H_2S_{(aq)} \rightarrow 2H_2O_{(l)} + S_{(s)}$

Redox Reactions (Section 5.3)

5.10   Assign oxidation numbers to the underlined element in each of the following formulas:

a) $\underline{C}lO_3^-$

b) $H_2\underline{S}O_4$

c) $Na\underline{N}O_3$

d) $\underline{N}_2O$

e) $K\underline{Mn}O_4$

f) $H\underline{C}lO_2$

Solution:

a) Cl = +5, (O = −2)

b) S = +6, (H = +1, O = −2)

c) N = +5, (Na = +1, O = −2)

d) N = 1, (O = −2)

e) Mn = +7, (K = +1, O = −2)

f) Cl = +3, (H = +1, O = −2)

5.12   Find the element with the highest oxidation number in each of the following formulas:

a) $Na_2Cr_2O_7$

b) $K_2S_2O_3$

c) HNO_3

d) $P_{10}O_5$

e) $Mg(ClO_4)_2$

f) $HClO_2$

Solution:

The charge on oxygen is normally 2−, and the net charge on a compound is zero (0).

a) $2Na + 2Cr + 7O; (2 \times 1+) + (2 \times 6+) + (7 \times 2-) = 0$

Chromium has the highest oxidation number at 6+ each.

b) $2K + 2S + 3O; (2 \times 1+) + (2 \times 2+) + (3 \times 2-) = 0$

Sulfur has the highest oxidation number at 2+ each.

c) $1H + 1N + 3O; (1 \times 1+) + (1 \times 5+) + (3 \times 2-) = 0$

Nitrogen has the highest oxidation number at 5+.

d) Phosphorous is positive and the oxygen is negative, therefore the phosphorous has the highest oxidation number.

e) $1Mg + 2Cl + 8O; (1 \times 2+) + (2 \times 7+) + (8 \times 2-) = 0$

Chlorine has the highest oxidation number, +7.

f) $1H + 1Cl + 2O; (1 \times 1+) + (1 \times 3+) + (2 \times 2-) = 0$

Chlorine, +3, has the highest oxidation number.

5.14   For each of the following equations, indicate whether the underlined element has been *oxidized, reduced,* or *neither oxidized nor reduced.*

a) $4Al_{(s)} + 3\underline{O}_{2(g)} \rightarrow 2Al_2O_{3(s)}$

b) $\underline{S}O_{2(g)} + H_2O_{(l)} \rightarrow H_2SO_{3(g)}$

c) $2K\underline{C}lO_{3(s)} \rightarrow 2KCl_{(s)} + 3O_{2(g)}$

d) $2\underline{C}O_{(g)} + O_{2(g)} \rightarrow 2CO_{2(g)}$
e) $2\underline{Na}_{(s)} + 2H_2O_{(l)} \rightarrow 2NaOH_{(aq)} + H_{2(g)}$

Solution:
a) Reduced. The oxidation number of oxygen changes from 0 to -2.
b) Neither. The oxidation number of sulfur does not change.
c) Reduced. The oxidation number of chlorine changes from $+5$ to -1.
d) Oxidized. The oxidation number of carbon changes from $+2$ to $+4$.
e) Oxidized. The oxidation number of sodium changes from 0 to $+1$.

5.16 Assign oxidation numbers to each element in the following equations and identify the oxidizing and reducing agents:
a) $2Cu_{(s)} + O_{2(g)} \rightarrow 2CuO_{(s)}$
b) $Cl_{2(aq)} + 2KI_{(aq)} \rightarrow 2KCl_{(aq)} + I_{2(aq)}$
c) $3MnO_{2(s)} + 4Al_{(s)} \rightarrow 2Al_2O_{3(s)} + 3Mn_{(s)}$
d) $2H^+_{(aq)} + 3SO_3^{2-}(aq) + 2NO_3^-(aq) \rightarrow 2NO_{(g)} + H_2O_{(l)} + 3SO_4^{2-}(aq)$
e) $Mg_{(s)} + 2HCl_{(aq)} \rightarrow MgCl_{2(aq)} + H_{2(g)}$
f) $4NO_{2(g)} + O_{2(g)} \rightarrow 2N_2O_{5(g)}$

Solution:
a) Left side: copper, zero; oxygen, zero
   Right side: copper, 2+; oxygen, 2–
   Cu = reducing agent; O_2 = oxidizing agent
b) Left side: chlorine, zero; potassium, 1+; iodine, 1–
   Right side: potassium, 1+; chlorine, 1–; iodine, zero
   Cl_2 = oxidizing agent; Ki = reducing agent
c) Left side: manganese, 4+; oxygen, 2–; aluminum, zero
   Right side: aluminum, 3+; oxygen, 2–; manganese, zero
   MnO_2 = oxidizing agent; Al = reducing agent
d) Left side: hydrogen, 1+; sulfur, 4+; oxygen, 2–; nitrogen, 5+
   Right side: nitrogen, 2+; oxygen, 2–; hydrogen, 1+; sulfur, 6+
   N^{5+} in NO_3^- = oxidizing agent; S^{4+} in SO_3^{2-} = reducing agent
e) Left side: magnesium, zero; hydrogen, 1+; chlorine, 1–
   Right side: magnesium, 2+; chlorine, 1–; hydrogen, zero
   Mg = reducing agent; H^+ in HCl = oxidizing agent
f) Left side: nitrogen, 4+; oxygen in NO_2, -2; oxygen as O_2; zero
   Right side: nitrogen, 5+; oxygen, 2–
   N^{4+} in NO_2 = reducing agent; O_2 = oxidizing agent

5.18 Aluminum metal reacts rapidly with highly basic solutions to liberate hydrogen gas and a large amount of heat. This reaction is utilized in a popular solid drain cleaner that is composed primarily of lye (sodium hydroxide) and aluminum granules. When wet, the mixture reacts as follows:
   $6NaOH_{(aq)} + 2Al_{(s)} \rightarrow 3H_{2(g)} + 2Na_3AlO_{3(aq)}$
The liberated H_2 provides agitation that, together with the heat, breaks the drain stoppage loose. What are the oxidizing and reducing agents in the reaction?

Solution:

$6NaOH_{(aq)} + 2Al_{(s)} \rightarrow 3H_{2(g)} + 2Na_3AlO_{3(aq)}$
  1 −2 1   0   0   1 3 −2

The oxidation number of Al changes from 0 to +3. Al is the reducing agent. The oxidation number of the hydrogen changes from +1 to 0. H in NaOH is the oxidizing agent.

Decomposition, Combination, and Replacement Reactions (Sections 5.4–5.6)

5.20   Classify each of the reactions represented by the following equations first as a *redox* or *nonredox* reaction. Then further classify each redox reaction as a *decomposition, single replacement,* or *combination* reaction; and each nonredox reaction as a *decomposition, double replacement,* or *combination* reaction.

a) $K_2CO_{3(s)} \rightarrow K_2O_{(s)} + CO_{2(g)}$
b) $Ca_{(s)} + 2H_2O_{(l)} \rightarrow Ca(OH)_{2(s)} + H_{2(g)}$
c) $BaCl_{2(aq)} + H_2SO_{4(aq)} \rightarrow BaSO_{4(s)} + 2HCl$
d) $SO_{2(g)} + H_2O_{(l)} \rightarrow H_2SO_{3(aq)}$
e) $2NO_{(g)} + O_{2(g)} \rightarrow 2NO_{2(g)}$
f) $2Zn_{(s)} + O_{2(g)} \rightarrow 2ZnO_{(s)}$

Solution:
a) nonredox, decomposition
b) redox, single replacement
c) nonredox, double replacement
d) nonredox, combination
e) redox, combination
f) redox, combination

5.22   Baking soda ($NaHCO_3$) can serve as an emergency fire extinguisher for grease fires in the kitchen. When heated, it liberates CO_2 which smothers the fire. The equation for the reaction is:

$2NaHCO_{3(s)} \xrightarrow{heat} Na_2CO_{3(s)} + H_2O_{(g)} + CO_{2(g)}$

Classify the reaction into the categories used in Exercise 5.20.

Solution:
nonredox, decomposition

5.24   Many homes are heated by the energy released when natural gas (represented by CH_4) reacts with oxygen. The equation for the reaction is:

$CH_{4(g)} + 2\,O_{2(g)} \rightarrow CO_{2(g)} + 2H_2O_{(g)}$

Classify the reaction as *redox* or *nonredox*.

Solution:
Redox

5.26   Chlorine, used to treat drinking water, undergoes the reaction represented by the following equation.

$Cl_{2(aq)} + H_2O_{(l)} \rightarrow HClO_{(aq)} + HCl_{(aq)}$

Classify the reaction as *redox* or *nonredox*.

Solution:
Redox. The oxidation number of the chlorine goes from 0 in Cl_2 to −1 in HCl and to +1 in HClO.

Ionic Equations (Section 5.7)

5.28   Consider all the following ionic substances to be water soluble and write the formulas of the ions that would be formed if the substances were dissolved in water. Table 4.6 will be helpful.

a) $LiNO_3$ b) Na_2HPO_4 c) $Ca(ClO_3)_2$

d) KOH e) $MgBr2$ f) $(NH_4)_2SO_4$

Solution:

a) Li^+ and NO_3^- b) Na^+ and HPO_4^{2-} c) Ca^{2+} and ClO_3^-

d) K^+ and OH^- e) Mg^{2+} and Br^+ f) NH_4^+ and SO_4^{2-}

5.30   Reactions represented by the following equations take place in water solutions. Write each molecular equation in total ionic form, then identify spectator ions and write the equations in net ionic form. Solids that do not dissolve are designated by (s); gases that do not dissolve are designated by (g); and substances that dissolve but do not dissociate are underlined.

a) $\underline{Cl_2}_{(aq)} + 2NaI_{(aq)} \rightarrow 2NaCl_{(aq)} + \underline{I_2}_{(aq)}$

b) $AgNO_{3(aq)} + NaCl_{(aq)} \rightarrow AgCl_{(s)} + NaNO_{3(aq)}$

c) $Zn(s) + 2HCl(aq) \rightarrow ZnCl_2(aq) + \underline{H_2}(g)$

d) $BaCl_{2(aq)} + H_2SO_{4(aq)} \rightarrow BaSO_{4(s)} + 2HCl_{(aq)}$

e) $\underline{SO_3}_{(aq)} + \underline{H_2O}_{(l)} \rightarrow H_2SO_{4(aq)}$

f) $2NaI_{(aq)} + 2H_2SO_{4(aq)} \rightarrow \underline{I_2}_{(aq)} + \underline{SO_2}_{(aq)} + Na_2SO_{4(aq)} + 2\underline{H_2O}_{(l)}$

Solution:

a) $\underline{Cl_2}_{(aq)} + 2Na^+_{(aq)} + 2I^-_{(aq)} \rightarrow 2Na^+_{(aq)} + 2Cl^-_{(aq)} + \underline{I_2}_{(aq)}$ (total ionic)

   The $2Na^+$ are spectator ions.

   $\underline{Cl_2}_{(aq)} + 2I^-_{(aq)} \rightarrow 2Cl^-_{(aq)} + \underline{I_2}_{(aq)}$ (net ionic)

b) $Ag^+_{(aq)} + NO^-_{3(aq)} + Na^+_{(aq)} + Cl^-_{(aq)} \rightarrow AgCl_{(s)} + Na^+_{(aq)} + NO^-_{3(aq)}$ (total ionic)

   The Na^+ and NO_3^- are spectator ions.

   $Ag^+_{(aq)} + Cl^-_{(aq)} \rightarrow AgCl_{(s)}$ (net ionic)

c) $Zn_{(s)} + 2H^+_{(aq)} + 2Cl^-_{(aq)} \rightarrow Zn^{2+}_{(aq)} + 2Cl^-_{(aq)} + \underline{H_2}_{(g)}$ (total ionic)

   The $2Cl^-$ are spectator ions.

   $Zn(s) + 2H^+(aq) \rightarrow Zn^{2+}(aq) + \underline{H_2}(g)$ (net ionic)

d) $Ba^{2+}_{(aq)} + 2Cl^-_{(aq)} + 2H^+_{(aq)} + SO^{2-}_{4(aq)} \rightarrow BaSO_{4(s)} + 2H^+_{(aq)} + 2Cl^-_{(aq)}$ (total ionic)

   The $2H^+$ and $2Cl^-$ are spectator ions.

   $Ba^{2+}_{(aq)} + SO^{2-}_{4(aq)} \rightarrow BaSO_{4(s)}$ (net ionic)

e) $\underline{SO_3}_{(aq)} + \underline{H_2O}_{(l)} \rightarrow 2H^+_{(aq)} + SO^{2-}_{4(aq)}$ (total ionic and net ionic)

   There are no spectator ions.

f) $2Na^+_{(aq)} + 2I^-_{(aq)} + 4H^+_{(aq)} + 2SO^{2-}_{4(aq)} \rightarrow \underline{I_2}_{(aq)} + \underline{SO_2}_{(aq)} + 2Na^+_{(aq)} + SO^{2-}_{4(aq)} + 2\underline{H_2O}_{(l)}$

   The two Na^+ ions and one of the two SO_4^{2-} are spectator ions.

   $2I^-_{(aq)} + 4H^+_{(aq)} + SO^{2-}_{4\ (aq)} \rightarrow \underline{I_2}_{(aq)} + \underline{SO_2}_{(aq)} + 2\underline{H_2O}_{(l)}$ (net ionic)

5.32   The following molecular equations all represent neutralization reactions of acids and bases. These reactions are discussed further in Chapter 9. Write each equation in total ionic form, identify the spectator ions, then write the net ionic equation. Water is the only substance that does not dissociate. What do you notice about all the net ionic equations?

a) $HNO_{3(aq)} + KOH_{(aq)} \rightarrow KNO_{3(aq)} + H_2O_{(l)}$

b) $H_3PO_{4(aq)} + 3NH_4OH_{(aq)} \rightarrow (NH_4)_3PO_{4(aq)} + 3H_2O_{(l)}$

c) $HI_{(aq)} + NaOH_{(aq)} \rightarrow NaI_{(aq)} + H_2O_{(l)}$

Solution:

a) $H^+ + NO_3^- + K^+ + OH^- \rightarrow K^+ + NO_3^- + H_2O$
The spectator ions are the potassium and nitrate ions.

b) $3H^+ + PO_4^{3-} + 3NH_4^+ + 3OH^- \rightarrow 3NH_4^+ + PO_4^{3-} + 3H_2O$
The spectator ions are the ammonium and phosphate ions.

c) $H^+ + I^- + Na^+ + OH^- \rightarrow Na^+ + I^- + H_2O$
The spectator ions are the sodium and iodide ions.

The simplified net ionic equation is the same for each of these acid/base reactions.
$H^+ + OH^- \rightarrow H_2O$

Energy and Reactions (Section 5.8)

5.34   In addition to emergency cold packs, emergency hot packs are available that heat up when water is added through an opening. Is the process that takes place in such packs *exothermic* or *endothermic*? Explain.

Solution:
Exothermic; Heat is released, making the solution warmer.

5.36   An individual wants to keep some food cold in a portable picnic cooler. A piece of ice is put into the cooler, but it is wrapped in a thick insulating blanket to slow its melting. Comment on the effectiveness of the cooler in terms of the direction of heat movement inside the cooler.

Solution:
The heat will move from the contents of the cooler into the ice and will melt it, but the rate will be very slow because of the insulating blanket. The purpose of putting the ice in the cooler is to absorb the heat from the contents of the cooler. The insulating blanket around the ice reduces the effectiveness of heat transfer from the contents of the cooler to the ice.

The Mole and Chemical Equations (Section 5.9)

5.38   For the reactions represented by the following equations, write statements equivalent to statements 1, 4, 5, and 6 given in the chapter (section 5.9).

a) $Ca_{(s)} + 2H_2O_{(l)} \rightarrow H_{2(g)} + Ca(OH)_{2(s)}$

b) $2NO_{(g)} + O_{2(g)} \rightarrow 2NO_{2(g)}$

c) $2C_2H_{6(g)} + 7O_{2(g)} \rightarrow 4CO_{2(g)} + 6H_2O_{(l)}$

d) $Zn_{(s)} + 2AgNO_{3(aq)} \rightarrow Zn(NO_3)_{2(aq)} + 2Ag_{(s)}$

e) $2HCl_{(aq)} + Mg(OH)_{2(s)} \rightarrow MgCl_{2(aq)} + 2H_2O_{(l)}$

Solution:

a) 1 atom Ca + 2 molecules $H_2O \rightarrow$ 1 molecule H_2 + 1 formula $Ca(OH)_2$
6.02×10^{23} atoms Ca + 12.04×10^{23} molecules $H_2O \rightarrow 6.02 \times 10^{23}$ molecules H_2 + 6.02×10^{23} formulas $Ca(OH)_2$
1 mol Ca + 2 mol $H_2O \rightarrow$ 1 mol H_2 + 1 mol $Ca(OH)_2$
40.08 g Ca + 36.02 g $H_2O \rightarrow$ 2.02 g H_2 + 74.10 g $Ca(OH)_2$

b) 2 molecules NO + 1 molecule $O_2 \rightarrow$ 2 molecules NO_2
12.04×10^{23} molecules NO + 6.02×10^{23} molecules $O_2 \rightarrow 12.04 \times 10^{23}$ molecules NO_2
2 mol NO + 1 mol $O_2 \rightarrow$ 2 mol NO_2
60.02 g NO + 32 g $O_2 \rightarrow$ 92.02 g NO_2

c) 2 molecules C_2H_6 + 7 molecules $O_2 \rightarrow$ 4 molecules CO_2 + 6 molecules H_2O
12.04×10^{23} molecules C_2H_6 + 42.14×10^{23} molecules $O_2 \rightarrow 24.08 \times 10^{23}$ molecules CO_2 + 36.12×10^{23} molecules H_2O
2 mol C_2H_6 + 7 mol $O_2 \rightarrow$ 4 mol CO_2 + 6 mol H_2O
60.14 g C_2H_6 + 224.0 g $O_2 \rightarrow$ 176.04 g CO_2 + 108.10 g H_2O

d) 1 atom Zn + 2 formulas $AgNO_3 \rightarrow$ 1 formula $Zn(NO_3)_2$ + 2 atoms Ag
6.02×10^{23} atoms Zn + 12.04×10^{23} formulas $AgNO_3 \rightarrow 6.02 \times 10^{23}$ formulas $Zn(NO_3)_2$ + 12.04×10^{23} atoms Ag
1 mol Zn + 2 mol $AgNO_3 \rightarrow$ 1 mol $Zn(NO_3)_2$ + 2 mol Ag
65.37 g Zn + 339.92 g $AgNO_3 \rightarrow$ 189.39 g $Zn(NO_3)_2$ + 215.8 g Ag

e) 2 molecules HCl + 1 formula $Mg(OH)_2 \rightarrow$ 1 formula $MgCl_2$ + 2 molecules H_2O
12.04×10^{23} molecules HCl + 6.02×10^{23} formulas $Mg(OH)_2 \rightarrow 6.02 \times 10^{23}$ formulas $MgCl_2$ + 12.04×10^{23} molecules H_2O
2 mol HCl + 1 mol $Mg(OH)_2 \rightarrow$ 1 mol $MgCl_2$ + 2 mol H_2O
72.92 g HCl + 58.33 g $Mg(OH)_2 \rightarrow$ 95.21 g $MgCl_2$ + 36.03 g H_2O

5.40 For the following equation, write statements equivalent to Statements 1, 4, 5, and 6 given in section 5.9). Then write at least six factors (including numbers and units) based on Figure 5.11 and the mole definition that could be used to solve problems by the factor-unit method.

$$2SO_{2(g)} + O_{2(g)} \rightarrow 2SO_{3(g)}$$

Solution:

2 molecules SO_2 + 1 molecule $O_2 \rightarrow$ 2 molecules SO_3
12.04×10^{23} molecules SO_2 + 6.02×10^{23} molecules $O_2 \rightarrow 12.04 \times 10^{23}$ molecules SO_3
2 mol SO_2 + 1 mol $O_2 \rightarrow$ 2 mol SO_3
128.12 g SO_2 + 32.0 g $O_2 \rightarrow$ 160.12 g SO_3

$$\frac{2\text{ mol }SO_2}{1\text{ mol }O_2} \quad \frac{2\text{ mol }SO_2}{2\text{ mol }SO_3} \quad \frac{1\text{ mol }O_2}{160.12\text{ g }SO_3} \quad \frac{32.0\text{ g }O_2}{128.12\text{ g }SO_2} \quad \frac{12.04 \times 10^{23}\text{ molecules }SO_2}{32.0\text{ g }O_2} \quad \frac{2\text{ mol }SO_3}{1\text{ mol }O_2}$$

5.42 Calculate the mass of limestone ($CaCO_3$) that must be decomposed to produce 500 g of lime (CaO). The equation for the reaction is:

$$CaCO_3(s) \rightarrow CaO(s) + CO_2(g)$$

Solution:

$$500\text{ g CaO} \times \frac{1\text{ mol CaO}}{56.1\text{ g CaO}} \times \frac{1\text{ mol }CaCO_3}{1\text{ mol CaO}} \times \frac{100.1\text{ g }CaCO_3}{1\text{ mol }CaCO_3} = 892\text{ g }CaCO_3$$

5.44 Calculate the grams of bromine (Br_2) needed to react with exactly 35.0 g of aluminum (Al). The equation for the reaction is:

$$2Al_{(s)} + 3Br_{2(l)} \rightarrow 2AlBr_{3(s)}$$

Solution:

$$35.0\text{ g Al} \times \frac{1\text{ mol Al}}{26.98\text{ g Al}} \times \frac{3\text{ mol }Br_2}{2\text{ mol Al}} \times \frac{479.4\text{ g }Br_2}{3\text{ mol }Br_2} = 311\text{ g }Br_2$$

5.46 How many grams of $AlBr_3$ are produced by the process in Exercise 5.44?

Solution:

$$35.0\text{ g Al} \times \frac{1\text{ mol Al}}{26.98\text{ g Al}} \times \frac{2\text{ mol }AlBr_3}{2\text{ mol Al}} \times \frac{266.68\text{ g }AlBr_3}{1\text{ mol }AlBr_3} = 346\text{ g }AlBr_3$$

5.48   Pure titanium metal is produced by reacting titanium(IV) chloride with magnesium metal. The equation for the reaction is:

$$TiCl_{4(s)} + 2Mg_{(s)} \rightarrow Ti_{(s)} + 2MgCl_{2(s)}$$

How many grams of Mg would be needed to produce 1.00 kg of pure titanium?

Solution:

$$1.00 \text{ kg Ti} \times \frac{1000 \text{ g Ti}}{1 \text{ kg Ti}} \times \frac{1 \text{ mol Ti}}{47.9 \text{ g Ti}} \times \frac{2 \text{ mol Mg}}{1 \text{ mol Ti}} \times \frac{24.3 \text{ g Mg}}{1 \text{ mol Mg}} = 1.01 \times 10^3 \text{ g Mg}$$

5.50   Caproic acid is oxidized in the body as follows:

$$C_6H_{12}O_{2(aq)} + 8O_{2(aq)} \rightarrow 6CO_{2(aq)} + 6H_2O_{(l)}$$

How many grams of oxygen are needed to oxidize 1.00 mol of caproic acid?

Solution:

$$1.00 \text{ mol } C_6H_{12}O_2 \times \frac{8 \text{ mol } O_2}{1 \text{ mol } C_6H_{12}O_2} \times \frac{32.0 \text{ g } O_2}{1 \text{ mol } O_2} = 256 \text{ g } O_2$$

The Limiting Reactant (Section 5.10)

5.52   Nitrogen and oxygen react as follows: $N_{2(g)} + 2O_{2(g)} \rightarrow 2NO_{2(g)}$

Suppose 1.24 mol N_2 and 50.0 g O_2 are mixed together.
a) Which one is the limiting reactant?
b) What is the maximum mass of NO_2 that can be produced from the mixture?

Solution:

a) If all the N_2 were to react, the mass of NO_2 produced would be:

$$1.24 \text{ mol } N_2 \times \frac{2 \text{ mol } NO_2}{1 \text{ mol } N_2} \times \frac{92.0 \text{ g } NO_2}{2 \text{ mol } NO_2} = 114 \text{ g } NO_2$$

If all the O_2 were to react, the mass of NO_2 produced would be:

$$50.0 \text{ g } O_2 \times \frac{2 \text{ mol } O_2}{64.0 \text{ g } O_2} \times \frac{2 \text{ mol } NO_2}{2 \text{ mol } O_2} \times \frac{92.0 \text{ g } NO_2}{2 \text{ mol } NO_2} = 71.9 \text{ g } NO_2$$

Since less NO_2 can be made using all of the O_2, the O_2 is the limiting reactant.
b) A maximum of 71.9 g NO_2 could be made.

5.54   Ammonia, carbon dioxide, and water vapor react to form ammonium bicarbonate as follows:

$$NH_{3(aq)} + CO_{2(aq)} + H_2O_{(l)} \rightarrow NH_4HCO_{3(aq)}$$

Suppose 50.0 g NH_3, 80.0 g CO_2, and 2.00 mol H_2O are reacted. What is the maximum number of grams of NH_4HCO_3 that can be produced?

Solution:

If all the NH_3 is used, the weight of NH_4HCO_3 produced could be:

$$34.0 \text{ g } NH_3 \times \frac{1 \text{ mol } NH_3}{17.0 \text{ g } NH_3} \times \frac{1 \text{ mol } NH_4HCO_3}{2 \text{ mol } NH_3} \times \frac{79.0 \text{ g } NH_4HCO_3}{1 \text{ mol } NH_4HCO_3} = 158 \text{ g } NH_4HCO_3$$

If all the CO_2 is used, the weight of NH_4HCO_3 produced could be:

$$3.00 \text{ mol } CO_2 \times \frac{1 \text{ mol } NH_4HCO_3}{1 \text{ mol } CO_2} \times \frac{79.0 \text{ g } NH_4HCO_3}{1 \text{ mol } NH_4HCO_3} = 237 \text{ g } NH_4HCO_3$$

If all the H_2O is used, the weight of NH_4HCO_3 produced could be:

$$1.50 \text{ mol } H_2O \times \frac{1 \text{ mol } NH_4HCO_3}{1 \text{ mol } H_2O} \times \frac{79.0 \text{ g } NH_4HCO_3}{1 \text{ mol } NH_4HCO_3} = 119 \text{ g } NH_4HCO_3$$

No more than 119 g NH_4HCO_3 could be made because that uses up all of the H_2O.

Reaction Yields (Section 5.11)

5.56   The actual yield of a reaction was 12.18 g of product, while the calculated theoretical yield was 15.93 g. What is the percentage yield?

Solution:

Percent Yield $= \frac{\text{Actual Yield}}{\text{Theoretical Yield}} \times 100$

Percent Yield $= \frac{12.18 \text{ g}}{15.93 \text{ g}} \times 100 = 76.46\%$

5.58   For a combination reaction, it was calculated that 7.59 g A would exactly react with 4.88 g B. These amounts were reacted and 9.04 g of the product was isolated. What was the percentage yield of the reaction?

Solution:

Percent Yield $= \frac{\text{Actual Yield}}{\text{Theoretical Yield}} \times 100$

Percent Yield $= \frac{9.04 \text{ g}}{7.59 \text{ g} + 4.88 \text{ g}} \times 100 = \frac{9.04 \text{ g}}{12.47 \text{ g}} \times 100 = 72.5\%$

5.60   Upon heating, mercury(II) oxide undergoes a decomposition reaction:

$$2HgO_{(s)} \rightarrow 2Hg_{(l)} + O_{2(g)}$$

A sample of HgO weighing 7.22 g was heated. The collected mercury weighed 5.95 g. What is the percentage yield of the reaction?

Solution:

The theoretical yield is calculated by

$7.22 \text{ g HgO} \times \frac{1 \text{ mol HgO}}{217.6 \text{ g HgO}} \times \frac{1 \text{ mol Hg}}{1 \text{ mol HgO}} \times \frac{201.6 \text{ g Hg}}{1 \text{ mol Hg}} = 6.69 \text{ g Hg}$

Percent Yield $= \frac{5.95 \text{ g Hg (actual)}}{6.69 \text{ g Hg (theoretical)}} \times 100 = 88.9\%$

PROGRAMMED REVIEW

Section 5.1 Chemical Equations

According to convention, a chemical equation is written with (a) _reactants_ on the left and (b) _products_ on the right. In a (c) _balanced_ equation the total number of each kind of atom is equal in the reactants and products.

Section 5.2 Types of Reactions

In this text reactions are first classified as (a) _redox_ or (b) _non redox_. Both of these types are further classified as (c) _combination_ or (d) _decomposition_. Only redox reactions are further classified as (e) _single replacement_ reactions, and only nonredox are further classified as (f) _double replacement_ reactions.

Section 5.3 Redox Reactions

One meaning of the term (a) _oxidation_ is to lose electrons. The term (b) _oxidizing_ can mean to combine with hydrogen. (c) _oxidation numbers_ are positive or negative numbers assigned to the elements in chemical formulas according to a set of rules. The oxidation number of an uncombined element is always (d) _zero_. In a redox reaction, the substance oxidized is called the (e) _reducing_ agent, and the substance reduced is called the (f) _oxidizing_ agent.

Section 5.4 Decomposition Reactions

In decomposition reactions, a single substance forms two or more (a) _simpler_ substances. Decomposition reactions can be either (b) _redox_ or (c) _non redox_

Section 5.5 Combination Reactions

Combination reactions are also called (a) _addition_ or (b) _synthesis_ reactions. In combination reactions, (c) _two_ or _more_ substances combine to form a (d) _single_ substance.

Section 5.6 Replacement Reactions

(a) _Single_ replacement or (b) _Substitution_ reactions are always redox reactions, but (c) _double_ replacement or (d) _metathesis_ reactions are not redox. Partner swapping is a characteristic of (e) _double_ replacement reactions.

Section 5.7 Ionic Equations

When ionic substances dissolve in water they (a) _break apart_ into their constituent (b) _ions_. (c) _Molecular_ equations contain no ions. In a (d) _total ionic_ equation all substances that form ions are written as ions. Ions that appear on both the reactant and product sides of equations are called (e) _spectator ions_, and are not shown in (f) _net ionic_ equations.

Section 5.8 Energy and Reactions

In addition to composition changes, (a) _energy_ changes accompany all chemical reactions. The energy of reactions can take the form of heat, and when heat is liberated the reaction is called (b) _exothermic_. When heat is absorbed the reaction is called (c) _endothermic_

Section 5.9 The Mole and Chemical Equations

Application of the (a) _mole_ concept to a balanced equation provides several (b) _statements_ from which (c) _factors_ can be obtained that are useful in solving problems using the (d) _factor_-_unit_ method.

Section 5.10 The Limiting Reactant

According to the (a) _limiting reactant_ principle, the maximum amount of product that can be obtained from a mixture of reactants is determined by the amount of (b) _limiting_ reactant present. As soon as the (c) _limiting_ reactant has all reacted, the reaction will (d) _stop._

Section 5.11 Reaction Yields

Reactions that do not give the desired products are called (a) _side_ reactions. Such reactions are one reason that the (b) _actual_ yield of a reaction is often less than the (c) _theoretical_ yield. The (d) _percentage_ yield of a reaction is the actual yield divided by the theoretical yield and multiplied by 100.

SELF-TEST QUESTIONS

Multiple Choice

1. Which of the following is a reactant in the reaction: $2H^+ + CaCO_3 \rightarrow H_2O + Ca^{2+} + CO_2$
    a) H_2O b) H^+ c) Ca^{2+} d) CO_2

2. What is the coefficient to the left of H_2 when the following equation is balanced? $Na + H_2O \rightarrow NaOH + H_2$
    a) 1   b) 2   c) 3   d) 4

3. A decomposition reaction can also be classified as
    a) a combination reaction   b) a double replacement reaction
    c) a single replacement reaction   d) a redox or nonredox reaction

4. The oxidation number of a monoatomic ion is always
    a) +1   b) −1
    c) 0   d) equal to the charge on the ion

5. In a redox reaction, the reducing agent
    a) is reduced
    b) is oxidized
    c) gains electrons
    d) contains an element whose oxidation number decreases

6. Identify the spectator ion in the following reaction: $Cl_2 + 2K^+ + 2Br^- \rightarrow 2K^+ + 2Cl^- + Br_2$
    a) Br^- b) Cl_2 c) K^+ d) Cl^-

7. Which of the following statements is consistent with the balanced equation: $2SO_2 + O_2 \rightarrow 2SO_3$
    a) one mol SO_2 reacts with one mol O_2 b) two mol SO_2 produces two mol SO_3
    c) 64.1 g SO_2 reacts with 32.0 g O_2 d) 32.0 g O_2 produces 80.1 g SO_3

8. According to the reaction $N_2 + 3H_2 \rightarrow 2NH_3$, how many grams of H_2 are needed to produce 4.0 moles of NH_3?
    a) 3.0   b) 6.0   c) 9.0   d) 12.0

9. For the reaction: $2H_2 + O_2 \rightarrow 2H_2O$ how many moles of H_2O could be obtained by reacting 64.0 grams of O_2 and 8.0 grams of H_2?
    a) 72   b) 18   c) 2.0   d) 4.0

10. Nitrogen and hydrogen react as follows to form ammonia: $N_2 + 3H_2 \rightarrow 2NH_3$ In a reaction mixture consisting of 1.50 g H_2 and 6.00 g N_2, it is true that
    a) H_2 is the limiting reactant
    b) N_2 is the limiting reactant
    c) H_2 is present in excess
    d) an exact reacting ratio of N_2 and H_2 is present

True-False

11. Oxidation numbers never change in combination reactions.

12. Oxidation numbers of some combined elements may be equal to zero.

13. The loss of hydrogen corresponds to an oxidation process.

14. A reaction that releases heat is classified as endothermic.

15. A balanced equation represents a statement of the law of conservation of matter.

16. The percentage yield of a reaction cannot exceed 100%, if the measured product is pure.

Matching
Match each of the reactions below with the correct category from the responses.

17. _____ $3Fe + 2O_2 \rightarrow Fe_3O_4$ a) decomposition

18. _____ $CuO + H_2 \rightarrow Cu + H_2O$ b) single replacement

19. _____ $CuCO_3 \rightarrow CuO + CO_2$ c) double replacement

20. _____ $KOH + HBr \rightarrow H_2O + KBr$ d) combination

21. _____ $2Ag_2O \rightarrow 4Ag + O_2$

Match the oxidation number of the underlined element to the correct value given as a response.

22. _____ $H_2\underline{C}O_3$

23. _____ $\underline{C}O_2$

24. _____ $\underline{Cs}_2O$

25. _____ $\underline{Cr}_2O_3$

26. _____ $\underline{S}O_3^{2-}$

27. _____ $\underline{Ba}^{2+}$

a) +1

b) +2

c) +3

d) +4

e) none of the above

SOLUTIONS

A. Answers to Programmed Review

5.1 a) reactants	b) products	c) balanced	
5.2 a) redox	b) nonredox	c) combination	d) decomposition
e) single replacement	f) double replacement		
5.3 a) oxidation	b) reduction	c) oxidation numbers	d) zero
e) reducing	f) oxidizing		
5.4 a) simpler	b) redox	c) nonredox	
5.5 a) addition	b) synthesis	c) two or more	d) single
5.6 a) single	b) substitution	c) double	d) metathesis
e) double			
5.7 a) break apart	b) ions	c) molecular	d) total ionic
e) spectator ions	f) net ionic		
5.8 a) energy	b) exothermic	c) endothermic	
5.9 a) mole	b) statements	c) factors	d) factor-unit
5.10 a) limiting reactant	b) limiting	c) limiting	d) stop
5.11 a) side	b) actual	c) theoretical	d) percentage

B. Answers to Self-Test Questions

1. b	10. b	19. a
2. a	11. F	20. c
3. d	12. T	21. a
4. d	13. T	22. d
5. b	14. F	23. d
6. c	15. T	24. a
7. b	16. T	25. c
8. d	17. d	26. d
9. d	18. b	27. b

Chapter 6

The States of Matter

CHAPTER OUTLINE

LEARNING OBJECTIVES

When you have completed your study of this chapter, you should be able to:
1. Do calculations based on the property of density.
2. Know the postulates of the kinetic molecular theory and use the theory to explain the properties of matter in different states.
3. Do calculations based on Boyle's law, Charles's law, the combined gas law, Dalton's law, and Graham's law.
4. Do calculations based on the ideal gas law.
5. Do calculations based on energy changes that accompany the heating, cooling, or changing of state of substances.

ANSWERS AND SOLUTIONS TO EVEN-NUMBERED PROBLEMS

Observed Properties of Matter (Section 6.1)

6.2   Calculate the volume of 125 g of the following liquids:
    a) sea water (d = 1.03 g/mL)
    b) methyl alcohol (d = 0.792 g/mL)
    c) concentrated sulfuric acid (d = 1.84 g/mL)

    Solution:
    a) $v = \frac{m}{d} = \frac{125 \text{ g}}{1.03 \text{ g/mL}} = 121$ mL
    b) $v = \frac{m}{d} = \frac{125 \text{ g}}{0.792 \text{ g/mL}} = 158$ mL
    c) $v = \frac{m}{d} = \frac{125 \text{ g}}{1.84 \text{ g/mL}} = 67.9$ mL

6.4   Liquid water has a density of 1.00 g/mL at 10.0°C and 0.996 g/mL at 30.0°C. Calculate the change in volume that occurs when a 500 mL of water is heated from 10.0°C to 30.0°C.

    Solution:
    At 10.0°C the mass of the water is:
        $m = d \times V = 1.00$ g/~~mL~~ $\times$ 500 ~~mL~~ $= 500$ g
    At 30.0°C, the volume of the water would be:
        $V = \frac{m}{d} = \frac{500 \text{ g}}{0.996 \text{ g/mL}} = 502$ mL
    The volume changed from 500 mL to 502 mL. The change in volume was 2 mL.

6.6   A 1.50-L rubber balloon is filled with nitrogen gas at a temperature of 0.00°C and a pressure of 1.00 atm. The density of nitrogen gas under these conditions is 1.25 g/L.
    a) Will the density of the nitrogen increase or decrease when the balloon is heated?
    b) At 50.0°C, the balloon has a volume of 1.78 L. Calculate the nitrogen density at this temperature.

    Solution:
    a) Gases tend to expand when heated. Since the mass is constant and the volume is increasing, the density decreases (Volume and density are inversely proportional.).
    b) The mass of the nitrogen is m = d × v = 1.25 g/L × 1.50 L = 1.875 g
        $d = \frac{m}{v} = \frac{1.875 \text{ g}}{1.78 \text{ L}} = 1.05 \frac{g}{L}$

The Kinetic Molecular Theory of Matter (Section 6.2)

6.8   Suppose a ball is thrown into the air such that the ball goes straight up, then the ball is caught by the person who threw it. Describe the changes in form of energy that occur for the ball from the time it is thrown until it is caught.

    Solution:
    As the ball travels upward, the kinetic energy decreases and the potential energy increases until the maximum height is reached. At that point, the kinetic energy equals 0 because the ball stops instantaneously. As the ball falls through the air, the kinetic energy increases and the potential energy decreases. If there were no air resistance, the ball would attain the same kinetic energy as when it was thrown.

6.10   At 25.0°C, He molecules (He) have an average velocity of 1.26×10^5 cm/s, and methane molecules (CH_4) have an average velocity of 6.30×10^4 cm/s. Calculate the kinetic energy of each type of molecule at 25.0°C and determine which is greater. Express molecular masses in u for this calculation.

Solution:

$KE = 1/2 \, mv^2$

for He: $KE = 1/2 \, mv^2 = 1/2(4.003 \text{ u})(1.26 \times 10^5 \text{ cm/s})^2 = 3.18 \times 10^{10} \text{ u cm}^2/\text{s}^2$

for CH_4: $KE = 1/2 \, mv^2 = 1/2(16.042 \text{ u})(6.30 \times 10^4 \text{ cm/s})^2 = 3.18 \times 10^{10} \text{ u cm}^2/\text{s}^2$

The result of calculations is the same kinetic energy. This result is expected as all gases have the same average kinetic energy at the same temperature according to the kinetic molecular theory.

The Solid, Liquid, and Gaseous States (Section 6.3–6.5)

6.12   Explain each of the following observations using the kinetic molecular theory of matter:
a) A liquid takes the shape, but not necessarily the volume, of its container.
b) Solids and liquids are practically incompressible.
c) A gas always exerts uniform pressure on all walls of its container.

Solution:
a) The particles of a liquid are relatively close to each other, are in constant motion, and readily move across each other. There is not enough kinetic energy for particles to separate far enough to completely fill a container if the volume of the liquid is less than the container. The aggregation of particles has a fluid action taking the shape of the container.
b) Solids and liquids are composed of particles that are close enough to each other that there is no room to be moved closer together.
c) The particles of a gas have sufficient kinetic energy to overcome the attraction between particles. Particles are free to move randomly providing for collisions with the container randomly and, therefore, evenly.

6.14   Discuss the differences in kinetic and potential energy of the constituent particles for a substance in the solid, liquid, and gaseous states.

Solution:
Heat must be added to a substance in order to convert it from the solid to the liquid state, and then from the liquid to the gaseous state. If the added heat causes no temperature difference between the two states, as when a solid melts to a liquid at the melting point, then the added heat increases the potential energy of the particles of the substance. Thus, at the melting point, the particles of both solid and liquid forms of a substance will have the same kinetic energy because they are at the same temperature. However, particles of the liquid would have more potential energy corresponding to the energy that had to be added to change the solid to a liquid. Similar arguments apply for a comparison of the liquid and gaseous states of a substance.

6.16   The following statements are best associated with the solid, liquid, or gaseous states of matter. Match the statements to the appropriate state of matter.
a) Temperature changes influence the volume of this state substantially.
b) In this state, constituent particles are less free to move about than in other states.
c) Pressure changes influence the volume of this state more than that of the other two states.
d) This state is characterized by an indefinite shape and a low density.

Solution:
a) Gases change volume most with temperature changes.
b) Solids have the least motion for the molecules.
c) Gases change volume most with pressure changes. They are very compressible.
d) Gases have very low density and an indefinite shape.

The Gas Laws (Section 6.6)

6.18   A weather reporter on TV reports the barometric pressure as 28.6 inches of mercury. Calculate this pressure in the following units:

    a) atm   b) torr   c) psi   d) bars

Solution:

a) $28.6 \text{ in} \times \frac{1 \text{ atm}}{29.9 \text{ in}} = 0.957$ atm

b) $28.6 \text{ in} \times \frac{1 \text{ atm}}{29.9 \text{ in}} \times \frac{760 \text{ torr}}{1 \text{ atm}} = 727$ torr

c) $28.6 \text{ in} \times \frac{1 \text{ atm}}{29.9 \text{ in}} \times \frac{14.7 \text{ psi}}{1 \text{ atm}} = 14.1$ psi

d) $28.6 \text{ in} \times \frac{1 \text{ atm}}{29.9 \text{ in}} \times \frac{1.01 \text{ bars}}{1 \text{ atm}} = 0.967$ bar

6.20   An engineer reads the pressure gauge of a boiler at 210 psi. Calculate this pressure in the following units:

    a) atm   b) bars   c) mm Hg   d) in. Hg

Solution:

a) $210 \text{ psi} \times \frac{1 \text{ atm}}{14.7 \text{ psi}} = 14.3$ atm

b) $210 \text{ psi} \times \frac{1 \text{ atm}}{14.7 \text{ psi}} \times \frac{1.01 \text{ bars}}{1 \text{ atm}} = 14.4$ bars

c) $210 \text{ psi} \times \frac{1 \text{ atm}}{14.7 \text{ psi}} \times \frac{760 \text{ mm Hg}}{1 \text{ atm}} = 10{,}900$ mm Hg $= 1.09 \times 10^4$ mm Hg

d) $210 \text{ psi} \times \frac{1 \text{ atm}}{14.7 \text{ psi}} \times \frac{29.9 \text{ in. Hg}}{1 \text{ atm}} = 427$ in. Hg

6.22   Convert each of the following temperatures from the unit given to the unit indicated.
    a) the melting point of potassium metal, 63.7°C, to kelvins
    b) the freezing point of liquid hydrogen, 14.1 K, to degrees Celsius
    c) the boiling point of liquid helium, −268.9°C, to kelvins

Solution:

$K = °C + 273$ $°C = K − 273$

a) $63.7°C + 273 = 337$ K

b) $14.1 \text{ K} − 273 = −259°C$

c) $−268.9°C + 273 = 4$ K

Pressure, Temperature, and Volume Relationships (Section 6.7)

6.24   Use the combined gas law (Equation 6.9) to calculate the unknown quantity for each gas sample described in the following table.

	Sample A	Sample B	Sample C
P_i	1.50 atm	2.35 atm	9.86 atm
V_i	2.00 L	1.97 L	11.7 L
T_i	300 K	293 K	500 K
P_f	?	1.09 atm	5.14 atm
V_f	3.00 L	?	9.90 L
T_f	450 K	310 K	?

Solution:

$$\frac{P_i V_i}{T_i} = \frac{P_f V_f}{T_f}$$

Gas A: $P_f = P_i \times \frac{V_i}{V_f} \times \frac{T_i}{T_f} = 1.50 \text{ atm} \times \frac{2.00 \text{ L}}{3.00 \text{ L}} \times \frac{450 \text{ K}}{300 \text{ K}} = 1.50$ atm

Gas B: $V_f = V_i \times \frac{P_i}{P_f} \times \frac{T_f}{T_i} = 1.97 \text{ L} \times \frac{2.35 \text{ atm}}{1.09 \text{ atm}} \times \frac{310 \text{ K}}{293 \text{ K}} = 4.49 \text{ L}$

Gas C: $T_f = T_i \times \frac{P_f}{P_i} \times \frac{V_f}{V_i} = 500 \text{ K} \times \frac{5.14 \text{ atm}}{9.86 \text{ atm}} \times \frac{9.90 \text{ L}}{11.7 \text{ L}} = 221 \text{ K}$

6.26   A 200-mL sample of nitrogen gas is collected at 45.0°C and a pressure of 610 torr. What volume will the gas occupy at STP (0°C and 760 torr)?

Solution:

$V_f = P_i \times \frac{V_i}{V_f} \times \frac{T_f}{T_i} = \frac{610 \text{ torr}}{760 \text{ torr/atm}} \times \frac{200 \text{ mL}}{1 \text{ atm}} \times \frac{273 \text{ K}}{318 \text{ K}} = 138 \text{ mL}$

6.28   A 2.50 L sample of neon at 0.00°C and 1.00 atm is compressed into a 0.75 L cylinder. What pressure will the gas exert in the cylinder at 30°C.

Solution:

Convert temperatures to K: $T_i = 273 \text{ K}$; $T_f = 30.0°\text{C} + 273 \text{ K} = 303 \text{ K}$

$P_f = P_i \times \frac{V_i}{P_f} \times \frac{T_f}{T_i} = 1.00 \text{ atm} \times \frac{2.50 \text{ L}}{0.75 \text{ L}} \times \frac{303 \text{ K}}{273 \text{ K}} = 3.7 \text{ atm (keep only 2 sig figs)}$

6.30   What volume (in liters) of air measured at 1.00 atm would have to be put into a car tire with a volume of 14.5 L if the pressure in the car tire is to be 32.0 psi? Assume the temperature of the gas remains constant.

Solution:

$V_i = ?$; $V_f = 14.5 \text{ L}$; $P_i = 1.0 \text{ atm}$; $P_f = 32.0 \text{ psi}$; $T_i = T_f$

P_i is converted to psi; one atmosphere is 14.7 psi.

$V_i = V_f \times \frac{P_f}{P_i} = 14.5 \text{ L} \times \frac{32.0 \text{ psi}}{14.7 \text{ psi}} = 31.6 \text{ L}$

6.32   A sample of gas has a volume of 750 mL at a pressure of 700 torr. What volume will the gas occupy at the same temperature but at the standard atmospheric pressure, 760 torr?

Solution:

$V_i = 750 \text{ mL}$; $V_f = ?$; $P_i = 700 \text{ torr}$; $P_f = 760 \text{ torr}$; $T_i = T_f$

$V_f = V_i \times \frac{P_i}{P_f} = 750 \text{ mL} \times \frac{700 \text{ torr}}{760 \text{ torr}} = 691 \text{ mL}$

6.34   A 3.8-L sample of gas at 1.0 atm and 20.0°C is heated to 75°C. Calculate the gas volume at the higher temperature if the pressure remains at 1.0 atm.

Solution:

$V_i = 3.8 \text{ L}$; $V_f = ?$; $P_i = P_f$; $T_i = 20°\text{C} + 273 = 293 \text{ K}$; $T_f = 75°\text{C} + 273 = 348 \text{ K}$

$V_f = V_i \times \frac{T_f}{T_i} = 3.8 \text{ L} \times \frac{348 \text{ K}}{293 \text{ K}} = 4.5 \text{ L}$

6.36   What volume of gas at 120°C must be cooled to 35°C if the gas volume, at a constant pressure and 35 °C, is to be 1.5 L?

Solution:

$V_i = ?$; $V_f = 1.5 \text{ L}$; $P_i = P_f$; $T_i = 120°\text{C} + 273 = 393 \text{ K}$; $T_f = 35°\text{C} + 273 = 308 \text{ K}$

$V_i = V_f \times \frac{T_i}{T_f} = 1.5 \text{ L} \times \frac{393 \text{ K}}{308 \text{ K}} = 1.9 \text{ L}$

6.38   A 2500 L sample of oxygen gas is produced at 1.00 atm pressure. It is to be compressed and stored in a 20.0 L steel cylinder. Assume it is produced and stored at the same temperature and calculate the pressure of the oxygen in the cylinder.

Solution:
$V_i = 2500$ L; $V_f = 20.0$ L; $P_i = 1.00$ atm; $P_f = ?$; $T_i = T_f$
$P_f = P_i \times \frac{V_i}{V_f} = 1.00$ atm $\times \frac{2500 \text{ L}}{20.0 \text{ L}} = 125$ atm

6.40   A helium balloon was partially filled with 8000 ft^3 of helium when the atmospheric pressure was 0.98 atm and the temperature was 23°C. The balloon rose to an altitude where the atmospheric pressure was 400 torr and the temperature was 5.3°C. What volume did the helium occupy at this altitude?

Solution:
$V_i = 8000$ ft^3; $V_f = ?$; $P_i = 0.98$ atm; $P_f = 400$ torr $\times$ (1 atm/760 torr) $= 0.526$ atm;
$T_i = 23°C + 273 = 296$ K; $T_f = 5.3°C + 273 = 278$ K
$V_f = V_i \times \frac{P_f}{P_i} \times \frac{T_f}{T_i} = 8000$ ft$^3 \times \frac{0.98 \text{ atm}}{0.526 \text{ atm}} \times \frac{278 \text{ K}}{296 \text{ K}} = 13,999$ (round to 2 sig figs) $= 1.4 \times 10^4$ ft^3

6.42   A gas occupies 250 mL at a pressure of 2.10 atm. What volume would it occupy at the same temperature and a pressure of 60.0 kPa?

Solution:
$V_i = 250$ mL; $V_f = ?$; $P_i = 2.10$ atm $\times$ (101 kPa/1 atm) $= 212$ kPa; $P_f = 60.0$ kPa; $T_i = T_f$
$V_f = V_i \times \frac{P_i}{P_f} = 250$ mL $\times \frac{212 \text{ kPa}}{60.0 \text{ kPa}} = 883$ mL

6.44   A 2.00 L sample of nitrogen gas at 760 torr and 0.0°C weighs 2.50 g. The pressure on the gas is increased to 3.00 atm at 0.0°C. Calculate the gas density at the new pressure in grams per liter.

Solution:
The density at the new pressure is (mass/V_f)
$V_i = 2.00$ L; $V_f = ?$; $P_i = 760$ torr $= 1$ atm; $P_f = 3.00$ atm; $T_i = T_f = 273$ K
$V_f = V_i \times \frac{P_i}{P_f} = 200$ L $\times \frac{1 \text{ atm}}{3.00 \text{ atm}} = 0.667$ L
density $= \frac{2.50 \text{ g}}{0.667 \text{ L}} = 3.75$ g/L

The Ideal Gas Law (Section 6.8)

6.46   Use the ideal gas law and calculate the following:
   a) The number of moles of argon in a gas sample that occupies a volume of 400 mL at a temperature of 90.0°C and has a pressure of 735 torr
   b) The pressure exerted by 0.738 mol of hydrogen gas confined to a volume of 2.60 L at 45°C
   c) The volume of a tank of nitrogen if 1.75 mol of the gas exerts a pressure of 4.32 atm at 25°C.

Solution:
PV = nRT where R = 0.0821 L atm/mol K
pressure must be in atm, volume in L and temperature in kelvins
   a) convert to K: 90°C + 273 = 363 K   convert to L: 400 mL $\times \frac{1 \text{ L}}{1000 \text{ mL}} = 0.400$ L
      convert to atm: 735 torr $\times \frac{1 \text{ atm}}{760 \text{ torr}} = 0.967$ atm
      $n = \frac{PV}{RT} = \frac{(0.967 \text{ atm})(0.400 \text{ L})}{(0.0821 \text{ L atm/mol K})(363 \text{ K})} = 0.0130$ mol

b) convert to K: $45°C + 273 = 318$ K

$P = \frac{nRT}{V} = \frac{(0.738 \text{ mol})(0.0821 \text{ L atm/mol K})(318 \text{ K})}{2.60 \text{ L}} = 7.41$ atm

c) convert to K: $25°C + 273 = 298$ K

$V = \frac{nRT}{P} = \frac{(1.75 \text{ mol})(0.0821 \text{ L atm/mol K})(298 \text{ K})}{4.32 \text{ atm}} = 9.91$ L

6.48   Suppose 10.0 g sulfur dioxide gas (SO_2) is compressed into a 0.750 L steel cylinder at a temperature of 27°C. What pressure would the compressed gas exert on the walls of the cylinder?

Solution:

convert to K: $27°C + 273 = 300$

convert to mol SO_2: $10.0 \text{ g } SO_2 \times \frac{1 \text{ mol } SO_2}{64.06 \text{ g } SO_2} = 0.156$ mol SO_2

$P = \frac{nRT}{V} = \frac{(0.156 \text{ mol})(0.0821 \text{ L atm/mol K})(300 \text{ K})}{0.750 \text{ L}} = 5.12$ atm

6.50   The pressure gauge of a steel cylinder of methane (CH_4) reads 380 psi. The cylinder has a volume of 0.500 L and is at a temperature of 30°C. How many grams of methane does the cylinder contain?

Solution:

convert to K: $30°C + 273 = 303$ K   convert to atm: $380 \text{ psi} \times \frac{1 \text{ atm}}{14.7 \text{ psi}} = 25.9$ atm

$n = \frac{PV}{RT} = \frac{(25.9 \text{ atm})(0.500 \text{ L})}{(0.0821 \text{ L atm/mol K})(303 \text{ K})} = 0.521$ mol

$0.521 \text{ mol } CH_4 \times \frac{16.042 \text{ g } CH_4}{1 \text{ mol } CH_4} = 8.36$ g CH_4

6.52   An experimental chamber has a volume of 60 L. How many moles of oxygen gas will be required to fill the chamber at STP?

Solution:

STP = 1.00 atm and 0°C (273 K)

$n = \frac{PV}{RT} = \frac{(1.00 \text{ atm})(6.00 \text{ L})}{(0.0821 \text{ L atm/mol K})(273 \text{ K})} = 2.68$ mol O_2

Alternate Method:

$60.0 \text{ L} \times \frac{1 \text{ mol } O_2}{22.4 \text{ L}} = 2.68$ mol O_2

6.54   A sample of methyl ether has a mass of 8.12 g and occupies a volume of 3.96 L at STP. What is the molecular weight of methyl ether?

Solution:

$MW = \frac{mRT}{PV}$ where m = mass and MW = molecular weight

$MW = \frac{mRT}{PV} = \frac{(8.12 \text{ g})(0.0821 \text{ L atm/mol K})(273 \text{ K})}{(1.00 \text{ atm})(3.96 \text{ L})} = 46.0$ g/mol

Alternate Method:

The molecular weight can be found by dividing the mass by the number of moles.

number of moles = $3.96 \text{ L} \times \frac{1 \text{ mol}}{22.4 \text{ L}} = 0.177$ mol

$\frac{8.12 \text{ g}}{0.177 \text{ mol}} = 45.9$ g/mole

Note: These answers differ because of rounding errors, but they have no significant difference.

6.56   A sample of gas weights 0.176 g and has a volume of 114.0 mL at a pressure and temperature of 640 torr and 20°C. Determine the molecular weight of the gas and identify is as CO, CO_2, or O_2.

Solution:

convert to K: 20°C + 273 = 293 K   convert to atm: 640 ~~torr~~ $\times \frac{1 \text{ atm}}{760 \text{ ~~torr~~}}$ = 0.842 atm

convert to L: 114.0 ~~mL~~ $\times \frac{1 \text{ L}}{1000 \text{ ~~mL~~}}$ = 0.114 L

$MW = \frac{mRT}{PV} = \frac{(0.176 \text{ g})(0.0821 \text{ ~~L~~ ~~atm~~/mol K})(293 \text{ K})}{(0.842 \text{ ~~atm~~})(0.114 \text{ ~~L~~})}$ = 44.1 g/mol

The gas is probably CO_2 (MW = 44).

Dalton's Law (Section 6.9)

6.58   A steel cylinder contains a mixture of nitrogen, oxygen, and carbon dioxide gases. The total pressure in the tank is 1800 torr. The pressure exerted by the nitrogen and oxygen is 750 torr and 810 torr respectively. What is the partial pressure of CO_2 in the mixture?

Solution:

$P_{total} = P_{nitrogen} + P_{oxygen} + P_{carbon \text{ dioxide}}$ or $P_{carbon \text{ dioxide}} = P_{total} - P_{nitrogen} - P_{oxygen}$

$P_{carbon \text{ dioxide}}$ = 1800 torr – 750 torr – 810 torr = 240 torr

Graham's Law (Section 6.10)

6.60   Hydrogen gas (H_2) is found to diffuse approximately four times as fast as oxygen gas (O_2). Use this information and determine how the masses of hydrogen molecules and oxygen molecules compare. How do they compare based on information in the periodic table?

Solution:

Based on Graham's Law, if the rate is four times faster, the mass of oxygen must be about 16 times as large.

$\frac{Rate_{hydrogen}}{Rate_{oxygen}} = \sqrt{\frac{M_{oxygen}}{M_{hydrogen}}}$ = 4 or $\frac{M_{oxygen}}{M_{hydrogen}} = \frac{(Rate_{hydrogen})^2}{(Rate_{oxygen})^2} = \frac{4^2}{1^2}$ = 16

This result is consistent with the atomic weights of H (1.008 u) and oxygen (16.00 u).

6.62   Two identical rubber balloons were filled with gas, one with helium and the other with nitrogen. After a time, it was noted that one of the balloons appeared to be going "flat." Which one do you think it was? Explain.

Solution:

The smaller molecule would diffuse more rapidly through the very small pores in the rubber of the balloon. The one that diffuses faster will "go flat" in less time. The smaller mass molecule will diffuse faster. The mass of He = 4.00 u; the mass of N_2 = 28.0 u. Thus, the helium-filled balloon will go "flat" faster.

Changes in State (Section 6.11)

6.64   Classify each of the following processes as *endothermic* or *exothermic*:

a) condensation   b) liquefaction   c) boiling

Solution:

a) exothermic   b) endothermic   c) endothermic

6.66   Discuss what is meant by a change in state.

Solution:
A change in state is a change from one of the states of matter (a solid, liquid, or gas) to a different state.

Evaporation and Vapor Pressure (Section 6.12)
6.68   Methylene chloride (CH_2Cl_2) was used at one time as a local anesthetic by dentists. It was sprayed onto the area to be anesthetized. Propose an explanation for how it worked.

Solution:
You may have noticed that the ability of the skin to feel pain decreases as the skin becomes cold. Methylene chloride is a very volatile liquid. As it is sprayed on the skin (or membranes of the mouth), it evaporates quickly. As it evaporates, the skin is cooled, creating a temporary anesthetic effect.

Boiling and the Boiling Point (Section 6.13)
6.70   Two glass containers each contain a clear, colorless, odorless liquid that has been heated until it is boiling. One liquid is water (H_2O) and the other is ethylene glycol ($C_2H_6O_2$). Explain how you could make one measurement of each boiling liquid, using the same device and tell which liquid was which.

Solution:
The two liquids would have different boiling points. (The ethylene glycol would be higher.) A simple thermometer could be used to measure the boiling point. The one with the higher boiling point would be the ethylene glycol.

6.72   Suppose you were on top of Mount Everest and wanted to cook a potato as quickly as possible. You left your microwave oven home, so you can either boil the potato in water or throw it into a campfire. Explain which method you would use and why.

Solution:
Because of the reduced air pressure at that altitude, the temperature of the boiling water would be much less than at lower altitudes. However, the temperature of a campfire would be the same at either location. (Assuming there is enough oxygen to build a fire.) Hence, the potato would cook faster in the higher temperature of the fire.

Energy and the States of Matter (Section 6.15)
6.74   Using the specific heat data of Table 6.8, calculate the amount of heat (in calories) needed to increase the temperature of the following:
a) 115 g of copper from 35°C to 75°C
b) 250 g of mercury from 110°C to 320°C
c) 5000 g of nitrogen from 200°C to 900°C

Solution:
a) heat = $(115\ g)(0.093\ cal/g\ °C)(75 - 35)°C = 4.3 \times 10^2$ cal
b) heat = $(250\ g)(0.033\ cal/g\ °C)(320 - 110)°C = 1.7 \times 10^3$ cal
c) heat = $(5000\ g)(0.25\ cal/g\ °C)(900 - 200)\ °C = 8.8 \times 10^5$ cal

6.76   Why wouldn't a solid such as K_2SO_4 (melting point = 1069°C, heat of fusion = 50.3 cal/g) be suitable for use in a solar heat storage system? (See Exercise 6.75)

Solution:

In order to be useful, the solid should have a melting point above room temperature but low enough for the sun to melt it. The melting point of K_2SO_4 is way too high.

6.78   Calculate the total amount of heat needed to change 500 g of ice at $-10°C$ into 500 g of steam at $120°C$. Do this by calculating the heat required for each of the following steps and adding to get the total:

Step 1. Ice $(-10°C) \rightarrow$ ice $(0°C)$
Step 2. Ice $(0°C) \rightarrow$ water $(0°C)$
Step 3. Water $(0°C) \rightarrow$ water $(100°C)$
Step 4. Water $(100°C) \rightarrow$ steam $(100°C)$
Step 5. Steam $(100°C) \rightarrow$ steam $(120°C)$

Solution:

Step 1. heat $= (500\text{ g})(0.51\text{ cal/g }°C)[0 - (-10)]\text{ }°C = 2.6 \times 10^3$ cal
Step 2. heat $= (500\text{ g})(80\text{ cal/g}) = 4.0 \times 10^4$ cal
Step 3. heat $= (500\text{ g})(1.00\text{ cal/g }°C)(100 - 0)\text{ }°C = 5.00 \times 10^4$ cal
Step 4. heat $= (500\text{ g})(540\text{ cal/g}) = 2.70 \times 10^5$ cal
Step 5. heat $= (500\text{ g})(0.48\text{ cal/g }°C)(120 - 100)\text{ }°C = 4.8 \times 10^3$ cal

Total $= 2.6 \times 10^3 + 4.0 \times 10^4 + 5.00 \times 10^4 + 2.70 \times 10^5 + 4.8 \times 10^3 = 3.67 \times 10^5$

PROGRAMMED REVIEW

Section 6.1 Observed Properties of Matter

The (a) <u>density</u> of a substance is defined as the mass of a sample divided by the volume of the sample. The change in volume resulting from a change in pressure is called the (b) <u>compressibility</u> and is quite high for matter in the (c) <u>gaseous</u> state. The change in volume resulting from a temperature change is called the (d) <u>thermal expansion</u> of matter.

Section 6.2 The Kinetic Molecular Theory of Matter

According to the kinetic molecular theory, matter is composed of tiny (a) <u>particles</u> that are in constant (b) <u>motion</u> and therefore possess (c) <u>kinetic</u> energy. The particles also possess (d) <u>potential</u> energy as a result of attracting or repelling each other. (e) <u>cohesive</u> forces are most closely associated with potential energy, while (f) <u>disruptive</u> forces are associated with kinetic energy of the particles.

Section 6.3 The Solid State

In the solid state, (a) <u>cohesive</u> forces are stronger than (b) <u>disruptive</u> forces. Solids generally have a (c) <u>high</u> density, (d) <u>definite</u> shape, (e) <u>small</u> compressibility, and (f) <u>very small</u> thermal expansion.

Section 6.4 The Liquid State

In the liquid state, (a) <u>cohesive</u> forces dominate slightly. Liquids are characterized by a (b) <u>high</u> density, (c) <u>indefinite</u> shape, (d) <u>small</u> compressibility, and (e) <u>small</u> thermal expansion.

Section 6.5 The Gaseous State

In the gaseous state, (a) <u>disruptive</u> forces completely overcome (b) <u>cohesive</u> forces. Gases are characterized by a (c) <u>low</u> density, (d) <u>torr</u> shape, (e) <u>large</u> compressibility, and (f) <u>moderate</u> thermal expansion.

Section 6.6 The Gas Laws

Mathematical expressions that describe the behavior of gases as they are mixed, subjected to pressure changes, etc. are called (a) _gas laws_. An important quantity in gas calculations is the (b) _preassure_ which is defined as a force per unit area. Two units of pressure are the standard (c) _atmosphere_ and the (d) _torr_ which are, respectively, the pressure needed to support 760 and 1 mm columns of mercury.

Section 6.7 Pressure, Temperature and Volume Relationships

Boyle's Law describes a relationship between the (a) _preassure_ and (b) _volume_ of a gas sample maintained at constant (c) _temperature_ A mathematical expression of Boyle's Law is (d) _PV=K_. Charles' Law describes the behavior of a gas in terms of the (e) _volume_ and (f) _temp._ of a sample maintained at constant (g) _pressure_ A mathematical expression of Charles' Law is (h) _V=K'T_. The temperature in this expression must be expressed in (i) _Kelvins_. Combination of (j) _Boyle's_ law and (k) _Charles_ law gives the (l) _combines_ gas law which provides a relationship between the (m) _pressure_ (n) _volume_ and (o) _temp._ of a gas sample. A mathematical expression of this law is (p) $\frac{P_i V_i}{T_i} = \frac{P_f V_f}{T_f}$.

Section 6.8 The Ideal Gas Law

According to Avogadro's law, (a) _equal_ volumes of gases at the same temperature and pressure contain (b) _equal_ numbers of molecules of gas. The standard conditions adopted for gas measurements are a pressure of (c) _one atmosphere_ and a temperature of (d) _273_ kelvins. A combination of Boyle's, Charles' and Avogadro's law leads to the (e) _ideal gas law_, which is written mathematically as (f) _PV=nRT_. The R found in the ideal gas law is called the (g) _universal_ gas constant.

Section 6.9 Dalton's Law

Dalton's Law is also called the law of (a) _partial pressures_. According to this law, the (b) _total_ pressure of a mixture of gases is equal to the sum of the (c) _partial pressure_ of the gases in the mixture.

Section 6.10 Graham's Law

Graham's Law describes the processes of (a) _effusion_ and (b) _diffusion_ for gases. According to this law, gases of lower molecular weight diffuse (c) _faster_ than gases of higher molecular weight.

Section 6.11 Changes in State

(a) _heating_ and (b) _cooling_ are the processes used most often to change matter from one state to another. Such changes are called (c) _exothermic_ when heat is released or taken away, and (d) _endothermic_ when heat is added to cause the change to occur.

Section 6.12 Evaporation and Vapor Pressure

(a) _Evaporation_, a process sometimes called (b) _vapourization_, is an (c) _endothermic_ process in which particles leave the surface of a liquid. The reverse process is called (d) _condensation_ The pressure exerted by a vapor in equilibrium with liquid is called the (e) _vapour pressure_ of the liquid.

Section 6.13 Boiling and the Boiling Point

The (a) _boiling point_ of a liquid is the (b) _temp._ at which the liquid vapor pressure is equal to the atmospheric pressure above the liquid. A (c) _normal_ or (d) _standard_ boiling point is the temperature at which the liquid vapor pressure is equal to one (e) _standard atmosphere_.

Section 6.14 Sublimation and Melting

The process in which a solid changes directly to a (a) _gas_ without first becoming a liquid is called (b) _sublimation_ The temperature at which solid and liquid forms of a substance have the same vapor pressure is called the (c) _melting point_ of the substance.

Section 6.15 Energy and the States of Matter

Substances not decomposed by heating undergo a temperature change or a change in state as heat is added. The amount of heat needed to change the temperature of a specified amount of substance by 1°C is called the (a) specific heat of the substance. The amount of heat needed to melt one gram of a solid to a liquid at constant temperature is called the (b) heat of fusion for the solid. The amount of heat needed to change one gram of liquid to a gas at constant temperature is called the (c) heat of vapourization

SELF-TEST QUESTIONS

Multiple Choice

1. The density of ether is 0.736 g/mL. How much would 20.0 mL of ether weigh?
   a) 14.7 g   b) 1.47 g   c) 2.72 g   d) 27.2 g

2. Calculate the density of a swimmer who weighs 40.0 kg and occupies a volume of 45.0 liters.
   a) 0.889 g/mL   b) 0.0088 g/mL   c) 1.12 g/mL   d) 0.0112 g/mL

3. As a pure liquid is heated, its temperature increases and it becomes less dense. Therefore, which of the following is true?
   a) its potential energy increases
   b) its kinetic energy increases
   c) both its potential and kinetic energy increase
   d) its potential energy increases while kinetic energy decreases

4. Which of the following is an endothermic process?
   a) freezing of water   b) condensation of steam
   c) melting of tin   d) solidification of liquid sulfur

5. Which of the following is a property of the gaseous state?
   a) it has a low density   b) it has a large degree of compressibility
   c) it has a moderate degree of thermal expansion   d) more than one response is correct.

6. When the temperature is 0°C, a balloon has a volume of 5 liters. If the temperature is changed to 50°C and if we assume that the pressure inside the balloon equals atmospheric pressure at all times, which of the following will be true?
   a) the new volume will be larger
   b) the new volume will be smaller
   c) there will be no change in volume
   d) the new volume could be calculated from Boyle's law

7. A steel cylinder of CO_2 gas has a pressure of 20 atmospheres (atm) at 20°C. If the cylinder is heated to 70°C, what will the pressure be?
   a) it will be greater than 20 atm   b) it will be less than 20 atm
   c) it will be exactly 20 atm   d) it cannot be calculated

8. A balloon is filled with exactly 5.0 liters of helium gas in a room where the temperature is 20°C. What volume will the balloon have when it is taken outside and cooled to 10°C?
   a) 2.5 liters   b) 4.8 liters   c) 5.2 liters   d) 10.0 liters

9. If the specific heat of water is 1.00 calorie per gram °C (cal/g °C) and the heat of vaporization is 540 cal/g, how much heat would it take to raise the temperature of 10.0 g of water 10°C?
    a) 10 cal   b) 100 cal   c) 5400 cal   d) 54,000 cal

10. The heat of fusion of water is 80.0 cal/g. Therefore, what happens when 1.00 g of ice is melted?
    a) 80 calories of heat would be absorbed   b) 80 calories of heat would be released
    c) no heat change would take place   d) what would happen is not predictable

11. Calculate the amount of heat necessary to convert 5.00 g of water at 100°C into steam at 100° C. The heat of vaporization of water is 540 cal/g. The specific heat of steam is 0.480 cal/g °C.
    a) 24.0 cal   b) 545 cal   c) 2700 cal   d) 2724 cal

12. What volume would 1 mole of gaseous CH_4 occupy at standard temperature and 0.82 atmospheres of pressure? Remember that R = 0.0821 L atm/mol K.
    a) 0.100 liter   b) 1.00 liter   c) 2.70 liters   d) 27.3 liters

13. A sample of gas is found to occupy a volume of 6.80 liters at a temperature of 30.0°C and a pressure of 640 torr. How many moles of gas are in the sample? (R = 0.0821 L atm/mol K)
    a) 0.230 mole   b) 4.33 moles   c) 1769 moles   d) 2.33 moles

14. A 1.20 L sample of gas is under a pressure of 15.0 atm. What volume would the sample occupy if the pressure were lowered to 5.20 atm and the temperature was kept constant?
    a) 0.289 L   b) 0.416 L   c) 3.46 L   d) 34.6 L

True-False

15. According to Graham's law, a gas with molecules four times as heavy as a second gas will diffuse two times faster than the second gas.

16. If the volume of a gas is held constant, increasing the temperature would result in no increase in pressure.

17. The total pressure of a sample of oxygen saturated with water vapor is equal to the partial pressure of the oxygen plus the partial pressure of the water vapor.

18. The state of matter in which disruptive forces predominate is the gaseous state.

19. Boiling points of liquids decrease as atmospheric pressure decreases.

20. When a liquid is placed in a closed container, evaporation continues for a time and then stops.

21. When the combined gas law is used in a calculation, the gas pressures must be expressed in atm.

Matching
Match the states of matter given as responses to the following descriptions.

22. _____ state has a high density   a) solid

23. _____ state has a large compressibility   b) liquid

24. _____ cohesive forces dominate slightly   c) gas

25. _____ cohesive forces predominate over disruptive forces   d) two or more states

26. _____ sample takes shape of its container

Match the terms given as responses to the following descriptions.

27. _____ molten steel changes to a solid   a) sublimation

28. _____ water in an open container disappears after a time   b) evaporation

29. _____ ice on a car windshield disappears without melting   c) freezing

SOLUTIONS

A. Answers to Programmed Review

6.1	a) density	b) compressibility	c) gaseous	d) thermal expansion

6.1   a) density   b) compressibility   c) gaseous   d) thermal expansion

6.2   a) particles   b) motion   c) kinetic   d) potential
      e) cohesive   f) disruptive

6.3   a) cohesive   b) disruptive   c) high   d) definite
      e) small   f) very small

6.4   a) cohesive   b) high   c) indefinite   d) small
      e) small

6.5   a) disruptive   b) cohesive   c) low   d) indefinite
      e) large   f) moderate

6.6   a) gas laws   b) pressure   c) atmosphere   d) torr

6.7   a) pressure   b) volume   c) temperature   d) $PV = k$ or $P = k/V$
      e) volume   f) temperature   g) pressure   h) $V = k'T$ or $V/T = k'$
      i) kelvins   j) Boyle's   k) Charles'   l) combined
      m) pressure   n) volume   o) temperature
      p) $PV/T = k''$ or $\frac{P_iV_i}{T_i} = \frac{P_fV_f}{T_f}$

6.8   a) equal   b) equal   c) one atmosphere d) 273
      e) ideal gas law   f) $PV = nRT$ g) universal

6.9   a) partial pressures b) total   c) partial pressures

6.10 a) effusion   b) diffusion   c) faster

6.11 a) heating   b) cooling   c) exothermic   d) endothermic

6.12 a) evaporation   b) vaporization   c) endothennic   d) condensation
      e) vapor pressure

6.13 a) boiling point   b) temperature   c) normal   d) standard
      e) standard atmosphere

6.14 a) gas   b) sublimation   c) melting point

6.15 a) specific heat   b) heat of fusion   c) heat of vaporization

B. Answers to Self-Test Questions

1. a   4. c   7. a
2. a   5. d   8. b
3. c   6. a   9. b

10. a	17. T	24. b
11. d	18. T	25. a
12. d	19. T	26. d
13. a	20. T	27. c
14. c	21. F	28. b
15. F	22. d	29. a
16. F	23. c	

Chapter 7

Solutions and Colloids

CHAPTER OUTLINE

7.1   Physical States of Solutions
7.2   Solubility
7.3   The Solution Process
7.4   Solution Concentrations
7.5   Solution Preparation

7.6   Solution Stoichiometry
7.7   Solution Properties
7.8   Colloids
7.9   Colloid Formation and Destruction
7.10 Dialysis

LEARNING OBJECTIVES

When you have completed your study of this chapter, you should be able to:
1. Predict in a general way the solubility of solutes in solvents on the basis of molecular polarity.
2. Describe the process of solution formation in terms of solute-solvent interactions.
3. Calculate solution concentrations in units of molarity, weight/weight percent, weight/volume percent, and volume/volume percent.
4. Do stoichiometric calculations based on solution concentrations.
5. Describe the preparation of solutions of specific concentration using pure solute and solvent or solutions of higher concentration than the one desired.
6. Do calculations based on the colligative solution properties of boiling point, freezing point, and osmotic pressure.
7. Describe the characteristics of colloids.
8. Describe the process of dialysis and contrast it to the process of osmosis.

ANSWERS AND SOLUTIONS TO EVEN-NUMBERED PROBLEMS

Physical States of Solutions (Section 7.1)

7.2   Many solutions are found in the home. Some are listed below, with the composition as printed on the label. When no percentage is indicated, components are usually given in order of decreasing amount. When water is present, it is often not mentioned on the label or it is listed in the inert ingredients. Identify the solvent and solutes of the following solutions:
   a) liquid laundry bleach: sodium hypochloride 5.25%, inert ingredients 94.75%

b) rubbing alcohol: isopropyl alcohol 70%
c) hydrogen peroxide: 3% hydrogen peroxide
d) after shave: SD alcohol, water, glycerin, fragrance, menthol, benzophenone, coloring

Solution:
a) Water is the solvent. It is the most prevalent inert ingredient. Sodium hypochlorite is the solute.
b) Isopropyl alcohol is the solvent. Water is the solute.
c) Water is the solvent. Hydrogen peroxide is the solute.
d) SD alcohol is the solvent. Everything else are solutes.

7.4   Classify the following as being a solution or not a solution. Explain your reasons when you classify one as *not* a solution. For the ones classified as solutions, identify the solvent and solute(s).
a) foggy air   b) tears   c) fresh squeezed orange juice
d) tea   e) hand lotion

Solution:
a) Foggy air is not a solution because a fog is not homogeneous.
b) Fresh squeezed orange juice is not a solution. There are solids included in the juice. The juice tends to separate on standing, even if the solids are strained out.
c) Tearing produces a solution composed of water as the solvent and solutes that include sodium chloride.
d) Tea is a complex solution of many solutes dissolved in water.
e) Hand lotion is not a solution. Part of the lotion is a water-based solution and part is the mixed in oils, lanolin, and other materials that do not dissolve in water; they are held in the water-based solution by an emulsifying agent.

Solubility (Section 7.2)
7.6   Use the terms *soluble, insoluble,* or *immiscible* to describe the behavior of the following pairs of substances when they are shaken together:
a) 25 mL of cooking oil and 25 mL of vinegar—the resulting mixture is cloudy and gradually separates
b) 25 mL of water and 10 mL of rubbing alcohol—the resulting mixture is clear and colorless
c) 25 mL of chloroform and 1 g of roofing tar—the resulting mixture is clear but dark brown in color

Solution:
a) The reason that separation occurs is that the two fluids are immiscible. This is not an example of a solution.
b) The mixture of rubbing alcohol and water is not cloudy and does not separate. The two liquids are miscible and produce a solution. Color is not a deciding factor when identifying a solution.
c) The mixing of chloroform and the tar produces a clear mixture indicating the formation of a solution. The tar is soluble. Color is not a deciding factor when identifying a solution.

7.8   Classify the following solutions as *unsaturated, saturated,* or *supersaturated:*
a) A solution to which a small piece of solute is added, and it dissolves.
b) A solution to which a small piece of solute is added, and much more solute comes out of solution.
c) The final solution resulting from the process in part (b).

Solution:
a) unsaturated; more solute could dissolve
b) supersaturated; too much was dissolved
c) saturated

7.10   Suppose you have a saturated solution that is at room temperature. Discuss how it could be changed into a supersaturated solution without using any additional solute.

Solution:
There would be more than one method. Since most solutes are more soluble at higher temperatures, one method would be to cool the solution. Another method would be to allow the solvent to evaporate.

7.12   Classify each of the following solutes into the approximate solubility categories of Table 7.3. The numbers in parentheses are the grams of solute that will dissolve in 100 g of water at the temperature indicated.
a) barium nitrate, $Ba(NO_3)_2$ (8.7 g at 20°C)
b) aluminum oxide, Al_2O_3 (9.8×10^{-5} g at 29°C)
c) calcium sulfate, $CaSO_4$ (0.21 g at 30°C)
d) manganese chloride, $MnCl_2$ (72.3 g at 27°C)
e) lead bromide, $PbBr_2$ (0.46 g at 0°C)

Solution:
a) soluble   b) insoluble   c) slightly soluble
d) soluble   e) slightly soluble

The Solution Process (Section 7.3)
7.14   Suppose you had a sample of white crystalline solid that was a mixture of barium chloride ($BaCl_2$) and barium sulfate ($BaSO_4$). Describe how you could treat the sample to isolate one of the solids in a pure state. Which solid would it be?

Solution:
Barium chloride is soluble in water and barium sulfate is not. The mixture is dissolved in water and filtered. The insoluble barium sulfate remains in the filter paper and can be recovered.

7.16   Indicate which of the following substances (with geometries shown in the text) would be soluble in water (a polar solvent) and in benzene (a nonpolar solvent):
a) CH_4 b) Ne   c) NH_3 d) BF_3

Solution:
a) CH_4 is nonpolar; it is not soluble in water, but is soluble in benzene.
b) Ne is nonpolar; it is not soluble in water, but is soluble in benzene.
c) NH_3 is polar; it is soluble in water, but is not soluble in benzene.
d) BF_3 is nonpolar; it is not soluble in water, but is soluble in benzene.

7.18   Freons are compounds formerly used in a variety of ways. Explain why Freon-114 was useful as a degreasing agent. The molecular structure is:

$$
\begin{array}{c}
\text{F}\quad\text{F}\\
|\quad\;\;|\\
\text{Cl}-\text{C}-\text{C}-\text{Cl}\\
|\quad\;\;|\\
\text{F}\quad\text{F}
\end{array}
$$

Solution:
Freon-114 is nonpolar and grease is nonpolar. Freon would dissolve grease.

Solution Concentrations (Section 7.4)

7.20   Calculate the molarity of the following solutions:
   a) 1.50 L of solution contains 0.294 mol of solute
   b) 200 mL of solution contains 0.151 mol of solute
   c) 0.335 mol solute is put into a container and enough distilled water is added to give 500 mL of solution

   Solution:
   $M = \frac{\text{moles solute}}{\text{L soln}}$
   a) $M = \frac{0.294 \text{ mol}}{1.50 \text{ L}} = 0.196$ M
   b) $M = \frac{0.151 \text{mol}}{0.200 \text{ L}} = 0.755$ M
   c) $M = \frac{0.335 \text{ mol}}{0.500 \text{ L}} = 0.670$ M

7.22   Calculate the molarity of the following solutions:
   a) A sample of solid NaOH weighing 4.00 g is put in enough distilled water to give 100 mL solution.
   b) 20.2 g sample of solid $CuCl_2$ is dissolved in enough water to give 1.00 L of solution.
   c) A 10.0 mL sample of solution is evaporated to dryness and leaves 0.51 g of solid residue that is identified as KNO_3.

   Solution:
   $M = \frac{\text{moles solute}}{\text{L soln}}$ convert g to mol and mL soln to L soln

   a) $M = \frac{4.00 \text{ g NaOH} \times \frac{1 \text{ mol NaOH}}{40.0 \text{ g NaOH}}}{0.100 \text{ L soln}} = 1.00$ M

   b) $M = \frac{20.2 \text{ g CuCl}_2 \times \frac{1 \text{ mol CuCl}_2}{134.5 \text{ g CuCl}_2}}{1.00 \text{ L soln}} = 0.150$ M

   c) $M = \frac{0.51 \text{ g KNO}_3 \times \frac{1 \text{ mol KNO}_3}{101 \text{ g KNO}_3}}{10.0 \text{ mL} \times \frac{1 \text{ L soln}}{1000 \text{ mL}}} = 0.50$ M

7.24   Calculate the quantity asked for.
   a) How many moles of solute is contained in 1.50 L of a 0.225 M solution?
   b) How many moles of solute is contained in 200 mL of 0.185 M solution?
   c) What volume of 0.452 M solution contains 0.200 mol solute?

   Solution:
   a) $M = \frac{\text{moles solute}}{\text{L soln}}$; moles solute = $M \times$ L soln = 0.255 M $\times$ 1.50 L = 0.383 mol
   b) Convert to L: $\frac{200 \text{ mL}}{1000 \text{ mL/L}} = 0.200$ L

      moles solute = $M \times$ L soln = 0.185 M $\times$ 0.200 L = 0.0370 mol
   c) L soln = $\frac{\text{moles solute}}{M} = \frac{0.200 \text{ mol solute}}{0.452 \text{ mol solute/L soln}} = 0.442$ L = 442 mL

7.26   Calculate the quantity asked for.
   a) How many grams of solid would be left behind if 20.0 mL of 0.550 M KCl were evaporated to dryness?
   b) What volume of 0.315 M HNO_3 solution is needed to provide 0.0410 mol of HNO_3?
   c) What volume of 1.21 M NH_4NO_3 contains 50.0 g of solute?

Solution:

a) mol KCl = M KCl × L soln = 0.550 mol KCl/~~L soln~~ × 0.020 ~~L soln~~ = 0.0110 mol KCl

  g solute = 0.0110 ~~mol KCl~~ × $\frac{74.6 \text{ g KCl}}{1 \text{ mol KCl}}$ = 0. 821 g KCl

  Note: If intermediate rounding is not done, the answer is 0.820 g KCl. This is not a significant difference.

b) L soln = $\frac{\text{mol solute}}{M}$ = $\frac{0.0410 \text{ mol HNO}_3}{0.315 \text{ mol HNO}_3/\text{L soln}}$ = 0.130 L soln

c) 50.0 ~~g NH₄NO₃~~ × $\frac{1 \text{ mol NH}_4\text{NO}_3}{80.0 \text{ g NH}_4\text{NO}_3}$ = 0.625 mol NH₄NO₃

  L soln = $\frac{\text{mol solute}}{M}$ = $\frac{0.625 \text{ mol NH}_4\text{NO}_3}{1.21 \text{ mol NH}_4\text{NO}_3/\text{L soln}}$ = 0.517 L soln

  Note: If intermediate rounding is not done, the answer is 0.516 L soln. This is not a significant difference.

7.28   Calculate the concentration in % (w/w) of the following solutions. Assume water has a density of 1.00 g/mL.

a) 6.5 g of sugar and 100 mL of water

b) 6.5 g of any solute and 100 mL of water

c) 6.5 g of any solute and 100 g of any solvent

Solution:

a) % (w/w) = $\frac{\text{wt of solute}}{\text{wt solution}}$ × 100 = $\frac{6.5 \text{ g sugar}}{106.5 \text{ g}}$ × 100 = 6.10%

b) % (w/w) = $\frac{\text{wt of solute}}{\text{wt solution}}$ × 100 = $\frac{6.5 \text{ g solute}}{106.5 \text{ g}}$ × 100 = 6.10%

c) % (w/w) = $\frac{\text{wt of solute}}{\text{wt solution}}$ × 100 = $\frac{6.5 \text{ g solute}}{106.5 \text{ g}}$ × 100 = 6.10%

7.30   Calculate the concentration in % (w/w) of the following solutions. Assume water has a density of 1.00 g/mL.

a) 20.0 g of salt is dissolved in 250 mL of water

b) 0.100 mol of solid glucose ($C_6H_{12}O_6$) is dissolved in 100 mL of water

c) 120 g of solid is dissolved in 100 mL of water

d) 10.0 mL of ethyl alcohol (density = 0.789 g/mL) is mixed with 10.0 mL of water.

Solution:

% (w/w) = $\frac{\text{wt solute}}{\text{wt soln}}$ × 100

a) wt soln = wt water + wt salt = 250 ~~mL H₂O~~ × (1.00 g H₂O/~~mL H₂O~~) + 20.0 g salt = 270 g soln

  % (w/w) = $\frac{20.0 \text{ g salt}}{270 \text{ g soln}}$ × 100 = 7.41%

b) wt solute = 0.100 mol C₆H₁₂O₆ × (180 g C₆H₁₂O₆/mol C₆H₁₂O₆) = 18.0 g C₆H₁₂O₆

  wt soln = 100 g H₂O + 18.0 g C₆H₁₂O₆ = 118 g soln

  % (w/w) = $\frac{18.0 \text{ g C}_6\text{H}_{12}\text{O}_6}{118 \text{ g soln}}$ × 100 = 15.3% (w/w)

c) wt soln = 100 ~~mL H₂O~~ × (1.00 g H₂O/~~mL H₂O~~) + 120 g solute = 220 g soln

  % (w/w) = $\frac{120 \text{ g solid}}{220 \text{ g soln}}$ × 100 = 54.5% (w/w)

d) wt alcohol = 10.0 mL × 0.789 g/mL = 7.89 g alcohol

  wt soln = 10.0 ~~mL H₂O~~ × (1.00 g/H₂O/~~mL H₂O~~) + 7.89 g alcohol = 17.9 g soln

  % (w/w) = $\frac{7.89 \text{ g alcohol}}{17.9 \text{ g soln}}$ × 100 = 44.1% (w/w)

7.32   Calculate the concentration in % (w/w) of the following solutious:
a) 20.0 g of solute is dissolved in water to give 150 mL of solution. The density of the resulting solution is 1.20 g/mL.
b) A 10.0-mL sample with a density of 1.10 g/mL leaves 1.18 g of solid residue when evaporated.
c) A 25.0-g sample of solution on evaporation leaves a 1.87-g residue of $MgCl_2$.

Solution:
a) $d = \frac{mass}{vol}$; mass = d × vol = 1.20 g/mL × 150 mL = 180 g

   % (w/w) = $\frac{\text{wt solute}}{\text{wt solution}}$ × 100 = $\frac{20.0 \text{ g solute}}{180 \text{ g soln}}$ × 100 = 11.1%

b) $d = \frac{mass}{vol}$; mass = d × vol = 1.10 g/mL × 10.0 mL = 11.0 g

   % (w/w) = $\frac{\text{wt solute}}{\text{wt solution}}$ × 100 = $\frac{1.18 \text{ g solute}}{11.0 \text{ g soln}}$ × 100 = 10.7%

c) % (w/w) = $\frac{\text{wt solute}}{\text{wt solution}}$ × 100 = $\frac{1.87 \text{ g solute}}{25.0 \text{ g soln}}$ × 100 = 7.48%

7.34   Calculate the concentration in % (v/v) of the following solutions:
a) 200 mL of solution contains 15 mL of alcohol
b) 200 mL of solution contains 15 mL of any soluble liquid solute
c) 8.0 fluid ounces of oil is added to 2.0 gallons (256 fluid ounces) of gasoline
d) a solution of alcohol and water is separated by distillation. A 200-mL sample give 85.9 mL of alcohol.

Solution:
a) % (v/v) = $\frac{\text{vol solute}}{\text{vol solution}}$ × 100 = $\frac{15 \text{ mL}}{200 \text{ mL}}$ × 100 = 7.5%

b) % (v/v) = $\frac{\text{vol solute}}{\text{vol solution}}$ × 100 = $\frac{15 \text{ mL}}{200 \text{ mL}}$ × 100 = 7.5%

c) % (v/v) = $\frac{\text{vol solute}}{\text{vol solution}}$ × 100 = $\frac{8.0 \text{ fl oz}}{256 \text{ fl oz}}$ × 100 = 3.1%

d) % (v/v) = $\frac{\text{vol solute}}{\text{vol solution}}$ × 100 = $\frac{85.9 \text{ mL}}{200 \text{ mL}}$ × 100 = 43.0%

7.36   Consider the blood volume of an adult to be 5.0 L. A blood alcohol level of 0.50% (v/v) can cause a coma. What volume of pure ethyl alcohol, if consumed in one long drink and assumed to be absorbed completely into the blood, would result in this critical blood alcohol level?

Solution:
$\frac{0.50 \text{ mL alcohol}}{100 \text{ mL blood}}$ × $\frac{1000 \text{ mL blood}}{1 \text{ L blood}}$ × 5.0 L blood = 25 mL alcohol

7.38   Calculate the concentration in % (w/v) of the following solutions:
a) 200 mL of solution contains 8.00 g of Na_2SO_4
b) 200 mL of solution contains 8.00 g of any solute
c) 750 mL of solution contains 58.7 g of solute

Solution:
a) % (w/v) = $\frac{\text{grams } Na_2SO_4}{\text{vol solution}}$ × 100 = $\frac{8.00 \text{ g } Na_2SO_4}{200 \text{ mL}}$ × 100 = 4.00% soln

b) % (w/v) = $\frac{\text{g solute}}{\text{vol solution}}$ × 100 = $\frac{8.00 \text{ g solute}}{200 \text{ mL}}$ × 100 = 4.00% soln

c) % (w/v) = $\frac{\text{g solute}}{\text{vol solution}}$ × 100 = $\frac{58.7 \text{ g solute}}{750 \text{ mL}}$ × 100 = 7.83% soln

7.40   A saturated solution of KBr in water is formed at 20.0°C. Consult Figure 7.3 and calculate the concentration of the solution in % (w/w).

Solution:

From Figure 7.3, the solubility of KBr is about 70 g/100 g H_2O

$\% \ (w/w) = \frac{wt \ solute}{wt \ solution} \times 100 = \frac{70 \ g \ solute}{170 \ g \ soln} \times 100 = 41.2\%$

Solution Preparation (Section 7.5)

7.42   Explain how you would prepare the following solutions using pure solute and water. Assume water
has a density of 1.00 g/mL.

moles =

a) 100 mL of 0.250 M Na_2SO_4 solution
b) 500 mL of 0.100 M $Zn(NO_3)_2$ solution
c) 250 g of 2.50% (w/w) NaCl solution
d) 100 mL of 0.55% (w/v) KCl solution

Solution:

a) $M = \frac{mol \ Na_2SO_4}{L \ soln}$; mol $Na_2SO_4 = 0.250 \ M \times 0.100 \ L \ = 0.0250$ mol Na_2SO_4

g $Na_2SO_4 = 0.0250$ mol $Na_2SO_4 \times \frac{142.04 \ g \ Na_2SO_4}{1 \ mol \ Na_2SO_4} = 3.55$ g Na_2SO_4

Dissolve 3.55 g Na_2SO_4 in sufficient distilled water to produce 100 mL of solution.

b) mol $Zn(NO_3)_2 = 0.100 \ M \times 0.500 \ L = 0.0500$ mol $Zn(NO_3)_2$

g $Zn(NO_3)_2 = 0.0500$ mol $Zn(NO_3)_2 \times \frac{189.39 \ g \ Zn(NO_3)_2}{1 \ mol \ Zn(NO_3)_2} = 9.47$ g $Zn(NO_3)_2$

Dissolve 9.47 g $Zn(NO_3)_2$ in sufficient distilled water to produce 500 mL of solution.

c) $\% \ (w/w) = \frac{wt \ solute}{wt \ solution} \times 100$

wt NaCl $= \frac{wt \ solution \times \% \ soln}{100} = \frac{250 \ g \times 2.50\%}{100} = 6.25$ g NaCl

And, using 250 g solution -6.25 g NaCl $= 243.75$ g water, the solution is produced by dissolving 6.25
g NaCl in 243.75 g water.

d) $\% \ (w/v) = \frac{g \ solute}{vol \ solution} \times 100$

wt KCl $= \frac{vol \ solution \times \% \ soln}{100} = \frac{100 \ mL \times 0.55\%}{100} = 0.55$ g KCl

Dissolving 0.55 g KCl in 100 mL of water produces this solution assuming no volume change.

7.44   A solution is prepared by mixing 45.0 g of water and 15.0 g of ethyl alcohol. The resulting solution has a
density of 0.952 g/mL. Express the solution concentration in % (w/w) ethyl alcohol and % (w/v) ethyl
alcohol.

Solution:

The solution has a total mass of 45.0 g H_2O + 15.0 g alcohol $= 60.0$ g soln

The volume of that solution $= \frac{60.0 \ g \ soln}{0.952 \ g \ soln/mL \ soln} = 63.3$ mL soln

$\% \ (w/w) = \frac{15.0 \ g \ alcohol}{60.0 \ g \ soln} \times 100 = 25.0\%$ alcohol (w/w)

$\% \ (w/v) = \frac{15.0 \ g \ alcohol}{63.3 \ mL \ soln} \times 100 = 23.8\%$ alcohol (w/v)

7.46   Calculate the following:

a) The number of grams of Li_2CO_3 in 500 mL of 2.50 M Li_2CO_3 solution
b) The number of moles of NH_3 in 100 mL of 6.00 M NH_3 solution
c) The number of milliliters of alcohol in 200 mL of 10.8% (v/v) solution
d) The number of grams of $CaCl_2$ in 20.0 mL of 3.15% (w/v) $CaCl_2$ solution

Solution:

a) mol Li_2CO_3 = 2.50 mol Li_2CO_3/L soln × 0.500 L soln = 1.25 mol Li_2CO_3

    1.25 mol Li_2CO_3 × $\frac{73.892 \text{ g } Li_2CO_3}{\text{mol } Li_2CO_3}$ = 92.4 g Li_2CO_3

b) mol NH_3 = 6.00 mol NH_3/L soln × 0.100 L soln = 0.600 mol NH_3

c) % (v/v) = $\frac{\text{vol solute}}{\text{vol solution}}$ × 100; vol solute = $\frac{\% \text{ soln} \times \text{vol soln}}{100}$

    vol alcohol = $\frac{10.8 \times 200 \text{ mL}}{100}$ = 21.6 mL alcohol

d) % (w/v) = $\frac{\text{g solute}}{\text{vol solution}}$ × 100; g $CaCl_2$ = $\frac{\% \text{ solute} \times \text{vol soln}}{100}$

    g $CaCl_2$ = $\frac{\% \text{ solute} \times \text{vol soln}}{100}$ = $\frac{3.15 \times 20.0}{100}$ = 0.63 g $CaCl_2$

7.48 Explain how you would prepare the following dilute solutions from the more concentrated ones:
a) 50.0 L of 6.00 M H_2SO_4 from 18.0 M H_2SO_4 solution
b) 250 mL of 0.500 M $CaCl_2$ from 3.00 M $CaCl_2$ solution
c) 200 mL of 1.50% (w/v) KBr from 10.0% (w/v) KBr solution
d) 500 mL of 10.0% (v/v) alcohol from 50.0% (v/v) alcohol

Solution:
Use Equation 7.9: $C_c \times V_c = C_d \times V_d$, where C can be M, % (w/v), or % (v/v).

$V_c = \frac{C_d \times V_d}{C_c}$

a) $V_c = \frac{6.00 \text{ M} \times 5.00 \text{ L}}{18.0 \text{ M}}$ = 1.67 L

    Add 1.67 L of 18.0 M H_2SO_4 to enough water to make 5.00 L of solution.
    Note: Always add the acid to the water. Adding the water to the acid is not safe. It can cause
    spattering.

b) $V_c = \frac{0.500 \text{ M} \times 250 \text{ mL}}{3.00 \text{ M}}$ = 41.7 mL

    Dilute 41.7 mL of 3.00 M $CaCl_2$ with enough water to make 250 mL of solution.

c) $V_c = \frac{1.50\% \text{ (w/v)} \times 200 \text{ mL}}{10.0\% \text{ (w/v)}}$ = 30.0 mL

    Dilute 30.0 mL of 10.0% KBr (w/v) with enough water to make 200 mL solution.

d) $V_c = \frac{10.0\% \text{ (v/v)} \times 500 \text{ mL}}{50.0\% \text{ (v/v)}}$ = 100 mL

    Dilute 100 mL of 50.0% (v/v) alcohol with enough water to make 500 mL solution.

7.50 What is the molarity of the solution prepared by diluting 50.0 mL of 0.195 M KBr to each of the following volumes?

a) 1.50 L   b) 200 mL   c) 500 mL   d) 700 mL

Solution:
By the equation $M_d V_d = M_c V_c$; $M_d = \frac{M_c \times V_c}{V_d}$ and 50.0 mL × $\frac{1 \text{ L}}{1000 \text{ mL}}$ = 0.0500 L

a) $M_d = \frac{0.195 \times 0.0500 \text{ L}}{1.50 \text{ L}}$ = 0.00650 M KBr

b) 200 mL × $\frac{1 \text{ L}}{1000 \text{ mL}}$ = 0.200 L; $M_d = \frac{0.195 \text{ M} \times 0.0500 \text{ L}}{.200 \text{ L}}$ = 0.0488 M KBr

c) 500 mL × $\frac{1 \text{ L}}{1000 \text{ mL}}$ = 0.500 L; $M_d = \frac{0.195 \text{ M} \times 0.0500 \text{ L}}{.500 \text{ L}}$ = 0.0195 M KBr

d) 700 mL × $\frac{1 \text{ L}}{1000 \text{ mL}}$ = 0.700 L; $M_d = \frac{0.195 \text{ M} \times 0.0500 \text{ L}}{.700 \text{ L}}$ = 0.0139 M KBr

Solution Stoichiometry (Section 7.6)

7.52   How many grams of solid Na_2CO_3 will exactly react with 250 mL of 1.25 M HCl solution?

$$Na_2CO_{3(s)} + 2HCl_{(aq)} \rightarrow 2NaCl_{(aq)} + CO_{2(g)} + H_2O_{(l)}$$

Solution:

$$250 \text{ mL} \times \frac{1 \text{ L}}{1000 \text{ mL}} \times \frac{1.25 \text{ mol HCl}}{1 \text{ L}} \times \frac{1 \text{ mol Na}_2\text{CO}_3}{2 \text{ mol HCl}} \times \frac{106 \text{ g Na}_2\text{CO}_3}{1 \text{ mol Na}_2\text{CO}_3} = 16.6 \text{ g Na}_2\text{CO}_3$$

7.54   How many mL of 0.250 M $AgNO_3$ will exactly react with 25.0 mL of 0.200 M NaCl solution?

$$NaCl_{(aq)} + AgNO_{3(aq)} \rightarrow NaNO_{3(aq)} + AgCl_{(s)}$$

Solution:

$$25.0 \text{ mL NaCl} \times \frac{1 \text{ L}}{1000 \text{ mL}} \times \frac{0.200 \text{ mol NaCl}}{1 \text{ L}} \times \frac{1 \text{ mol AgNO}_3}{1 \text{ mol NaCl}} \times \frac{1 \text{ L AgNO}_3 \text{ soln}}{0.250 \text{ mol AgNO}_3} \times \frac{1000 \text{ mL}}{1 \text{ L}} = 20.0 \text{ mL AgNO}_3 \text{ soln}$$

7.56   How many milliliters of 0.150 M NH_3 solution will exactly react with 30.5 mL of 0.109 M H_2SO_4 solution?

$$2NH_{3(aq)} + H_2SO_{4(aq)} \rightarrow (NH_4)_2SO_{4(aq)}$$

Solution:

$$30.5 \text{ mL} \times \frac{1 \text{ L}}{1000 \text{ mL}} \times \frac{0.109 \text{ mol H}_2\text{SO}_4}{1 \text{ L}} \times \frac{2 \text{ mol NH}_3}{1 \text{ mol H}_2\text{SO}_4} \times \frac{1 \text{ L NH}_3}{0.150 \text{ mol NH}_3} \times \frac{1000 \text{ mL}}{1 \text{ L}} = 44.3 \text{ mL of ammonia solution.}$$

7.58   How many milliliters of 0.124 M NaOH solution will exactly react with 25.0 mL of 0.210 M HCl solution?

$$NaOH_{(aq)} + H_2SO_{4(aq)} \rightarrow NaCl_{(aq)} + H_2O_{(l)}$$

Solution:

$$25.0 \text{ mL} \times \frac{1 \text{ L}}{1000 \text{ mL}} \times \frac{0.210 \text{ mol HCl}}{1 \text{ L}} \times \frac{1 \text{ mol NaOH}}{1 \text{ mol HCl}} \times \frac{1 \text{ L NaOH}}{0.124 \text{ mol NaOH}} \times \frac{1000 \text{ mL}}{1 \text{ L}} = 42.3 \text{ mL NaOH}$$

7.60   Stomach acid is essentially 0.10 M HCl. An active ingredient found in a number of popular antacids is calcium carbonate, $CaCO_3$. Calculate the number of grams of $CaCO_3$ needed to react exactly with 250 mL of stomach acid.

$$CaCO_3(s) + 2HCl(aq) \rightarrow CO_2(g) + CaCl_2(aq) + H_2O(l)$$

Solution:

$$25.0 \text{ mL acid} \times \frac{1 \text{ L acid}}{1000 \text{ mL acid}} \times \frac{0.10 \text{ mol HCl}}{1 \text{ L acid}} \times \frac{1 \text{ mol CaCO}_3}{2 \text{ mol HCl}} \times \frac{100 \text{ g CaCO}_3}{1 \text{ mol CaCO}_3} = 1.3 \text{ g CaCO}_3$$

Solution Properties (Section 7.7)

7.62   Before it is frozen, ice cream is essentially a solution of sugar, flavorings, etc. dissolved in water. Use the idea of colligative solution properties and explain why a mixture of ice (and water) and salt is used to freeze homemade ice cream. Why won't just ice work?

Solution:

The freezing point of the ice cream mixture is lower than the freezing point of water. In order to freeze the ice cream, the ice, salt, and water must be at a lower temperature than the freezing point of the ice cream. If the salt water, a strong electrolyte, is more concentrated than the ice cream mixture, the salt water will have a lower freezing point than the ice cream mixture, and the ice cream can freeze.

7.64   Calculate the boiling and freezing points of water solutions that are 1.00 M in the following solutes:
a) NH_4Cl, a strong electrolyte   b) glycerol, a nonelectrolyte
c) $(NH_4)_2SO_4$, a strong electrolyte   d) $Al(NO_3)_3$, a strong electrolyte

Solution:
Strong electrolytes are assumed to disassociate to release all component ions (n)

$\Delta t_b = nK_bM$ $\Delta t_f = nK_fM$

a) $\Delta t_b = 2 \times 0.52°C/M \times 1.00$ M $\Delta t_f = 2 \times 1.86°C/M \times 1.00$ M
   $\Delta t_b = 1.04°C$ $\Delta t_f = 3.72°C$
   $BP = 100°C + \Delta t_b$ $FP = 0°C - \Delta t_f$
   $BP = 101.04°C$ $FP = -3.72°C$

b) $\Delta t_b = 1 \times 0.52°C/M \times 1.00$ M $\Delta t_f = 1 \times 1.86°C/M \times 1.00$ M
   $\Delta t_b = 0.52°C$ $\Delta t_f = 1.86°C$
   $BP = 100°C + \Delta t_b$ $FP = 0°C - \Delta t_f$
   $BP = 100.52°C$ $FP = -1.86°C$

c) $\Delta t_b = 3 \times 0.52°C/M \times 1.00$ M $\Delta t_f = 3 \times 1.86°C/M \times 1.00$ M
   $\Delta t_b = 1.56°C$ $\Delta t_f = 5.58°C$
   $BP = 100°C + \Delta t_b$ $FP = 0°C - t_f$
   $BP = 101.56°C$ $FP = -5.58°C$

d) $\Delta t_b = 4 \times 0.52°C/M \times 1.00$ M $\Delta t_f = 4 \times 1.86°C/M \times 1.00$ M
   $\Delta t_b = 2.08°C$ $\Delta t_f = 7.44°C$
   $BP = 100°C + \Delta t_b$ $FP = 0°C - \Delta t_f$
   $BP = 102.08°C$ $FP = -7.44°C$

7.66   Calculate the boiling and freezing points of the following solutions. Water is the solvent unless otherwise indicated.
a) a 0.50 M solution of urea, a nonelectrolyte
b) a 0.250 M solution of $CaCl_2$, a strong electrolyte
c) a solution containing 100 g of ethylene glycol ($C_2H_6O_2$), a nonelectrolyte, per 250 mL

Solution:
Use equations 7.13 ($\Delta t_b = nK_bM$) and 7.14 ($\Delta t_f = nK_fM$); for water, $K_b = 0.52$; $K_f = 1.86$
a) For urea, n = 1; M = 0.50
   $\Delta t_b = 1 \times 0.52 \times 0.50 = 0.26°C$ higher; Boiling point $= 100.26°C$
   $\Delta t_f = 1 \times 1.86 \times 0.50 = 0.93°C$ lower; Freezing point $= -0.93°C$
b) For $CaCl_2$, n = 3; M = 0.250
   $\Delta t_b = 3 \times 0.52 \times 0.250 = 0.39°C$ higher; Boiling point $= 100.39°C$
   $\Delta t_f = 3 \times 1.86 \times 0.250 = 1.40°C$ lower; Freezing point $= -1.40°C$
c) For ethylene glycol, n = 1

   $\text{mol } C_2H_6O_2 = 100 \text{ g } C_2H_6O_2 \times \frac{1 \text{ mol } C_2H_6O_2}{62.0 \text{ g } C_2H_6O_2} = 1.61 \text{ mol } C_2H_6O_2$

   $M = \frac{1.61 \text{ mol } C_2H_6O_2}{0.250 \text{ L soln}} = 6.44$ M

   $\Delta t_b = 1 \times 0.52 \times 6.44 = 3.35°C$ higher; Boiling point $= 103.35°C$
   $\Delta t_f = 1 \times 1.86 \times 6.44 = 12.0°C$ lower; Freezing point $= -12.0°C$

7.68   Calculate the osmolarity for the following solutions:
a) a 0.15 M solution of glycerol, a nonelectrolyte
b) a 0.15 solution of $(NH_4)_2SO_4$, a strong electrolyte
c) a solution containing 25.3 g of LiCl (a strong electrolyte) per liter

Solution:

Osmolarity $= n \times M$

a) for glycerol, $n = 1$

osmolarity $= 1 \times 0.15 = 0.15$

b) for $(NH_4)_2SO_4$, $n = 3$

osmolarity $= 3 \times 0.15 = 0.45$

c) for LiCl, $n = 2$

osmolarity $= \dfrac{2 \times 25.3 \text{ g LiCl} \times \frac{1 \text{ mol LiCl}}{42.39 \text{ g LiCl}}}{1 \text{ L}} = 1.19$

Note: In Exercises 7.70–7.78, assume temperature of 25.0°C (298 K), and express your answer in torr, mm Hg, and atm.

7.70 Calculate the osmotic pressure of any solution with an osmolarity of 0.300.

Solution:

$\pi = nMRT = $ where $n \times M = $ osmolarity

0.300 mol/L $\times$ 0.0821 L atm/mol K $\times$ 298 K $= 7.34$ atm

7.34 atm $\times \frac{760 \text{ torr}}{1 \text{ atm}} = 5.58 \times 10^3$ torr $= 5.58 \times 10^3$ mm Hg

7.72 Calculate the osmotic pressure of a 0.125 M solution of Na_2SO_4, a strong electrolyte.

Solution:

$\pi = nMRT$; $n = 3$

$\pi = 3 \times 0.125$ mol/L $\times$ 0.0821 L atm/mol K $\times$ 298 K $= 9.17$ atm

9.17 atm $\times$ 760 torr/atm $= 6.97 \times 10^3$ torr $= 6.97 \times 10^3$ mm Hg

7.74 Calculate the osmotic pressure of a solution that contains 35.0 g of the nonelectrolyte urea, CH_4N_2O, per 100 mL of solution.

Solution:

$\pi = nMRT$; $n = 1.0$

mol $CH_4N_2O = 35$ g $CH_4N_2O \times \frac{1.00 \text{ mol } CH_4N_2O}{60.04 \text{ 35 g } CH_4N_2O} = 0.583$ mol CH_4N_2O

$M = \frac{0.583 \text{ mol/L}}{0.100 \text{ L soln}} = 5.83$

$\pi = 1 \times 5.83$ mol/L $\times$ 0.0821 L atm/mol K $\times$ 298 K $= 143$ atm

143 atm $\times \frac{760 \text{ torr}}{1 \text{ atm}} = 1.09 \times 10^5$ torr $= 1.09 \times 10^5$ mm Hg

7.76 Calculate the osmotic pressure of a solution that has a freezing point of −0.35°C.

Solution:

The freezing point depression provides the osmolarity through the equation for the change in freezing point, $\Delta t = nMK$

$nM = \frac{\Delta t_f}{K_f} = \frac{0.35}{1.86} = 0.19$

$\pi = nMRT = 0.19$ mol/L $\times$ 0.0821 L atm/mol K $\times$ 298 K $= 4.7$ atm

4.7 atm $= \frac{760 \text{ torr}}{1 \text{ atm}} = 3.6 \times 10^3$ torr $= 3.6 \times 10^3$ mmHg

7.78   Calculate the osmotic pressure of a solution that contains 5.30 g of NaCl and 8.20 g of KCl per 750 mL.

Solution:

$\pi = nMRT$

For NaCl, n = 2

mol NaCl = 5.30 g̶ ̶N̶a̶C̶l̶ $\times \frac{1 \text{ mol NaCl}}{58.5 \text{ g̶ ̶N̶a̶C̶l̶}}$ = 0.0906 mol NaCl

$M = \frac{0.0906 \text{ mol NaCl}}{0.750 \text{ L soln}}$ = 0.121 M NaCl

osmolarity of NaCl = 2 × 0.121 M = 0.242

For KCl, n = 2

mol KCl = 8.20 g̶ ̶K̶C̶l̶ $\times \frac{1 \text{ mol KCl}}{74.6 \text{ g̶ ̶K̶C̶l̶}}$ = 0.110 mol KCl

$M = \frac{0.110 \text{ mol KCl}}{0.750 \text{ L soln}}$ = 0.147

osmolarity of KCl = 2 × 0.147 M = 0.294

total osmolarity = 0.294 + 0.242 = 0.536

$\pi = nMRT$ = 0.536 m̶o̶l̶/̶L̶ × 0.0821 L̶ atm/m̶o̶l̶ K × 298 K = 13.1 atm

13.1 a̶t̶m̶ $\times \frac{760 \text{ torr}}{1 \text{ a̶t̶m̶}}$ = 9.96 × 10³ torr = 9.96 × 10³ mm Hg

Colloids and Dialysis (Section 7.8–7.10)

7.80   Suppose you have a bag made of a membrane like that in Figure 7.18. Inside the bag is a solution containing water and dissolved small molecules. Describe the behavior of the system when the bag functions as an osmotic membrane and when it functions as a dialysis membrane.

Solution:

When the membrane is functioning as an osmotic membrane, water molecules may move through the membrane. As a dialysis membrane, both the water and small molecules may move through it.

7.82   Explain how the following behave in a colloidal suspension: dispersing medium, dispersed phase, and colloid emulsifying agent.

Solution:

The dispersing medium is present in larger amount, analogous to the solvent in a solution. It is the phase in which the colloid is dispersed. The dispersed phase, analogous to the solute in solutions, is the large (compared to molecular sizes) particles spread throughout the dispersing medium. The emulsifying agent stabilizes the dispersed colloidal particles, preventing them from coagulating to make large particles that would settle out under the influence of gravity.

PROGRAMMED REVIEW

Section 7.1 Physical States of Solutions

Solutions are (a) homogenous mixtures in which the components are present as (b) atoms, (c) molecules, or (d) ions. The most abundant substance in a solution is called the (e) solvent, and any other substances are called (f) solutes.

Section 7.2 Solubility

Substances that dissolve to a significant extent in a solvent are called (a) soluble substances. (b) insoluble substances do not dissolve significantly in a solvent. The maximum amount of solute that can be dissolved in a specific amount of solvent under specific conditions is called the (c) solubility of the solute. A (d) saturated solution contains the maximum amount of dissolved solute that is stable under the prevailing conditions. An unstable solution that contains more dissolved solute than the solute solubility is called a (e) _Super saturated_ solution.

Section 7.3 The Solution Process

An ion that has broken away from a solid lattice and is in solution surrounded by water molecules is called a (a) _hydrated_ ion. A saturated solution represents an (b) _equilibrium_ between dissolved and undissolved solute in which undissolved solute enters solution at the same rate dissolved solute leaves solution. A good solubility generalization that applies well to nonionic compounds is (c) _like disolves like_. One solubility rule for ionic compounds in water is that all nitrates are (d) _soluble_. The rate of dissolving should not be confused with a solute's (e) _solubility_. The rate of dissolving is influenced by a number of factors such as the temperature of the (f) _solvent_.

Section 7.4 Solution Concentrations

Relationships between the quantities of solute and solvent in solutions are called (a) _concentrations_ concentration given as a (b) _molarity_ relates the number of moles of solute in each (c) _liter_ of solution. In general, a concentration in (d) _percent_ gives the number of parts of solute contained in 100 parts of solution. The mass of solute in 100 mass units of solution is called a (e) _W_ / W _____ percent. A weight/volume percent gives the grams of solute in (f) _100_ mL of solution. A volume/volume percent is a useful concentration unit used when both the solvent and solute are either (g) _liquid_ or (h) _gases_.

Section 7.5 Solution Preparation

Solutions are usually prepared by mixing together proper amounts of (a) _solute_ and (b) _solvent_, or by diluting a (c) _concentrated_ solution with (d) _solvent_ to produce a solution of lower concentration. When a specified volume of solution is desired, a (e) _volumetric flask_ is used, which when filled to mark holds an accurately known volume. A solution made by mixing 5.00 g NaCl and 195.0 gram water would have a % (w/w) of (f) _2.5%_. If 25.0 mL of 3.00 M H_2SO_4 is diluted with enough water to give 500 mL of diluted H_2SO_4 the molarity of the diluted H_2SO_4 is (g) _0.150M_

Section 7.6 Solution Stoichiometry

In 25.0 mL of 0.200 M HCl, there are (a) _5.00×10⁻³_ moles HCl. It would require (b) _29.9_ mL of 0.167 M NaOH to react exactly with that HCl.

Section 7.7 Solution Properties

Solutes that produce water solutions that conduct (a) _electricity_ are called (b) _electrolytes_. Solutes that produce nonconductive solutions are called (c) _non electrolyte_. Solution properties that depend only on the concentration of solute particles are called (d) _colligative_ properties. Three of these properties that are related are solution (e) _vapour_ pressure, (f) _boiling_ point and (g) _freezing_ point. A fourth property that involves (h) _semipermeable_ membranes is called (i) _osmotic_ pressure.

Section 7.8 Colloids

Colloids are (a) _homogenous_ mixtures of two or more components in which the terms (b) _dispering_ medium and (c) _dispersed_ phase are used in a manner analagous to the terms solvent and solute for solutions. The (d) _Tyndall_ effect is a property of colloids in which the path of a light beam passing through the colloid is visible. Colloids are usually differentiated according to the physical (e) _state_ of the dispersing medium and dispersed phase. In a colloid called a foam, the dispersing medium is a (f) _liquid_ and the dispersed phase is a (g) _gas_.

Section 7.9 Colloid Formation and Destruction

Much of the interest in colloids is related to their (a) _formation_ or (b) _destruction_. Some colloids are stabilized by substances called (c) _emulsifying_ or (d) _stabalizing_ agents.

Section 7.10 Dialysis

(a) di_____ membranes are semipermeable membranes that hold back (b) _____ particles and large (c) _____, but allow solvent, hydrated ions and small molecules to pass through. The passage of the ions and small molecules is called (d) _____, a process used to cleanse the blood of people suffering (e) _____ malfunction.

SELF-TEST QUESTIONS

Multiple Choice

1. A solution is prepared by dissolving a small amount of sugar in a large amount of water. In this case sugar would be the
    a) filtrate   b) solute   c) precipitate   d) solvent

2. The freezing point of a solution
    a) is lower than that of pure solvent   b) cannot be measured
    c) is higher than that of pure solvent   d) is the same as that of pure solvent

3. Which of the following would show the Tyndall effect?
    a) a solution of salt in water
    b) a solution of sugar in water
    c) a solution of CO_2 gas in water
    d) a colloidal suspension

4. A crystal of solid magnesium sulfate is placed into a solution of magnesium sulfate in water. It is observed that the crystal dissolves slightly. The original solution was
    a) saturated
    b) unsaturated
    c) supersaturated
    d) cannot determine from the data given

5. The weight/weight percent of sugar in a solution containing 25 grams of sugar and 75 grams of water would be:
    a) 25%   b) 33%   c) 75%   d) 80%

6. How many moles of Mg(NO,), is contained in 500 mL of 0.400 M solution?
    a) 0.400   b) 0.200   c) 0.800   d) 2.00

7. How many grams of Mg(NO,), must be dissolved in water to give 500 mL of 0.400 M solution?
    a) 29.7   b) 17.3   c) 59.3   d) 172.6

8. Which of the following aqueous solutions would be expected to have the highest boiling point at 1 atm pressure?
    a) 1 M NaCl   b) 1 M $C_{12}H_{22}O_{11}$ (sucrose)
    c) pure water   d) 1 M AlF_3

9. Calculate the freezing point of a water solution that contains 1.60 grams of methyl alcohol (a nonelectrolyte), CH_3OH, in each 100.0 mL. The K_f for water is 1.86°C/M.
    a) 0.93°C   b) −0.93°C   c) −1.86°C   d) −3.72°

10. How many moles of solute would be needed to prepare 250 mL of 0.150 M solution?
    a) 37.5   b) 0.150   c) 0.0375   d) 0.600

11. What volume of 0.200 M silver nitrate solution ($AgNO_3$) would have to be diluted to form 500 mL of 0.050 M solution?

    a) 200 mL   b) 2.0×10^{-5} mL   c) 80.0 mL   d) 125 mL

12. Calculate the osmolarity of a 0.400 M NaCl solution.

    a) 0.400   b) 0.800   c) 0.200   d) 0.100

13. What is the osmotic pressure (in torr) of a 0.0015 M NaCl solution at 25°C? R = 62.4 L torr/K mol

    a) 4.68   b) 2.34   c) 27.9   d) 55.8

True-False

14. Paint is a colloid.

15. A solution can contain only one solute.

16. A solution is a homogeneous mixture.

17. A colligative solution property is used to prevent winter freeze-up of cars.

18. Water is a good solvent for both ionic and polar covalent materials.

19. A nonpolar solute would be relatively insoluble in water.

20. On the basis of their solubility in each other, water and gasoline molecules have similar polarities.

21. Potassium nitrate is soluble in water.

22. Soluble solutes always dissolve rapidly.

Matching

Match the colloid names given as responses to the descriptions below.

23. _____ a liquid dispersed in a gas   a) foam

24. _____ a liquid dispersed in a liquid   b) emulsion

25. _____ mayonnaise is an example   c) aerosol

26. _____ a solid dispersed in a liquid   d) sol

27. _____ gelatin dessert is an example

SOLUTIONS

A. Answers to Programmed Review

7.1   a) homogeneous   b) atoms   c) molecules   d) ions
      e) solvent   f) solutes

7.2   a) soluble   b) insoluble   c) solubility   d) saturated
      e) supersaturated

7.3   a) hydrated   b) equilibrium   c) like dissolves like   d) soluble
      e) solubility   f) solvent

7.4   a) concentrations   b) molarity   c) liter   d) percent
      e) weight/weight   f) 100   g) liquids   h) gases

7.5   a) solute   b) solvent   c) concentrated   d) solvent
      e) volumetric flask   f) 2.50% (w/w)   g) 0.150 M

7.6   a) 5.00×10^{-3} b) 29.9

7.7   a) electricity   b) electrolytes   c) non-electrolytes   d) colligative
      e) vapor   f) boiling   g) freezing   h) semipermeable
      i) osmotic

7.8   a) homogeneous   b) dispersing   c) dispersed   d) Tyndall
      e) state   f) liquid   g) gas

7.9   a) formation   b) destruction   c) emulsifying   d) stabilizing

7.10   a) dialyzing   b) colloid   c) molecules   d) dialysis
      e) kidney

B. Answers to Self-Test Questions

1. b	10. c	19. T
2. a	11. d	20. F
3. d	12. b	21. T
4. b	13. d	22. F
5. a	14. T	23. c
6. b	15. F	24. b
7. a	16. T	25. b
8. d	17. T	26. d
9. b	18. T	27. d

Chapter 8

Reaction Rates and Equilibrium

CHAPTER OUTLINE

8.1   Spontaneous and Non-
      spontaneous Processes
8.2   Reaction Rates
8.3   Molecular Collisions
8.4   Energy Diagrams

8.5   Factors That Influence Reaction Rates
8.6   Chemical Equilibrium
8.7   The Position of Equilibrium
8.8   Factors That Influence Equilibrium
      Position

LEARNING OBJECTIVES

When you have completed your study of this chapter, you should be able to:
1. Use energy and entropy considerations to predict the spontaneity of processes and reactions.
2. Calculate reaction rates from experimental results.
3. Use the basic assumptions of reaction mechanisms to explain reaction characteristics.
4. Represent and interpret the energy relationships for reactions by using energy diagrams.
5. Explain how factors such as concentration, temperature, and catalysts influence reaction rates.
6. Do calculations based on equilibrium expressions for reactions.
7. Explain the concept of equilibrium and use Le Chatelier's principle to predict the influences of
    concentrations of reactants, concentrations of products, catalysts, and reaction temperature on the
    position of equilibrium.

ANSWERS AND SOLUTIONS TO EVEN-NUMBERED PROBLEMS

Spontaneous and Nonspontaneous Changes (Section 8.1)

8.2   Classify the following processes as *spontaneous* or *nonspontaneous*. Explain your answers in terms of
      whether energy must be continually supplied to keep the process going.
      a) The space shuttle leaves its pad and goes into orbit.
      b) The fuel in a booster rocket of the space shuttle burns.
      c) Water boils at 100°C and 1 atm pressure.

d) Water temperature increases to 100°C at 1 atm pressure.

e) Your bedroom becomes orderly.

Solution:

a) Nonspontaneous; energy must be continually supplied.

b) Spontaneous; only an initial energy is needed to start the burning.

c) Nonspontaneous; energy must be continually supplied to keep it boiling. It should be pointed out that the evaporation of water from the surface is spontaneous, but boiling is not spontaneous.

d) Nonspontaneous; energy must be continually supplied.

e) Nonspontaneous; work must be done to straighten it.

8.4   Classify the following processes as *exergonic* or *endergonic*. Explain your answers.

a) an automobile being pushed up a slight hill (from point of view of the one pushing)

b) ice melting (from point of view of ice)

c) ice melting (from point of view of surroundings of the ice)

d) steam condensing to liquid water (from point of view of steam)

e) steam condensing to liquid water (from point of view of surroundings of the steam)

Solution:

a) Exergonic; the "pusher" gives up energy to the automobile.

b) Endergonic; ice absorbs energy from surroundings to melt.

c) Exergonic; the surroundings give up energy to the ice.

d) Exergonic; condensing steam gives energy to surroundings.

e) Endergonic; the surroundings absorb energy from the steam.

8.6   Describe the energy and entropy changes that occur in the following processes and indicate whether the processes are spontaneous under the conditions stated:

a) On a cold day, water freezes.

b) A container of water at 40°C cools to room temperature.

c) The odor from an open bottle of perfume spreads throughout a room.

Solution:

a) The energy of the ice decreases and the entropy decreases. It is spontaneous if the temperature is below the freezing point.

b) The energy of the water decreases and the entropy stays constant. Spontaneous.

c) There are two processes involved. As the perfume evaporates, the energy slightly increases and the entropy greatly increases. This process is spontaneous. As the perfume vapors diffuse through the room, the energy does not change and the entropy increases. This process is spontaneous. The whole process is spontaneous.

8.8   Pick the example with the highest entropy from each of the following sets. Explain your answers.

a) two opposing football teams just before the ball is snapped, two opposing football teams 1 second after the ball is snapped, two opposing football teams when the whistle is blown, ending the play

b) a 10% copper/gold alloy, a 2% copper/gold alloy, pure gold

c) a purse on which the strap just broke, a purse just hitting the ground, a purse on the ground with contents scattered

d) coins in a piggy bank, coins in piles containing same type of coins, coins in stacks of same type of coins

e) a dozen loose pearls in a box, a dozen pearls randomly strung on a string, a dozen pearls strung on a string in order of decreasing size

Solution:

Entropy measures disorder: The more disorder, the higher the entropy.

a) The teams have the highest entropy when the whistle blows and the players are scattered on the field.

b) (1) From the gold atoms' point of view, the gold atoms have the most disorder in the mixture with the most copper atoms, (or the fewest gold atoms) That would be the 10% Cu mixture (90% gold). (2) From the copper atoms' point of view, the copper atoms have the most disorder in the mixture with the most (but not pure) gold atoms (the fewest copper atoms). The 2% copper (98% gold) has the lowest % copper, so it has the highest entropy.

c) The purse with the scattered contents has the highest entropy.

d) The coins in the piggy bank have the highest entropy.

e) The loose pearls in the box have the highest entropy.

Reaction Rates (Section 8.2)

8.10　Classify the following processes according to their rates as *very slow, slow,* or *fast.*

a) the souring of milk stored in a refrigerator

b) the cooking of an egg in boiling water

c) the ripening of a banana stored at room temperature

d) the rising of bread dough in a warm room

e) the melting of butter put into a hot pan

Solution:

a) Very slow; it takes several days or weeks.

b) Slow; it takes a few minutes.

c) Very slow; it takes a few days.

d) Slow; it takes an hour or so.

e) Fast; the butter melts on contact.

8.12　Describe the observation or measurements that could be made to allow you to follow the rate of the following processes:

a) the melting of a block of ice

b) the setting (hardening) of concrete

c) the burning of a candle

Solution:

a) visual estimate of the size of the ice block; weighing the water produced or the remaining ice; measuring the volume of the water or the remaining ice.

b) the ability of an object to penetrate the surface; the force required to break a specific sized piece.

c) length of unburned candle; weight of candle.

8.14　Consider the following hypothetical reaction: $A + B \rightarrow C$

Calculate the average rate of the reaction on the basis of the following information:

a) Pure A and B are mixed, and after 12.0 minutes the measured concentration of C is 0.396 mol/L.

b) Pure A, B, and C are mixed together at equal concentrations of 0.300 M. After 8.00 minutes, the concentration of C is found to be 0.455 M.

Solution:

The average rate can be calculated by: $rate = \frac{\Delta C}{time}$

a) $\Delta C = 0.396 \ M - 0 \ M = 0.396 \ M$

$rate = \frac{0.396 \ M}{12 \ min} = 0.033 \ M/min$

b) $\Delta C = 0.455 \text{ M} - 0.033 \text{ M} = 0.155 \text{ M}$

rate $= \frac{0.155 \text{ M}}{8 \text{ min}} = 0.0194 \text{ M/min}$

8.16 A reaction generates carbon dioxide gas (CO_2) as a product. The reactants are mixed in a sealed 500-mL vessel. After 25.0 minutes, 1.93×10^{-3} mol CO_2 has been generated. Calculate the average rate of the reaction.

Solution:

$C_0 = 0 \text{ M}; C_t = \frac{1.93 \times 10^{-3} \text{ mol}}{0.5 \text{ L}} = 3.86 \times 10^{-3} \text{ M}; \Delta C = 3.86 \times 10^{-3} \text{ M}$

The average rate can be calculated by:

rate $= \frac{\Delta C}{\text{time}} = \frac{3.86 \times 10^{-3} \text{ M}}{25.0 \text{ min}} = 1.54 \times 10^{-4} \text{ M/min}$

8.18 The ammonium and nitrite ions react in solution to form nitrogen gas:

$$NH_4^+{}_{(aq)} + NO_2^-(aq) \rightarrow N_{2(g)} + H_2O_{(l)}$$

A reaction is run, and the liberated N_2 gas is collected in a previously evacuated 500-mL container. After the reaction has gone on for 750 seconds, the pressure of N_2 in the 500-mL container is 2.77×10^{-2} atm, and the temperature of the N_2 is 25.0°C. Use the ideal gas law (Equation 6.9) to calculate the number of moles of N_2 liberated. Then calculate the average rate of the reaction.

Solution:

$n = \frac{PV}{RT} = \frac{2.77 \times 10^{-2} \text{ atm} \times 0.500 \text{ L}}{0.0821 \text{ L atm/mol K} \times 298 \text{ K}} = 5.66 \times 10^{-4} \text{ mol}$

$C_0 = 0 \text{ M}; C_t = \frac{5.66 \times 10^{-4} \text{ mol}}{0.5 \text{ L}} = 1.13 \times 10^{-3} \text{ M}; \Delta C = 1.13 \times 10^{-3} \text{ M}$

rate $= \frac{1.13 \times 10^{-3} \text{ M}}{750 \text{ sec}} = 1.51 \times 10^{-6} \text{ M/s}$

Molecular Collisions (Section 8.3)

8.20 In each of the following, which reaction mechanism assumption is apparently being violated? Explain your answers.

a) A reaction takes place more rapidly when the concentration of reactants is decreased.

b) A reaction takes place more rapidly when the reaction mixture is cooled.

c) The reaction rate of $A + B \rightarrow A - B$ increases as the concentration of A is increased but does not change as the concentration of B is increased.

Solution:

a) Decreasing the concentration of the reactants would be expected to decrease the number of collisions and decrease the rate at which the molecules react. The observation apparently violates assumption #1.

b) Assumption #2 requires a collision between molecules having minimum combined kinetic energy in order to react. Cooling the mixture will decrease the number with that minimum energy; R × n rate should decrease. The observation violates assumption #2.

c) Increasing the concentration of either reactant would be expected to increase the number of collisions and increase the rate of reaction. (Assumption #1) The observation violates assumption #1.

8.22   Describe two ways by which an increase in temperature increases a reaction rate.

Solution:
Increasing the temperature makes the molecules move faster, increasing the number of collisions per second; and increasing temperature increases the average kinetic energy of the molecules, increasing the probability that a collision will be effective.

Energy Diagrams (Section 8.4)

8.24   Sketch energy diagrams to represent each of the following. Label the diagrams completely and tell how they are similar to each other and how they are different.
a) Endothermic (endergonic) reaction with activation energy
b) Endothermic (endergonic) reaction without activation energy

Solution:
a)   b)

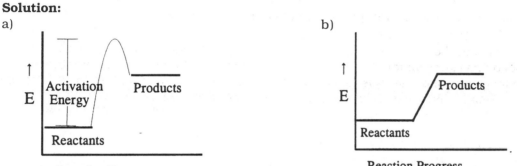

Both reactions have the same net energy difference between reactants and product. They differ in the activation energy required.

8.26   One reaction occurs at room temperature and liberates 500 kJ/mol of reactant. Another reaction does not take place until the reaction mixture is heated to 150°C. However, it also liberates 500 kJ/mol of reactant. Draw an energy diagram for each reaction and indicate the similarities and differences between the two.

Solution:

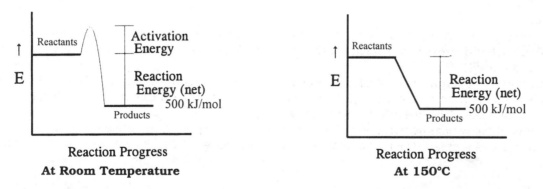

The difference between the two reactions is that one does not require additional energy above that of room temperature; the other reaction requires the input of the activation energy (to 150°C) before the reaction progresses. The similarity between the reactions is that the amount of energy given off is the same. Both are exothermic.

Factors That Influence Reaction Rates (Section 8.5)

8.28   The following reactions are proposed. Make a rough estimate of the rate of each one: *rapid, slow, won't react.* Explain each answer.

a) $CaO_{(s)} + 2HCl_{(g)} \rightarrow CaCl_{2(s)} + H_2O_{(l)}$

b) $2KI_{(s)} + Pb(NO_3)_{2(s)} \rightarrow PbI_{2(s)} + 2KNO_{3(s)}$

c) $Cl^-_{(aq)} + I^-_{(aq)} \rightarrow ICl^{2-}_{(aq)}$

d) $I_{2(aq)} + I^-_{(aq)} \rightarrow I^-_{3(aq)}$

Solution:
a) Slow to rapid. It will depend on the surface area of the solid CaO exposed to the gas.
b) Slow. The reaction limited to the surface contact between solids.
c) Won't react. The negative ions repel each other and do not collide.
d) Rapid. Collisions occur frequently in solutions.

8.30   Suppose you are running a reaction and you want to speed it up. Describe three things you might try to do this.

Solution:
Raising the temperature, increasing the concentration of the reactants, or adding a catalyst will speed up a reaction.

8.32   A reaction is run at 10°C and takes 3.7 hours to go to completion. How long would it take to complete the reaction at 30°C.

Solution:
Using the rule of thumb of doubling the rate for every 10°C change, heating it from 10°C to 20°C would double the rate, and then from 20°C to 30°C would double it again. Overall, it would react 4 times more quickly:

$$\frac{3.7 \text{ hrs}}{4} = 0.93 \text{ hours}$$

8.34   Describe two ways catalysts might speed up a reaction.

Solution:
A catalyst can lower the activation energy of a reaction, increasing the chance of an effective collision; or it can hold one of the reactants in a favorable orientation.

Chemical Equilibrium (Section 8.6)

8.36   Describe the observation or measurement result that would indicate when each of the following had reached equilibrium:

a) H_2 + I_2 $\rightleftarrows$ $2HI$
     colorless gas   violet gas   colorless gas

b) solid sugar + water $\rightleftarrows$ sugar solution

c) N_2 + $2O_2$ $\rightleftarrows$ $2NO_2$
     colorless gas   colorless gas   red-brown gas

Solution:
When the system is at equilibrium, there is no net change in the amounts of reactants and products.
a) The intensity of the color does not change.
b) The amount of undissolved sugar is constant.
c) The intensity of the color does not change. Or the total pressure does not change.

Position of Equilibrium (Section 8.7)

8.38   Write an equilibrium expression for each of the following gaseous reactions:
   a) $CH_4 + 2O_2 \Leftrightarrow CO_2 + 2H_2O$
   b) $2O_3 \Leftrightarrow 3O_2$
   c) $H_2 + CO_2 \Leftrightarrow H_2O + CO$
   d) $4NH_3 + 7O_2 \Leftrightarrow 4NO_2 + 6H_2O$
   e) $CO + 3H_2 \Leftrightarrow CH_4 + H_2O$

   Solution:

   a) $K = \frac{[CO_2][H_2O]^2}{[CH_4][O_2]^2}$ b) $K = \frac{[O_2]^3}{[O_3]^2}$ c) $K = \frac{[H_2O][CO]}{[H_2][CO_2]}$

   d) $K = \frac{[NO_2]^4[H_2O]^6}{[NH_3]^4[O_2]^7}$ e) $K = \frac{[CH_4][H_2O]}{[CO][H_2]^3}$

8.40   The following equilibria are established in water solutions. Write an equilibrium expression for each reaction.
   a) $Au^+ + 2CN^- \Leftrightarrow Au(CN)_2^-$
   b) $Pt^{2+} + 4Cl^- \Leftrightarrow PtCl_4^{2-}$
   c) $Co^{2+} + 6NH_3 \Leftrightarrow Co(NH_3)_6^{2+}$

   Solution:

   a) $K = \frac{[Au(CN)_2^-]}{[Au^+][CN^-]^2}$ b) $K = \frac{[PtCl_4^{2-}]}{[Pt^{2+}][Cl^-]^4}$ c) $K = \frac{[Co(NH_3)_6^{2+}]}{[Co^{2+}][NH_3]^6}$

8.42   Write an equation that corresponds to each of the following equilibrium expressions:
   a) $K = \frac{[CO_2]^2}{[CO]^2[O_2]}$ b) $K = \frac{[COCl_2]}{[CO][Cl_2]}$ c) $K = \frac{[H_2O]^2}{[H_2]^2[O_2]}$

   d) $K = \frac{[PCl_3][Cl_2]}{[PCl_5]}$

   Solution:

   The equilibrium expression is the product of the products divided by the product of the reactants.
   a) $2CO + O_2 \Leftrightarrow 2CO_2$ b) $CO + Cl_2 \Leftrightarrow COCl_2$
   c) $2H_2 + O_2 \Leftrightarrow 2H_2O$ d) $PCl_5 \Leftrightarrow PCl_3 + Cl_2$

8.44   A sample of gaseous BrCl is allowed to decompose in a closed container at 25°C;
   $2BrCl_{(g)} \Leftrightarrow Br_{2(g)} + Cl_{2(g)}$
   When the reaction reaches equilibrium, the following concentrations are measured:
   $[BrCl] = 0.38$ M, $[Cl_2] = 0.26$ M, $[Br_2] = 0.26$ M.
   Evaluate the equilibrium constant for the reaction at 25°C.

   Solution:

   $K = \frac{[Br_2][Cl_2]}{[BrCl]^2} = \frac{[0.26][0.26]}{[0.38]^2} = 0.47$

8.46   A mixture of the gases NOCl, Cl_2, and NO is allowed to reach equilibrium at 25°C. The measured equilibrium concentrations are $[NO] = 0.92$ mol/L, $[NOCl] = 1.31$ mol/L, and $[Cl_2] = 0.20$ mol/L. What is the value of the equilibrium constant at 25°C for the reaction
   $2NOCl_{(g)} \Leftrightarrow 2NO_{(g)} + Cl_{2(g)}$

   Solution:

   $K = \frac{[NO]^2[Cl]}{[NOCl]^2} = \frac{[0.92]^2[0.20]}{[1.31]^2} = 0.099$

8.48   Consider the following equilibrium constants. Describe how you would expect the equilibrium concentrations of reactants and products to compare with each other (larger than, smaller than, etc.) for each case.
a) $K = 4.4 \times 10^{-8}$ b) $K = 12.8$ c) $K = 3.5 \times 10^5$ d) $K = 0.000086$

Solution:
If K is greater than 1, the numerator is larger than the denominator.
If K is less than 1, the numerator is smaller than the denominator.
a) K is much less than 1. The concentrations of the reactants are greater than that of the products.
b) K is larger than 1. The concentrations of the reactants are less than that of the products.
c) K is much larger than 1. The concentrations of the reactants are less than that of the reactants.
d) K is much less than 1. The concentrations of the reactants are greater than that of the products.

Factors That Influence Equilibrium Position (Section 8.8)

8.50   Use Le Châtelier's principle to predict the direction of equilibrium shift in the following equilibria when the indicated stress is applied:
a) $Ag^+_{(aq)} + Cl^-_{(aq)} \Leftrightarrow AgCl_{(s)}$; some Cl^- is added.
b) $2HBr_{(g)} + heat \Leftrightarrow H_{2(g)} + Br_{2(g)}$; the system is cooled.
c) $6Cu_{(s)} + N_{2(g)} + heat \Leftrightarrow 2Cu_3N_{(s)}$; the system is heated and some N_2 is added.

Solution:
a) The reaction will shift to the right to relieve the stress to the left side.
b) The reaction will shift to the left due to the loss of heat.
c) The reaction will shift to the right due to the stress of the additions.

8.52   Using Le Châtelier's principle, predict the direction of equilibrium shift and the changes that will be observed (color, amount of precipitate, etc.) in the following equilibria when the indicated stress is applied:
a) $Cu^{2+}_{(aq)}$ + $4NH_{3(aq)}$ $\Leftrightarrow$ $Cu(NH_3)^{2+}_{4(aq)}$:
blue   colorless   dark purple
Some NH_3 is added to the equilibrium mixture.
b) $Pb^{2+}_{(aq)}$ + $2Cl^-_{(aq)}$ $\Leftrightarrow$ $PbCl_{2(s)}$ +   heat;
colorless   colorless   white solid
The equilibrium mixture is cooled.
c) C_2H_4 + I_2 $\Leftrightarrow$ $C_2H_4I_2$ +   heat;
colorless gas   violet gas   colorless gas
Some $C_2H_4I_2$ is removed from the equilibrium mixture.
d) C_2H_4 + I_2 $\Leftrightarrow$ $C_2H_4I_2$ +   heat;
colorless gas   violet gas   colorless gas
the equilibrium mixture is cooled.
e) heat   + $4NO_2$ + $6H_2O$ $\Leftrightarrow$ $7O_2$ + $4NH_3$;
              brown gas   colorless gas   colorless gas   colorless gas
A catalyst is added, and NH_3 is added to the equilibrium mixture.

Solution:
a) The addition of ammonia will shift the reaction to the right deepening the purple color.
b) Cooling the mixture shifts the reaction to the right increasing the amount of precipitate.
c) The removal of ethylene iodide shifts the reaction to the right; the violet color of the gas mixture will become less intense.
d) Cooling this mixture shifts the reaction to the right decreasing the violet color.

e) A catalyst will not stress the equilibrium as it would affect the forward and reverse reactions to the same extent; however, the addition of the ammonia would cause stress that can only be relieved by shifting to the left. The effect would be to deepen the brown color.

8.54   Tell what will happen to each equilibrium concentration in the following when the indicated stress is applied and a new equilibrium position is established:
a) $LiOH_{(s)} + CO_{2(g)} \Leftrightarrow LiHCO_{3(s)} + heat$; CO_2 is removed.
b) $2NaHCO_{3(s)} + heat \Leftrightarrow Na_2O_{(s)} + 2CO_{2(g)} + H_2O_{(g)}$; the system is cooled.
c) $CaCO_{3(s)} + heat \Leftrightarrow CaO_{(s)} + CO_{2(g)}$; the system is cooled

Solution:
a) Equilibrium shifts left. Amount of solid $LiOH$ increases and amount of solid $LiHCO_3$ decreases. Amount of CO_2 will be less than in original equilibrium, but greater than it was at the instant it was removed.
b) Equilibrium shifts left. Amount of $NaHCO_3$ increases, amounts of N_2O, CO_2 and H_2O decrease.
c) Equilibrium shifts left. Amount of $CaCO_3$ increases, amounts of CAO and CO_2 decrease.

8.56   The gaseous reaction $N_{2(g)} + O_{2(g)} \Leftrightarrow 2NO_{(g)}$ is exothermic. Tell which direction the equilibrium will shift for each of the following:
a) Some N_2 is removed.
b) The temperature is decreased.
c) Some NO is added.
d) Some O_2 is removed.
e) A catalyst is added.
f) The temperature is increased, and some O_2 is removed.

Solution:
a) Removal of N_2 shifts the reaction to the left.
b) Cooling shifts the reaction to the right.
c) Addition of NO shifts the reaction to the left.
d) Removing oxygen shifts the reaction to the left.
e) A catalyst has no effect on the position of equilibrium, but does affect the rate of reaction (establishing equilibrium). No shift.
f) The removal of oxygen shifts the reaction to the left, as does the increase in temperature.

PROGRAMMED REVIEW

Section 8.1 Spontaneous and Nonspontaneous Changes

Processes that take place naturally with no apparent cause or stimulus are called (a) _Spontaneous_ processes. (b) _Exergonic_ processes give up energy as they occur, while (c) _undergoni_ processes gain or accept energy. (d) _entropy_ is a measurement or indication of the disorder of a system. Substances that do not undergo spontaneous changes are said to be (e) _Stable_ .

Section 8.2 Reaction Rates

The (a) _speed_ of a reaction is called the reaction rate. A reaction rate is determined experimentally as a change in (b) _concentr_ of a (c) _reactant_ or (d) _product_ divided by the time required for the change. This measured rate is an (e) _average_ rate for the reaction.

Section 8.3 Molecular Collisions

A detailed explanation of how a reaction takes place is called a (a) _reaction mechanism_. Reactions between particles are assumed not to take place unless the particles (b) _collide_ with each other. The (c) _internal energy_ of molecules is the energy associated with vibrations within the molecules. The rubbing of a match head against a rough surface provides (d) _activation_ energy.

Section 8.4 Energy Diagrams

In an energy diagram, the vertical axis represents (a) _energy_, and the horizontal axis represents the (b) _reaction_ progress. In an energy diagram, the energy of products is lower than that of reactants for an (c) _exothermic_ reaction. The energy hump on an energy diagram between the energy of reactants and products represents (d) _activation_ energy for the reaction.

Section 8.5 Factors that Influence Reaction Rates

Four factors that influence reaction rates are the (a) _nature_ of the reactants, the (b) _concentration_ of the reactants, the (c) _temperature_ of the reactants, and the presence of (d) _catalysts_. Collisions between molecules that have the potential to cause a reaction to occur are called (e) _effective collisions_. (f) _catalysts_ are substances that change reaction rates without being used up in the reaction. (g) _effective collisions_ are ions or molecules uniformly dispersed throughout a reaction mixture, while (h) _heterogeneous catalysts_ are used in the form of solids.

Section 8.6 Chemical Equilibrium

In principle, all reactions can occur in (a) _both_ _directions_. When the (b) _forward_ and (c) _reverse_ reaction rates are equal, the reaction is in a state of (d) _equilibrium_. The concentrations of reactants and products in this state are called (e) _equilibrium_ concentrations.

Section 8.7 Position of Equilibrium

The position of equilibrium for a reaction is an indication of the relative amounts of (a) _reactants_ and (b) _products_ present at equilibrium. When the equilibrium position is described as being far to the right, the concentration of (c) _products_ is much higher than that of (d) _reactants_.

Section 8.8 Factors that Influence Equilibrium Position

The influence of a number of factors on the position of equilibrium can be predicted by using (a) _Le Chatelier_ principle. According to this principle, the addition of a reactant to an equilibrium mixture will shift the equilibrium toward the (b) _right (prod)_, and heating an equilibrium mixture of an exothermic reaction will shift the equilibrium toward the (c) _left (reactants)_.

SELF-TEST QUESTIONS

Multiple Choice

1. A catalyst
    a) is not used up in a reaction
    b) changes the rate of a reaction
    c) affects the forward reaction the same as it affects the reverse reaction
    d) all of the above

2. Which of the following responses correctly arranges the states of matter for a pure substance in the order of decreasing entropy?
    a) gas, liquid, solid   b) liquid, solid, gas   c) solid, liquid, gas   d) solid, gas, liquid

3. Four processes occur as the following changes take place in energy and entropy. Which process is definitely nonspontaneous?
    a) energy decrease and entropy increase
    b) energy decrease and entropy decrease
    c) energy increase and entropy increase
    d) energy increase and entropy decrease

4. A carrot cooks in 15 minutes in boiling water (100°C). How long will it take to cook a carrot inside a pressure cooker where the temperature is 10°C greater (110°C)?
    a) 3.7 minutes   b) 5.0 minutes   c) 7.5 minutes   d) 30 minutes

Questions 5 and 6 refer to the following reaction, which is assumed to be at equilibrium:
    heat $+2NO + O_2 \rightleftarrows 2NO_2$
In each case choose the response which best indicates the effects resulting from the described change in conditions.

5. The reaction mixture is heated.
    a) equilibrium shifts left   b) equilibrium shifts right
    c) equilibrium does not shift   d) the effect cannot be predicted

6. A catalyst is added to the reaction mixture.
    a) equilibrium shifts left   b) equilibrium shifts right
    c) equilibrium does not shift   d) the effect cannot be predicted

7. In the equilibrium constant expression for the reaction: $2N_2O_5 \rightleftarrows 4NO_2 + O_2$, the exponent on the concentration of N_2O_5 is
    a) 1
    b) 2
    c) 0
    d) can't be determined from the information given

8. A sample of ICl is placed in a container and equilibrium is established according to the reaction: $2ICl \rightleftarrows I_2 + Cl_2$ where all the materials are gases. Analysis of the equilibrium mixture gave the following molar concentrations:
       $[ICl] = 0.26$, $[I_2] = [Cl_2] = 0.09$
    What is the value of K, the equilibrium constant for the reaction?
    a) 0.031   b) 0.12   c) 0.35   d) 14.8

True-False

9. A reaction rate can be thought of as the speed of a reaction.

10. Effective molecular collisions are those that allow molecules to collide but not react.

11. Catalysts that slow reactions are called inhibitors.

12. In a reaction at equilibrium, the forward and reverse reactions have both stopped.

13. If an endothermic reaction is spontaneous, then entropy must have decreased.

14. Catalysts act by lowering the activation energy.

15. The entropy of a cluttered room is higher than that of an orderly room.

Matching
For each of the following processes choose the appropriate response from those on the right.

16. _____ a match burns

17. _____ perspiration evaporates

18. _____ melted lead becomes a solid

19. _____ an explosive detonates

a) both entropy and energy increase

b) both entropy and energy decrease

c) entropy increases; energy decreases

d) entropy decreases; energy increases

Three liquid fuels are to be tested. A 1.0 gram sample of each fuel is weighed out and heated to its ignition temperature. When the fuel burns, the total heat liberated is measured. The results of this experiment are given in the table below. For Questions 20–25 choose the answer from the column on the right that best fits each statement on the left.

Fuel	Ignition temperature	Heat liberated
X	210°C	1680 cal
Y	110°C	1410 cal
Z	285°C	1206 cal

20. _____ it has the highest activation energy

21. _____ it has the second highest activation energy

22. _____ it has the lowest activation energy

23. _____ it has the smallest energy difference between reactants and products

24. _____ it has the largest energy difference between reactants and products

25. _____ the described reaction is exothermic (exergonic)

a) fuel X

b) fuel Y

c) fuel Z

d) two or more reactants fuels fit this category

Use the following equilibrium expression and match the effects on the equilibrium from the right with the changes made to the equilibrium system listed on the left.

$$heat + CO + 2H_2 \rightleftarrows CH_3OH$$

26. _____ add CO

27. _____ add H_2

28. _____ remove some CH_3OH

29. _____ heat the system

30. _____ add a catalyst

a) shift left

b) shift right

c) no effect on equilibrium

d) cannot be determined from the information given

SOLUTIONS

A. Answers to Programmed Review

8.1   a) spontaneous   b) exergonic   c) endergonic   d) entropy
      e) stable

8.2   a) speed   b) concentration   c) reactant   d) product
      e) average

8.3   a) reaction mechanism   b) collide   c) internal energy   d) activation

8.4   a) energy   b) reaction   c) exothermic   d) activation

8.5   a) nature   b) concentration
      c) temperature   d) catalysts
      e) effective collisions   f) catalysts
      g) homogeneous catalysts   h) heterogeneous catalysts

8.6   a) both directions   b) forward   c) reverse   d) equilibrium
      e) equilibrium

8.7   a) reactants   b) products   c) products   d) reactants

8.8   a) Le Châtelier's   b) right (or products) c) left (or reactants)

B. Answers to Self-Test Questions

1. d	11. T	21. a
2. a	12. F	22. b
3. d	13. F	23. c
4. c	14. T	24. a
5. b	15. T	25. d
6. c	16. c	26. b
7. b	17. a	27. b
8. b	18. b	28. b
9. T	19. c	29. b
10. F	20. c	30. c

Chapter 9

Acids, Bases, and Salts

CHAPTER OUTLINE

9.1 The Arrhenius Theory
9.2 The Brønsted Theory
9.3 Naming Acids
9.4 The Self-Ionization of Water
9.5 The pH Concept
9.6 Properties of Acids
9.7 Properties of Bases

9.8 Salts
9.9 Strengths of Acids and Bases
9.10 Analysis of Acids and Bases
9.11 Titration Calculations
9.12 Hydrolysis Reactions of Salts
9.13 Buffers

LEARNING OBJECTIVES

When you have completed your study of this chapter, you should be able to:

1. Write equations that illustrate Arrhenius and Brønsted acid-base behavior.
2. Identify Brønsted acids and bases from written equations.
3. Name common acids based on their formulas.
4. Do calculations involving the pH concept.
5. Write equations that illustrate the characteristic reactions of acids.
6. Write equations that illustrate various ways to prepare salts.
7. Write equations that represent neutralization reactions between acids and bases.
8. Write equations that illustrate the stepwise dissociation and reaction of polyprotic acids.
9. Describe the strengths of acids and bases in a qualitative way (no calculations).
10. Do problems related to the analysis of acids and bases by titration.
11. Explain and write equations that illustrate the hydrolysis of salts.
12. Explain and write equations that illustrate the action of buffers.
13. Calculate buffer pH using the Henderson-Hasselbalch equation.

ANSWERS AND SOLUTIONS TO EVEN-NUMBERED PROBLEMS

The Arrhenius Theory (Section 9.1)

9.2   Write the dissociation equations for the following that emphasize their behavior as Arrhenius acids:

a) HF

b) $HClO_3$

c) H_3BO_3 (show only 1st H)

d) HSe^-

Solution:

a) $HF(aq) \rightarrow H^+(aq) + F^-(aq)$

b) $HClO_3(aq) \rightarrow H^+(aq) + ClO_3^-(aq)$

c) $H_3BO_3(aq) \rightarrow H^+(aq) + H_2BO_3^-(aq)$

d) $HSe^-(aq) \rightarrow H^+(aq) + Se^{2-}(aq)$

9.4   Each of the following produces a basic solution when dissolved in water. Identify those that behave as Arrhenius bases and write dissociation equations to illustrate that behavior.

a) $NaNH_2$

b) RbOH

c) $C_3H_7NH_2$

d) $Ba(OH)_2$

Solution:

a) not an Arrhenius base

b) $RbOH(aq) \rightarrow Rb^+(aq) + OH^-(aq)$

c) not an Arrhenius base

d) $Ba(OH)_2(aq) \rightarrow Ba^{2+}(aq) + 2OH^-(aq)$

The Brønsted Theory (Section 9.2)

9.6   Identify each Brønsted acid and base in the following equations. Note that the reactions are assumed to be reversible.

a) $OCl^-(aq) + H_2O(l) \Leftrightarrow HOCl(aq) + OH^-(aq)$

b) $H_2C_2O_4(aq) + H_2O(l) \Leftrightarrow H_3O^+(l) + HC_2O_4^-(aq)$

c) $HPO_4^{2-}(aq) + H_2O(l) \Leftrightarrow H_3O^+(aq) + PO_4^{3-}(aq)$

d) $C_2O_4^{2-}(aq) + H_2O(l) \Leftrightarrow HC_2O_4^-(aq) + OH^-(aq)$

e) $H_3AsO_4(aq) + H_2O(aq) \Leftrightarrow H_3O^+(aq) + H_2AsO_4^-(aq)$

Solution:

a) $OCl^-(aq) + H_2O(l) \Leftrightarrow HOCl(aq) + OH^-(aq)$
    base   acid   acid   base

b) $H_2C_2O_4(aq) + H_2O(l) \Leftrightarrow H_3O^+(l) + HC_2O_4^-(aq)$
    acid   base   acid   base

c) $HPO_4^{2-}(aq) + H_2O(l) \Leftrightarrow H_3O^+(aq) + PO_4^{3-}(aq)$
    acid   base   acid   base

d) $C_2O_4^{2-}(aq) + H_2O(l) \Leftrightarrow HC_2O_4^-(aq) + OH^-(aq)$
    base   acid   acid   base

e) $H_3AsO_4(aq) + H_2O(l) \Leftrightarrow H_3O^+(aq) + H_2AsO_4^-(aq)$
    acid   base   acid   base

9.8   Identify each conjugate acid-base pair in the equations you wrote for Exercise 9.6.

Solution:

	Acid	Conjugate Base	Base	Conjugate Acid
a)	H_2O	OH^-	OCl^-	$HOCl$
b)	$H_2C_2O_4$	$H_2C_2O_4^-$	H_2O	H_3O^+
c)	HPO_4^{2-}	PO_4^{3-}	H_2O	H_3O^+
d)	H_2O	OH^-	$C_2O_4^{2-}$	$HC_2O_4^-$
e)	H_3AsO_4	$H_2AsO_4^-$	H_2O	H_3O^+

9.10   Write a formula for the conjugate base formed when each of the following behaves as a Brønsted acid:
a) $H_2BO_3^-$ b) $C_6H_5NH_3^+$ c) HS^- d) $HC_2O_4^-$
e) $HClO_4$

Solution:
a) HBO_3^{2-} b) $C_6H_5NH_2$ c) S^{2-} d) $C_2O_4^{2-}$
e) ClO_4^-

9.12   Write a formula for the conjugate acid formed when each of the following behaves as a Brønsted base:
a) $C_6H_5NH_3^+$ b) $S_2O_3^{2-}$ c) CN^- d) $HAsO_4^{2-}$
e) F^-

Solution:
a) $C_6H_5NH_3^-$ b) $HS_2O_3^-$ c) HCN d) $H_2AsO_4^-$
e) HF

9.14   The following reactions illustrate Brønsted acid-base behavior. Complete each equation.
a) $H_2AsO_4^-(aq) + NH_3(aq) \rightarrow NH_4^+ + ?$
b) $? + H_2O(l) \rightarrow C_6H_5NH_3^+(aq) + OH^-(aq)$
c) $S^{2-}(aq) + ? \rightarrow HS^-(aq) + OH^-(aq)$
d) $? + HBr \rightarrow (CH_3)_2NH_2^+(aq) + Br^-(aq)$
e) $CH_3NH_2(aq) + HCl(aq) \rightarrow ? + Cl^-(aq)$

Solution:
The substance needed to complete the equation is underlined.
a) $H_2AsO_4^-(aq) + NH_3(aq) \rightarrow NH_4^+(aq) + \underline{HAsO_4^{2-}}(aq)$
b) $\underline{C_6H_5NH_2}(aq) + H_2O(l) \rightarrow C_6H_5NH_3^+(aq) + OH^-(aq)$
c) $S^{2-}(aq) + \underline{H_2O}(l) \rightarrow HS^-(aq) + OH^-(aq)$
d) $\underline{(CH_3)_2NH} + HBr(aq) \rightarrow (CH_3)_2NH_2^+(aq) + Br^-(aq)$
e) $CH_3NH_2(aq) + HCl(aq) \rightarrow \underline{CH_3NH_3^+}(aq) + Cl^-(aq)$

9.16   Write equations to illustrate the acid-base reaction of each of the following pairs of Brønsted acids and bases:

	Acid	Base
a)	H_3O^+	NH_2^-
b)	$H_2PO_4^-$	NH_3
c)	$HS_2O_3^-$	OCl^-
d)	H_2O	ClO_4^-
e)	H_2O	NH_3

Solution:
a) $H_3O^+(aq) + NH_2^-(aq) \Leftrightarrow H_2O(l) + NH_3(aq)$
b) $H_2PO_4^-(aq) + NH_3(aq) \Leftrightarrow HPO_4^{2-}(aq) + NH_4^+(aq)$
c) $HS_2O_3^-(aq) + OCl^-(aq) \Leftrightarrow S_2O_3^{2-}(aq) + HOCl(aq)$
d) $H_2O(l) + ClO_4^-(aq) \Leftrightarrow OH^-(aq) + HClO_4(aq)$
e) $H_2O(l) + NH_3(aq) \Leftrightarrow OH^-(aq) + NH_4^+(aq)$

Naming Acids (Section 9.3)

9.18   Hydrogen cyanide, HCN, behaves in water solution very much like the binary covalent compounds of hydrogen, but it liberates the cyanide ion, CN^-. Name the acidic water solution by following the rules for binary covalent compounds of hydrogen.

Solution:
hydrocyanic acid

9.20   Name the following acids. Refer to Table 4.7 as needed.

a) $H_2Te(aq)$ b) HClO   c) H_2SO_3 d) HNO_2

Solution:
a) hydrotelluric acid   b) hypochlorous acid   c) sulfurous acid   d) nitrous acid

9.22   The acid $H_2C_4H_4O_4$ forms the succinate ion, $C_4H_4O_4^{2-}$, when both hydrogens are removed. This acid is involved in the same energy-storing process as the acid of Exercise 9.21. Name $H_2C_4H_4O_4$ as an acid.

Solution:
succinic acid

9.24   Refer to Table 4.7 and write the formula for permanganic acid.

Solution:
$HMnO_4$

The Self-Ionization of Water (Section 9.4)

9.26   Calculate the molar concentration of OH^- in water solutions with the following H_3O^+ molar concentrations:

a) 1.2×10^{-5} b) 0.27 c) 0.031 d) 3.6×10^{-9}
e) 5.3×10^{-2}

Solution:

$K_w = 1.0 \times 10^{-14} = [H_3O^+][OH^-]$; rearranged to $[OH^-] = \frac{1.0 \times 10^{-14}}{[H_3O^+]}$

a) $[OH^-] = \frac{1.0 \times 10^{-14}}{1.2 \times 10^{-5}} = 8.3 \times 10^{-10}$ M

b) $[OH^-] = \frac{1.0 \times 10^{-14}}{2.7 \times 10^{-1}} = 3.7 \times 10^{-14}$ M

c) $[OH^-] = \frac{1.0 \times 10^{-14}}{3.1 \times 10^{-2}} = 3.2 \times 10^{-13}$ M

d) $[OH^-] = \frac{1.0 \times 10^{-14}}{3.6 \times 10^{-9}} = 2.8 \times 10^{-6}$ M

e) $[OH^-] = \frac{1.0 \times 10^{-14}}{5.3 \times 10^{-2}} = 1.9 \times 10^{-13}$ M

9.28   Calculate the molar concentration of H_3O^+ in water solutions with the following OH^- molar concentrations:

a) 0.0071 M   b) 4.2×10^{-4} M   c) 2.8 M   d) 7.9×10^{-10} M
e) 9.1×10^{-6} M

Solution:

a) $[H_3O^+] = \frac{1.0 \times 10^{-14}}{7.1 \times 10^{-3}} = 1.4 \times 10^{-12}$ M

b) $[H_3O^+] = \frac{1.0 \times 10^{-14}}{4.2 \times 10^{-4}} = 2.4 \times 10^{-11}$ M

c) $[H_3O^+] = \frac{1.0 \times 10^{-14}}{2.8} = 3.6 \times 10^{-15}$ M

d) $[H_3O^+] = \frac{1.0 \times 10^{-14}}{7.9 \times 10^{-10}} = 1.3 \times 10^{-5}$ M

e) $[H_3O^+] = \frac{1.0 \times 10^{-14}}{9.1 \times 10^{-6}} = 1.1 \times 10^{-9}$ M

9.30   Classify the solutions represented in Exercises 9.26 and 9.28 as *acidic, basic,* or *neutral.*

Solution:

#9.26 a) acidic	b) acidic	c) acidic
d) basic	e) acidic	
#9.28 a) basic	b) basic	c) basic
d) acidic	e) basic	

9.32   Explain what is wrong with each of the following statements. Water is the solvent.
a) An acidic solution is prepared in which $[H_3O^+]$ is greater than $[OH^-]$. The $[H_3O^+] = 4.4 \times 10^{-10}$ mol/L.
b) A solution is prepared so the ratio $[H_3O^+]/[OH^-] = 1.1 \times 10^6$ and $[H_3O^+] = 3.3 \times 10^{-4}$ mol/L.
c) A neutral solution is prepared in which $[H_3O^+]$ is twice as great as the $[OH^-]$.

Solution:
a) If the $[H_3O^+]$ really equals 4.4×10^{-10} M, the solution is not acidic. It would be basic.
$[OH^-] = 2.3 \times 10^{-5}$
b) If the $[H_3O^+]$ really equals 3.3×10^{-4} M, then:
$[OH^-] = \frac{1.0 \times 10^{-14}}{3.3 \times 10^{-4}} = 3.0 \times 10^{-11}$
The ratio $\frac{[H_3O^+]}{[OH^-]}$ would then be $\frac{3.3 \times 10^{-4}}{3.0 \times 10^{-11}} = 1.1 \times 10^7$ not 1.1×10^6
c) In a neutral water solution, the $[H_3O^+]$ is equal to the $[OH^-]$ and they both must be 1.0×10^{-7} M.

The pH Concept (Section 9.5)

9.34   Classify solution with the following characteristics as *acidic, basic,* or *neutral.*
a) pH = 2.8   b) pH = 8   c) pII = 6.9   d) pII = 12

Solution:
a) acidic   b) basic   c) acidic   d) basic

9.36   Determine the pH of water solutions with the following characteristics. Classify each solution as *acidic, basic,* or *neutral.*
a) $[H^+] = 3.0 \times 10^{-8}$ M   b) $[OH^-] = 7.0 \times 10^{-3}$ M
c) $[H^+] = 10[OH^-]$ M   d) $[OH^-] = 8.0 \times 10^{-11}$ M
e) $[H^+] = 5.0 \times 10^{-2}$ M

Solution:
Use equation 9.11 to get pH from $[H^+]$.   pH = $-\log[H^+]$
a) pH = 7.5; basic
b) $[H^+] = \frac{1.0 \times 10^{-14}}{7.0 \times 10^{-3}} = 1.4 \times 10^{-12}$ M; pH = 11.8 or 12 (two significant figures; basic
c) $[H^+][OH^-] = 1 \times 10^{-14}$ M; substitute $10[OH^-]$ for $[H^+]$
$10[OH^-] \times [OH^-] = 1 \times 10^{-14}$ M or $[OH^-]^2 = \frac{1.0 \times 10^{-14}}{10} = 1.0 \times 10^{-15}$ M
take the square root; $[OH^-] = 3.2 \times 10^{-8}$ M; $[H^+] = \frac{1.0 \times 10^{-14}}{3.2 \times 10^{-8}} = 3.1 \times 10^{-7}$ M; pH = 6.5; acidic
d) $[H^+] = \frac{1.0 \times 10^{-14}}{8.0 \times 10^{-11}} = 1.3 \times 10^{-4}$ M; pH = 3.9; acidic
e) pH = 1.3; acidic

9.38   Determine the pH of water solutions with the following characteristics. Classify each solution as *acidic*, *basic*, or *neutral*.
   a) $[H^+] = 3.9 \times 10^{-6}$ M
   b) $[H^+] = 6.2 \times 10^{-3}$ M
   c) $[H^+] = 1.8 \times 10^{-11}$ M
   d) $[OH^-] = 4.7 \times 10^{-9}$ M
   e) $[OH^-] = 8.4 \times 10^{-4}$ M

   Solution:
   a) pH = 5.4; acidic
   b) pH = 2.2; acidic
   c) pH = 10.7; basic
   d) $[H^+] = \frac{1.0 \times 10^{-14}}{4.7 \times 10^{-9}} = 2.1 \times 10^{-6}$ M; pH = 5.7; acidic
   e) $[H^+] = \frac{1.0 \times 10^{-14}}{8.4 \times 10^{-4}} = 1.2 \times 10^{-11}$ M; pH = 10.9; basic

9.40   Determine the $[H^+]$ value for solutions with the following characteristics:
   a) pH = 8.19
   b) pH = 3.72
   c) pH = 11.58

   Solution:
   Use calculator to get $[H^+]$; $[H^+] = 10^{-pH}$
   a) 6.5×10^{-9}
   b) 1.9×10^{-4}
   c) 2.6×10^{-12}

9.42   Convert the following pH values into both $[H^+]$ and $[OH^-]$ values:
   a) pH = 5.00
   b) pH = 2.88
   c) pH = 10.79

   Solution:
   a) $[H^+] = 1.0 \times 10^{-5}$; $[OH^-] = \frac{1.00 \times 10^{-14}}{1.0 \times 10^{-5}} = 1.0 \times 10^{-9}$ M
   b) $[H^+] = 1.3 \times 10^{-3}$; $[OH^-] = \frac{1.0 \times 10^{-14}}{1.3 \times 10^{-3}} = 7.6 \times 10^{-12}$ M
   c) $[H^+] = 1.6 \times 10^{-11}$; $[OH^-] = \frac{1.0 \times 10^{-14}}{1.6 \times 10^{-11}} = 6.2 \times 10^{-4}$ M

9.44   The pH values listed in Table 9.1 are generally the average values for the listed materials. Most natural materials such as body fluids and fruit juices have pH values that cover a range for different samples. Some measured pH values for specific body fluid samples are given below. Convert each one to $[H^+]$ and classify the fluid as *acidic*, *basic*, or *neutral*.
   a) bile, pH = 8.05
   b) vaginal fluid, pH = 3.93
   c) semen, pH = 7.38
   d) cerebrospinal fluid, pH = 7.40
   e) perspiration, pH = 6.23

   Solution:
   a) $[H^+] = 8.9 \times 10^{-9}$ M; basic
   b) $[H^+] = 1.2 \times 10^{-4}$ M; acidic
   c) $[H^+] = 4.2 \times 10^{-8}$ M; basic
   d) $[H^+] = 4.0 \times 10^{-8}$ M; basic
   e) $[H^+] = 5.9 \times 10^{-7}$ M; acidic

9.46   The pH values of specific samples of food items are listed below. Convert each value to $[H^+]$ and classify the sample as *acidic*, *basic*, or *neutral*.
   a) soft drink, pH = 2.91
   b) tomato juice, pH = 4.11
   c) lemon juice, pH = 2.32
   d) grapefruit juice, pH = 3.07

   Solution:
   a) $[H^+] = 1.2 \times 10^{-3}$ M; acidic
   b) $[H^+] = 7.8 \times 10^{-5}$ M; acidic
   c) $[H^+] = 4.8 \times 10^{-3}$ M; acidic
   d) $[H^+] = 8.5 \times 10^{-4}$ M; acidic

Properties of Acids (Section 9.6)

9.48   Use the information in Table 9.4 and describe how you would prepare each of the following solutions.
   a) about 2 L of 3.0 M HNO_3 from dilute nitric acid solution
   b) about 500 mL of 1.5 M aqueous ammonia from concentrated aqueous ammonia solution
   c) about 5 L of 0.2 M HCl from concentrated hydrochloric acid solution

   Solution:
   Solve equation 7.9 for V_c $V_c = V_d \times \frac{C_d}{C_c}$

   a) $V_c = 2\ L \times \frac{3.0\ M}{6\ M} = 1\ L$
      Measure 1 L of the dilute (6 M) HNO_3 and add 1 L water
   b) $V_c = 500\ mL \times \frac{1.5\ M}{15\ M} = 50\ mL$
      Measure 50 mL of the concentrated (15 M) ammonia solution and add 450 mL water
   c) $V_c = 5\ L \times \frac{0.2\ M}{12\ M} = 0.083\ L = 83\ mL$
      Measure 83 mL of the concentrated HCl (12 M) and add 4917 mL of water

9.50   Write balanced molecular equations to illustrate the following characteristic reactions of acids, using sulfuric acid (H_2SO_4).
   a) reaction with water to form hydronium ions   b) reaction with the solid oxide CaO
   c) reaction with the solid hydroxide, $Mg(OH)_2$ d) reaction with the solid carbonate, $CuCO_3$
   e) reaction with the solid bicarbonate, $KHCO_3$ f) reaction with Mg metal

   Solution:
   a) $H_2SO_4(aq) + 2H_2O(l) \rightarrow 2H_3O^+(aq) + SO_4^{2-}(aq)$
   b) $CaO(s) + H_2SO_4(aq) \rightarrow CaSO_4(aq) + H_2O(l)$
   c) $Mg(OH)_2(s) + H_2SO_4(aq) \rightarrow MgSO_4(aq) + 2H_2O(l)$
   d) $CuCO_3(s) + H_2SO_4(aq) \rightarrow CuSO_4(aq) + H_2O(l) + CO_2(g)$
   c) $2KHCO_3(s) + H_2SO_4(aq) \rightarrow K_2SO_4(aq) + 2H_2O(aq) + 2CO_2(g)$
   f) $Mg(s) + H_2SO_4(aq) \rightarrow MgSO_4(aq) + H_2(g)$

9.52   Write each molecular equation of Exercise 9.50 in total ionic and net ionic form. Use Table 7.4 to decide which products will be soluble.

   Solution:
   a) total ionic: $2H^+(aq) + SO_4^{2-}(aq) + 2H_2O(l) \rightarrow 2H_3O^+(aq) + SO_4^{2-}$
      net ionic: $2H^+(aq) + 2H_2O(l) \rightarrow 2H_3O^+(aq)$
      simplified: $H^+(aq) + H_2O(l) \rightarrow H_3O^+(aq)$
   b) total ionic: $CaO(s) + 2H^+(aq) + SO_4^{2-}(aq) \rightarrow Ca^{2+}(aq) + SO_4^{2-}(aq) + H_2O(l)$
      net ionic: $CaO(s) + 2H^+(aq) \rightarrow Ca^{2+}(aq) + H_2O(l)$
   c) total ionic: $Mg(OH)_2(s) + 2H^+(aq) + SO_4^{2-}(aq) \rightarrow Mg^{2+}(aq) + SO_4^{2-}(aq) + H_2O(l)$
      net ionic: $Mg(OH)_2(s) + 2H^+(aq) \rightarrow Mg^{2+}(aq) + 2H_2O(l)$
   d) total ionic: $CuCO_3(s) + 2H^+(aq) + SO_4^{2-}(aq) \rightarrow Cu^{2+}(aq) + SO_4^{2-}(aq) + H_2O(l) + CO_2(g)$
      net ionic: $CuCO_3(s) + 2H^+(aq) \rightarrow Cu^{2+}(aq) + H_2O(l) + CO_2(g)$
   e) total ionic: $2KHCO_3(s) + 2H^+(aq) + SO_4^{2-}(aq) \rightarrow 2K^+(aq) + SO_4^{2-}(aq) + 2H_2O(l) + 2CO_2(g)$
      net ionic: $2KHCO_3(s) + 2H^+(aq) \rightarrow 2K^+(aq) + 2H_2O(l) + 2CO_2(g)$
      simplified: $KHCO_3(s) + H^+(aq) \rightarrow K^+(aq) + H_2O(l) + CO_2(g)$
   f) total ionic: $Mg(s) + 2H^+(aq) + SO_4^{2-}(aq) \rightarrow Mg^{2+}(aq) + SO_4^{2-}(aq) + H_2(g)$
      net ionic: $Mg(s) + 2H^+(aq) \rightarrow Mg^{2+}(aq) + H_2(g)$

9.54   Write the balanced molecular equations to illustrate five different reactions that could be used to prepare $SrCl_2$ from hydrochloric acid (HCl) and other appropriate substances.

Solution:
a) $SrO(s) + 2HCl(aq) \rightarrow SrCl_2(aq) + H_2O\ (l)$
b) $Sr(OH)_2(s) + 2HCl(aq) \rightarrow SrCl_2(aq) + H_2O\ (l)$
c) $SrCO_3(s) + 2HCl(aq) \rightarrow SrCl_2(aq) + H_2O\ (l) + CO_2(g)$
d) $Sr(HCO_3)_2(s) + 2HCl(aq) \rightarrow SrCl_2(aq) + 2H_2O\ (l) + 2CO_2(g)$
e) $Sr(s) + 2HCl(aq) \rightarrow SrCl_2(aq) + H_2(g)$

9.56   Write balanced molecular, total ionic, and net ionic equations to illustrate each of the following reactions. All of the metals form 2+ ions.
a) tin with H_2SO_3 b) magnesium with H_3PO_4 c) calcium with HBr

Solution:
a) molecular: $Sn(s) + H_2SO_3(aq) \rightarrow SnSO_3(aq) + H_2(g)$
  total: $Sn(s) + 2H^+(aq) + SO_3^{2-}(aq) \rightarrow Sn^{2+}(aq) + SO_3^{2-}(aq) + H_2(g)$
  net: $Sn(s) + H_2^+(aq) \rightarrow Sn^{2+}(aq) + H_2(g)$
b) molecular: $3Mg(s) + 2H_3PO_4(aq) \rightarrow Mg_3(PO_4)_2(aq) + 3H_2(g)$
  total: $3Mg(s) + 6H^+(aq) + 2PO_4^{3-}(aq) \rightarrow 3Mg^{2+}(aq) + 2PO_4^{3-}(aq) + 3H_2(g)$
  net: $3Mg(s) + 6H^+(aq) \rightarrow 3Mg^{2+}(aq) + 3H_2(g)$
  simplified: $Mg(s) + 2H^+(aq) \rightarrow Mg^{2+}(aq) + H_2(g)$
c) molecular: $Ca(s) + 2HBr(aq) \rightarrow CaBr_2(aq) + H_2(g)$
  total: $Ca(s) + 2H^+(aq) + 2Br^-(aq) \rightarrow Ca^{2+}(aq) + 2Br^-(aq) + H_2(g)$
  net: $Ca(s) + 2H^+(aq) \rightarrow Ca^{2+}(aq) + H_2(g)$

Properties of Bases (Section 9.7)
9.58   Write balanced molecular, total ionic, and net ionic equations to represent neutralization reactions between RbOH and the following acids. Use all H's possible for each acid.
a) H_3PO_4 b) $H_2C_2O_4$ (oxalic acid)   c) $HC_2H_3O_2$

Solution:
a) molecular: $H_3PO_4(aq) + 3RbOH(aq) \rightarrow Rb_3PO_4(aq) + 3H_2O(l)$
  total: $3H^+(aq) + PO_4^{3-}(aq) + 3Rb^+(aq) + 3OH^-(aq) \rightarrow 3Rb^+(aq) + PO_4^{3-}(aq) + 3H_2O(l)$
  net: $3H^+(aq) + 3OH^-(aq) \rightarrow 3H_2O(l)$
  simplified: $H^+(aq) + OH^-(aq) \rightarrow H_2O(l)$
b) molecular: $H_2C_2O_4(aq) + 2RbOH(aq) \rightarrow Rb_2C_2O_4(aq) + 2H_2O(l)$
  total: $2H^+(aq) + C_2O_4^{2-}(aq) + 2Rb^+(aq) + 2OH^-(aq) \rightarrow 2Rb^+(aq) + C_2O_4^{2-}(aq) + 2H_2O(l)$
  net: $2H^+(aq) + 2OH^-(aq) \rightarrow 2H_2O(l)$
  simplified: $H^+(aq) + OH^-(aq) \rightarrow H_2O(l)$
c) molecular: $HC_2H_3O_2(aq) + RbOH(aq) \rightarrow RbC_2H_3O_2(aq) + H_2O(l)$
  total: $H^+(aq) + C_2H_3O_2^-(aq) + Rb^+(aq) + OH^-(aq) \rightarrow Rb^+(aq) + C_2H_3O_2^-(aq) + H_2O(l)$
  net: $H^+(aq) + OH^-(aq) \rightarrow H_2O(l)$

9.60   Some polyprotic acids can form more than one salt depending on the number of H's that react with base. Write balanced molecular, total ionic, and net ionic equations to represent the following neutralization reactions between KOH and
a) H_3PO_4 (react 2 H's)   b) H_3PO_4 (react 3 H's)   c) $H_2C_2O_4$ (react one H)

Solution:
a) molecular: $H_3PO_4(aq) + 2KOH(aq) \rightarrow K_2HPO_4(aq) + 2H_2O(l)$
     total: $2H^+(aq) + HPO_4^{2-}(aq) + 2K^+(aq) + 2OH^-(aq) \rightarrow 2K^+(aq) + HPO_4^{2-}(aq) + 2H_2O(l)$
     net: $2H^+(aq) + 2OH^-(aq) \rightarrow 2H_2O(l)$
     simplified: $H^+(aq) + OH^-(aq) \rightarrow H_2O(l)$
b) molecular: $H_3PO_4(aq) + 3KOH(aq) \rightarrow K_3PO_4(aq) + 3H_2O(l)$
     total: $3H^+(aq) + PO_4^{3-}(aq) + 3K^+(aq) + 3OH^-(aq) \rightarrow 3K^+(aq) + HPO_4^{3-}(aq) + 3H_2O(l)$
     net: $3H^+(aq) + 3OH^-(aq) \rightarrow +3H_2O(l)$
     simplified: $H^+(aq) + OH^-(aq) \rightarrow H_2O(l)$
c) molecular: $H_2C_2O_4(aq) + KOH(aq) \rightarrow KHC_2O_4(aq) + H_2O(l)$
     total: $H^+(aq) + HC_2O_4^-(aq) + K^+(aq) + OH^-(aq) \rightarrow K^+(aq) + HC_2O_4^-(aq) + H_2O(l)$
     net: $H^+(aq) + OH^-(aq) \rightarrow H_2O(l)$

Salts (Section 9.8)

9.62   Identify with ionic formulas the cations and anions of the following salts:
     a) NH_4NO_3 b) $CaCl_2$ c) $Mg(HCO_3)_2$ d) $KC_2H_3O_2$
     e) $LiHSO_3$

Solution:
a) cation: NH_4^+, anion: NO_3^-
b) cation: Ca^{2+}, anion: Cl^-
c) cation: Mg^{2+}, anion: HCO_3^-
d) cation: K^+, anion: $C_2H_3O_2^-$
e) cation: Li^+, anion: HSO_3^-

9.64   Identify with formulas the acid and base from which the anion and cation of each salt in Exercise 9.62 was derived. Pay special attention to salts derived from polyprotic acids and be sure to list the acid formula with all H's.

Solution:
a) acid: HNO_3, base: NH_3 (or NH_4OH)   b) acid: HCl, base: $Ca(OH)_2$
c) acid: H_2CO_3, base: $Mg(OH)_2$ d) acid: $HC_2H_3O_2$, base: KOH
e) acid: H_2SO_3 base: $LiOH$

9.66   Calculate the mass of water that would be released if the water of hydration were completely driven off 1.0 mol of
     a) epsom salts   b) borax
     (See Table 9.6) How would the products of these reactions compare?

Solution:
From Table 9.6, epsom salts is $MgSO_4 \cdot 7H_2O$; and borax is $Na_2B_4O_7 \cdot 10H_2O$

a) $1 \; \cancel{\text{mol } MgSO_4 \cdot 7H_2O} \times \frac{7 \; \cancel{\text{mol } H_2O}}{1 \; \cancel{\text{mol } MgSO_4 \cdot 7H_2O}} \times \frac{18 \text{ g } H_2O}{\cancel{\text{mol } H_2O}} = 126 \text{ g } H_2O$

b) $1 \; \cancel{\text{mol } Na_2B_4O_7 \cdot 10H_2O} \times \frac{10 \; \cancel{\text{mol } H_2O}}{1 \; \cancel{\text{mol } Na_2B_4O_7 \cdot 10H_2O}} \times \frac{18 \text{ g } H_2O}{\cancel{\text{mol } H_2O}} = 180 \text{ g } H_2O$

The products would be the "anhydrous" $MgSO_4$ and $Na_2B_4O_7$ respectively.

9.68   Write formulas for the acid and indicated solid that could be used to prepare each of the following salts:
     a) KNO_3 (solid is a bicarbonate)

b) $ZnCl_2$ (solid is a metal)
c) LiBr (solid is an oxide)

Solution:
a) acid: HNO_3 and solid: $KHCO_3$
b) acid: HCl and solid: Zn
c) acid: HBr and solid: Li_2O

9.70   Write balanced molecular equations to illustrate each salt preparation described in Exercise 9.68.

Solution:
a) $KHCO_3(s) + HNO_3(aq) \rightarrow KNO_3(aq) + CO_2(g) + H_2O\,(l)$
b) $2HCl(aq) + Zn(s) \rightarrow ZnCl_2(aq) + H_2(g)$
c) $2HBr(aq) + Li_2O(s) \rightarrow 2LiBr(aq) + H_2O(l)$

9.72   Determine the number of moles of each of the following salts that would equal 1 eq of salt:
a) $MgCO_3$ b) $Zn(HCO_3)_2$ c) $FeCl_3$

Solution:
a) 1 mol of $MgCO_3$ gives 2 mol (+) charge; 1 eq $MgCO_3$ = 1/2 mol $MgCO_3$
b) 1 mol of $Zn(HCO_3)_2$ gives 2 mol (+) charge; 1 eq $Zn(HCO_3)_2$ = 1/2 mol $Zn(HCO_3)_2$
c) 1 mol of $FeCl_3$ gives 3 mol (+) charge; 1 eq $FeCl_3$ = 1/3 mol $FeCl_3$

9.74   Determine the number of equivalents and milliequivalents in each of the following:
a) 0.22 mol $ZnCl_2$ b) 0.45 mol CsCl   c) 3.12×10^{-2} mol $Fe(NO_3)_2$

Solution:
a) $0.22 \text{ mol } ZnCl_2 \times \frac{2 \text{ eq}}{1 \text{ mol } ZnCl_2} = 0.44$ eq $ZnCl_2$ = 440 meq $ZnCl_2$

b) $0.45 \text{ mol } CsCl \times \frac{1 \text{ eq}}{1 \text{ mol } CsCl} = 0.45$ eq CsCl = 450 meq CsCl

c) $3.12 \times 10^{-2} \text{ mol } Fe(NO_3)_2 \times \frac{2 \text{ eq}}{1 \text{ mol } Fe(NO_3)_2} = 6.24 \times 10^{-2}$ eq $Fe(NO_3)_2$ = 62.4 meq $Fe(NO_3)_2$

9.76   Determine the number of equivalents and milliequivalents in 5.00 g of each of the following salts. Include any waters of hydration given in the salt formula when you calculate salt formula weights.
a) $Na_2CO_3 \cdot 10H_2O$ b) $CuSO_4 \cdot 5H_2O$ c) Li_2CO_3 d) NaH_2PO_4

Solution:
a) $5.00 \text{ g } Na_2CO_3 \times 10H_2O \times \frac{1 \text{ mol } Na_2CO_3 \cdot 10H_2O}{286.15 \text{ g } Na_2CO_3 \cdot 10H_2O} \times \frac{2 \text{ eq}}{1 \text{ mol } Na_2CO_3 \cdot 10H_2O} = 0.0349$ eq = 34.9 meq

b) $5.00 \text{ g } CuSO_4 \times 5H_2O \times \frac{1 \text{ mol } CuSO_4 \cdot 5H_2O}{249.7 \text{ g } CuSO_4 \cdot 5H_2O} \times \frac{2 \text{ eq}}{1 \text{ mol } CuSO_4 \cdot 5H_2O} = 4.0 \times 10^{-2}$ eq = 40.0 meq

c) $5.00 \text{ g } Li_2CO_3 \times \frac{1 \text{ mol } Li_2CO_3}{73.892 \text{ g } Li_2CO_3} \times \frac{2 \text{ eq}}{1 \text{ mol } Li_2CO_3} = 0.135$ eq = 135 meq

d) $5.00 \text{ g } NaH_2PO_4 \times \frac{1 \text{ mol } NaH_2PO_4}{120 \text{ g } NaH_2PO_4} \times \frac{1 \text{ eq}}{1 \text{ mol } NaH_2PO_4} = 4.2 \times 10^{-2}$ eq = 41.7 meq

9.78   A sample of intracellular fluid contains 133 meq/L of K^+ ion. Assume the K^+ comes from dissolved K_2SO_4 and calculate the number of moles and number of grams of K_2SO_4 that would be found in 150 mL of the intracellular fluid.

Solution:
$\frac{133 \text{ meq}}{\text{L soln}} \times 0.150 \text{ L soln} \times \frac{1 \text{ eq}}{1000 \text{ meq}} \times \frac{1 \text{ mol } K_2SO_4}{2 \text{ eq}} = 9.98 \times 10^{-3}$ mol K_2SO_4

$9.98 \times 10^{-3} \text{ mol } K_2SO_4 \times \frac{174 \text{ g } K_2SO_4}{1 \text{ mol } K_2SO_4} = 1.74$ g K_2SO_4

Strengths of Acids and Bases (Section 9.9)

9.80   The K_a values have been determined for four acids and are listed below. Arrange the acids in order of increasing acid strength (weakest first, strongest last).
        acid A ($K_a = 5.6 \times 10^{-5}$)
        acid B ($K_a = 1.8 \times 10^{-5}$)
        acid C ($K_a = 1.3 \times 10^{-4}$)
        acid D ($K_a = 1.1 \times 10^{-3}$)

Solution:
The acid with the smallest K_a will be the weakest and should be listed first:
acid B ($K_a = 1.8 \times 10^{-5}$), acid A ($K_a = 5.6 \times 10^{-5}$), acid C ($K_a = 1.3 \times 10^{-4}$), acid D ($K_a = 1.1 \times 10^{-3}$)

9.82 K_a values for four weak acids are given below:
        acid A ($K_a = 2.6 \times 10^{-4}$)
        acid B ($K_a = 3.7 \times 10^{-5}$)
        acid C ($K_a = 5.8 \times 10^{-4}$)
        acid D ($K_a = 1.5 \times 10^{-3}$)

   a) Arrange the four acids in order of increasing acid strength (weakest first, strongest last).
   b) Arrange the conjugate bases of the acids (identify as base A etc.) in order of increasing base strength (weakest base first, strongest last).

Solution:
   a) The acid with the smallest K_a will be the weakest and should be listed first:
      acid B ($K_a = 3.7 \times 10^{-5}$), acid A ($K_a = 2.6 \times 10^{-4}$), acid C ($K_a = 5.8 \times 10^{-4}$), acid D ($K_a = 1.5 \times 10^{-3}$)
   b) The stronger the acid, the weaker the conjugate base:
      base D, base C, base A, base B

9.84   Write dissociation reactions and K_a expressions for the following weak acids:
   a) hydrogen selenide ion, HSe^-
   b) dihydrogen borate ion, $H_2BO_3^-$ (1st H only)
   c) hydrogen borate ion, HBO_3^{2-}
   d) hydrogen arsenate ion, $HAsO_4^{2-}$
   e) hypochlorous acid, $HClO$

Solution:
   a) $HSe^-(aq) \rightarrow H^+(aq) + Se^{2-}(aq)$ $K_a = \frac{[H^+][Se^{2-}]}{[HSe^-]}$

   b) $H_2BO_3^-(aq) \rightarrow H^+(aq) + HBO_3^{2-}(aq)$ $K_a = \frac{[H^+][HBO_3^{2-}]}{[H_2BO_3^-]}$

   c) $HBO_3^{2-}(aq) \rightarrow H^+(aq) + BO_3^{3-}(aq)$ $K_a = \frac{[H^+][BO_3^{3-}]}{[HBO_3^{2-}]}$

   d) $HAsO_4^{2-}(aq) \rightarrow H^+(aq) + AsO_4^{3-}(aq)$ $K_a = \frac{[H^+][AsO_4^{3-}]}{[HAsO_4^{2-}]}$

   e) $HClO(aq) \leftrightarrow H^+(aq) + ClO^-(aq)$ $K_a = \frac{[H^+][ClO^-]}{[HClO]}$

9.86   If someone asked you for a weak acid solution, which of the following would you provide according to the definitions in this chapter? If the individual really wanted the other solution, what term should have been used instead of weak?
   a) 0.05 M HCl   b) 20% acetic acid

Solution:
The weak acid is the 20% acetic acid. The 0.05 M HCl is a dilute (strong) acid.

Analysis of Acids and Bases (Section 9.10)

9.88   Explain the purpose of doing a titration.

Solution:
A titration is done to measure the volume of titrant required to react with an analyte, one of which has a known number of moles or a known concentration.

9.90   Suppose a student is going to titrate an acidic solution with a base and just picks an indicator at random. Under what circumstances will
a) the endpoint and equivalence point be the same?
b) the endpoint and equivalence point be different?

Solution:
a) If the pH of the color change equals the pH of the equivalence point, the endpoint and equivalence point will be the same. The pH of the equivalence point is the pH of the solution of salt produced in the titration.
b) If not, the endpoint will be different from the equivalence point.
Note: Remember that if two measurements differ by less than the error in the measurement, the measurements are equal.

9.92   Determine the number of moles of NaOH that could be neutralized by each of the following:
a) 500 mL of 0.300 M $HC_2H_3O_2$ (monoprotic)
b) 1.50 L of 0.200 M HBr

Solution:
a) $0.500 \text{ L soln} \times \frac{0.300 \text{ mol } HC_2H_3O_2}{\text{L soln}} \times \frac{1 \text{ mol NaOH}}{1 \text{ mol } HC_2H_3O_2} = 0.150$ mol NaOH

b) $1.50 \text{ L soln} \times \frac{0.200 \text{ mol HBr}}{\text{L soln}} \times \frac{1 \text{ mol NaOH}}{1 \text{ mol HBr}} = 0.300$ mol NaOH

Titration Calculations (Section 9.11)

9.94   Write a balanced molecular equation to represent the neutralization reaction between NaOH and each of the following acids:
a) trichloroacetic acid, $HC_2O_2Cl_3$
b) dithionic acid, $H_2S_2O_6$
c) hypophosphorous acid, $H_4P_2O_6$

Solution:
a) $HC_2O_2Cl_3 + NaOH \rightarrow NaC_2O_2Cl_3 + H_2O$
b) $H_2S_2O_6 + 2NaOH \rightarrow Na_2S_2O_6 + 2H_2O$
c) $H_4P_2O_6 + 4NaOH \rightarrow Na_4P_2O_6 + 4H_2O$

9.96   Write a balanced molecular equation to represent the neutralization reaction between HCl and each of the following bases:
a) $Zn(OH)_2$ b) $Tl(OH)_3$ c) CsOH

Solution:
a) $Zn(OH)_2 + 2HCl \rightarrow ZnCl_2 + 2H_2O$
b) $Tl(OH)_3 + 3HCl \rightarrow TlCl_3 + 3H_2O$
c) $CsOH + HCl \rightarrow CsCl + H_2O$

9.98   A 25.0 mL sample of H_2SO_4 solution required 34.7 mL of 0.0400 M NaOH solution to titrate to the equivalence point. Calculate the molarity of the H_2SO_4 solution.

Solution:
$$H_2SO_4 + 2NaOH \rightarrow Na_2SO_4 + 2H_2O$$
$$34.7 \text{ mL NaOH} \times \frac{1 \text{ L NaOH}}{1000 \text{ mL NaOH}} \times \frac{0.0400 \text{ mol NaOH}}{1 \text{ L NaOH}} \times \frac{1 \text{ mol } H_2SO_4}{2 \text{ mol NaOH}} = 6.94 \times 10^{-4} \text{ mol } H_2SO_4$$
$$M = \frac{\text{mol}}{\text{L soln}} = \frac{6.94 \times 10^{-4} \text{ mol } H_2SO_4}{25.0 \text{ mL}} \times \frac{1000 \text{ mL}}{L} = 2.78 \times 10^{-2} \text{ M } H_2SO_4$$

9.100 A 20.00 mL sample of each of the following acid solutions is to be titrated to the equivalence point using 0.120 M NaOH solution. Determine the milliliters of NaOH solution that will be needed for each acid sample.
a) 0.200 M $HClO_4$
b) 0.125 M H_2SO_4
c) 0.150 M $H_4P_2O_6$
d) 0.120 mol H_3PO_4 in 500 mL of solution
e) 6.25 g H_2SO_4 in 250 mL of solution
f) 0.500 mol $HClO_3$ in 1.00 L of solution

Solution:
The equivalence point is reached when the number of moles of hydrogen (hydronium) ions and hydroxide ions reacted are the same. A simple variation on Equation 7.9, $V_aM_b = V_aM_b$, can be used to calculate the unknown volume when one volume and the concentrations of the $[H^+]$ from the acid and the $[OH^-]$ from the base are known.

a) $V_b = V_a \times \frac{M_a}{M_b} = 20.00 \text{ mL } HClO_4 \times \frac{0.200 \text{ M } H^+}{0.120 \text{ M } OH^-} = 33.3 \text{ mL NaOH}$

b) $V_b = V_a \times \frac{M_a}{M_b} = 20.00 \text{ mL } H_2SO_4 \times \frac{0.250 \text{ M } H^+}{0.120 \text{ M } OH^-} = 41.7 \text{ mL NaOH}$

c) $V_b = V_a \times \frac{M_a}{M_b} = 20.00 \text{ mL } H_4P_2O_6 \times \frac{0.600 \text{ M } H^+}{0.120 \text{ M } OH^-} = 100 \text{ mL NaOH}$

d) 0.120 mol H_3PO_4 in 500 mL of solution (.5 L) yields 0.240 M H_3PO_4.

   $V_b = V_a \times \frac{M_a}{M_b} = 20.00 \text{ mL } H_3PO_4 \times \frac{0.720 \text{ M } H^+}{0.120 \text{ M } OH^-} = 120.0 \text{ mL NaOH}$

e) 6.25 g of H_2SO_4 (6.25 g / 98.076 g/mol) is 0.0637 mol H_2SO_4. 0.0637 mol H_2SO_4 in 250 mL of solution (0.0637 mol H_2SO_4 / 0.250 L soln) is 0.255 M H_2SO_4.

   $V_b = V_a \times \frac{M_a}{M_b} = 20.00 \text{ mL } H_2SO_4 \times \frac{0.510 \text{ M } H^+}{0.120 \text{ M } OH^-} = 85.0 \text{ mL NaOH}$

f) 0.500 mol $HClO_3$ in 1.00 L of solution is a 0.500 M $HClO_3$ solution.

   $V_b = V_a \times \frac{M_a}{M_b} = 20.00 \text{ mL } HClO_3 \times \frac{0.500 \text{ M } H^+}{0.120 \text{ M } OH^-} = 83.3 \text{ mL NaOH}$

9.102 The following acid solutions were titrated to the equivalence point with the base listed. Use the titration data to calculate the molarity of each acid solution.
a) 5.00 mL of dilute H_2SO_4 required 29.88 mL of 1.17 M NaOH solution.
b) 10.00 mL of vinegar (acetic acid) required 35.62 mL of 0.250 M KOH solution.
c) 10.00 mL of muriatic acid (HCl) used to clean brick and cement required 20.63 mL of 6.00 M NaOH solution.

Solution:

The equivalence point reached when the number of moles of hydrogen (hydronium) ions and hydroxide ions reacted are the same. A simple variation on Equation 7.9, $V_aM_b = V_aM_b$, can be used to calculate the number of moles of acid reacted when the molarity and volume of base reacted are known.

$V_aM_b = V_aM_b$ is rearranged to become $M_a = V_b \times \frac{M_b}{V_a}$.

a) $29.88 \text{ mL NaOH} \times \frac{1 \text{ L}}{1000 \text{ mL}} \times \frac{1.17 \text{ mol NaOH}}{\text{L NaOH}} \times \frac{1 \text{ mol H}_2\text{SO}_4}{2 \text{ mol NaOH}} = 0.0175 \text{ mol H}_2\text{SO}_4$

molarity $H_2SO_4 = \frac{\text{mol solute}}{\text{L soln}} = \frac{0.0175 \text{ mol H}_2\text{SO}_4}{5.00 \text{ mL}} \times \frac{1000 \text{ mL}}{1 \text{ L}} = 3.50 \text{ M H}_2\text{SO}_4$

b) $35.62 \text{ mL KOH} \times \frac{1 \text{ L}}{1000 \text{ mL}} \times \frac{0.250 \text{ mol KOH}}{\text{L KOH}} \times \frac{1 \text{ mol HC}_2\text{H}_3\text{O}_2}{1 \text{ mol KOH}} = 0.00891 \text{ mol HC}_2\text{H}_3\text{O}_2$

molarity $HC_2H_3O_2 = \frac{\text{mol solute}}{\text{L soln}} = \frac{0.00891 \text{ mol HC}_2\text{H}_3\text{O}_2}{10.00 \text{ mL}} \times \frac{1000 \text{ mL}}{1 \text{ L}} = 0.891 \text{ M HClO}_4$

c) $20.63 \text{ mL NaOH} \times \frac{1 \text{ L}}{1000 \text{ mL}} \times \frac{6.00 \text{ mol NaOH}}{\text{L NaOH}} \times \frac{1 \text{ mol HCl}}{1 \text{ mol NaOH}} = 0.124 \text{ mol HCl}$

molarity $HCl = \frac{\text{mol solute}}{\text{L soln}} = \frac{0.0124 \text{ mol HCl}}{10.00 \text{ mL}} \times \frac{1000 \text{ mL}}{1 \text{ L}} = 12.4 \text{ M HCl}$

9.104 A sample of monoprotic benzoic acid weighing 0.5823 g is dissolved in about 25 mL of water. The solution is titrated to the equivalence point using 0.1021 M NaOH. The volume of base required was 46.75 mL. Calculate the molecular weight of the solid acid.

Solution:

Let HBz represent benzoic acid.

mol HBz $= 46.75 \text{ mL NaOH} \times \frac{1 \text{ L}}{1000 \text{ mL}} \times \frac{0.1021 \text{ mol NaOH}}{\text{L NaOH}} \times \frac{1 \text{ mol HBz}}{1 \text{ mol NaOH}} = 4.773 \times 10^{-3} \text{ mol HBz}$

Molecular weight can be obtained by dividing the weight of the HBz by the number of moles of HBz.

MW $= \frac{0.5823 \text{ g HBz}}{4.773 \times 10^{-3} \text{ mol HBz}} = 122.0 \text{ g/mol HBz}$

Hydrolysis Reactions of Salts (Section 9.12)

9.106 A solution of solid Na_3PO_4 in pure water is basic (the pH is greater than 7). Explain.

Solution:

The PO_4^{3-} ion is a Brønsted base. In water, it reacts according to the equation:

$$PO_4^{3-} + H_2O \rightleftarrows HPO_4^{2-} + OH^-$$

The OH^- formed makes the solution basic.

9.108 Predict the relative pH (*greater than 7, less than 7*, etc.) for water solutions of the following salts. Table 9.9 may be useful. For each solution in which the pH is greater or less than 7, explain why and write a net ionic equation to justify your answer.
a) sodium hypochlorite, $NaClO$ ($HClO$ is a weak acid)
b) sodium formate, $NaCHO_2$
c) potassium nitrate, KNO_3
d) sodium phosphate, Na_3PO_4

Solution:

a) Greater than 7. ClO^- is a weak base. $ClO^- + H_2O \rightleftarrows HClO + OH^-$
b) Greater than 7. CHO_2^- is a weak base. $CHO_2^- + H_2O \rightleftarrows HCH_2O + OH^-$
c) Equal to 7. No hydrolysis occurs. KOH is a strong base, and HNO_3 is a strong acid.
d) Greater than 7. The PO_4^{3-} is a weak base. $PO_4^{3-} + H_2O \rightleftarrows HPO_4^{2-} + OH^-$

9.110 Explain why the hydrolysis of salts makes it necessary to have available in a laboratory more than one acid-base indicator for use in titrations.

Solution:
The pH at the equivalence point of a titration depends on the degree of hydrolysis of the salt formed in the titration. Since the color of the indicator must change at the approximate pH of the equivalence point, different indicators would be needed to titrate different acids and bases.

Buffers (Section 9.13)

9.112 Write equations similar to Equations 9.48 and 9.49 in the text to illustrate how a mixture of sodium hydrogen phosphate (Na_2HPO_4) and sodium dihydrogen phosphate (NaH_2PO_4) could function as a buffer when dissolved in water. Remember that phosphoric acid (H_3PO_4) ionizes in three steps.

Solution:
The HPO_4^{2-} could react with added acid: $HPO_4^{2-} + H^+ \rightarrow H_2PO_4^-$
and the $H_2PO_4^-$ could react with added base: $H_2PO_4^- + OH^- \rightarrow HPO_4^{2-} + H_2O$

9.114 Some illnesses lead to a condition of excess acid (acidosis) in the body fluids. An accepted treatment is to inject solutions containing bicarbonate ions (HCO_3^-) directly into the bloodstream. Write an equation to show how this treatment would help combat the acidosis.

Solution:
The HCO_3^- could react with the excess acid. The equation is: $HCO_3^- + H^+ \rightarrow H_2CO_3$
Note: See Chapter 25 for a discussion of how the H_2CO_3 is removed from the blood.

9.116 a) Calculate the pH of a buffer that is 0.1 M lactic acid ($C_2H_4(OH)COOH$) and 0.1 M sodium lactate, $C_2H_4(OH)COONa$.
    b) What is the pH of a buffer that is 1 M in lactic acid and 1 M in sodium lactate?
    c) What is the difference between the buffers described in parts (a) and (b)?

Solution:
From Table 9.9, $pK_a = 3.85$ Use Equation 9.54, $pH = pK_a + \log \frac{[B^-]}{[HB]}$

a) $pH = 3.85 + \log \frac{0.1}{0.1} = 3.85$

b) $pH = 3.85 + \log \frac{1}{1} = 3.85$

c) The only difference is that the buffer in (b) has much greater capacity to resist change of pH because of the higher concentration of both the lactic acid and lactate ion.

9.118 Calculate the pH of the buffers with the acid and conjugate base concentrations listed below.
    a) $[CH_3COOH] = 0.40$ M, $[CH_3COO^-] = 0.25$ M
    b) $[H_2PO_4^-] = 0.10$ M, $[HPO_4^{2-}] = 0.40$ M
    c) $[HSO_3^-] = 1.50$ M, $[SO_3^{2-}] = 0.20$ M

Solution:
These calculations are performed by applying the Henderson-Hasselbalch equation:

$pH = pK_a + \log \frac{[A^-]}{[HA]}$

a) pK_a for CH_3COOH ($HC_2H_3O_2$) = 4.74

   $pH = 4.74 + \log \frac{0.25}{0.40} = 4.54$

b) pK_a for $H_2PO_4^- = 7.21$

   $pH = 7.21 + \log \frac{0.40}{0.10} = 7.81$

c) pK_a for $HSO_3^- = 7.00$

   $pH = 7.00 + \log \frac{0.20}{1.50} = 6.12$

9.120 What ratio of concentrations of NaH_2PO_4 and Na_2HPO_4 in solution would give a buffer with pH = 7.65?

Solution:

The calculation is performed by applying the Henderson-Hasselbalch equation, as in problem 9.118.

$$pH = pK_a + \log\frac{[HPO_4^{2-}]}{[H_2PO_4^-]}$$

rearrange to $\log\frac{[HPO_4^{2-}]}{[H_2PO_4^-]} = pH - pK_a$; $\log\frac{[HPO_4^{2-}]}{[H_2PO_4^-]} = 7.65 - 7.21 = 0.44$

Taking the antilog ($10^{0.44}$) yields the ratio required, 2.75. The concentration of HPO_4^{2-} must be 2.75 times greater than that of the $H_2PO_4^-$.

PROGRAMMED REVIEW

Section 9.1 The Arrhenius Theory

Svante Arrhenius defined an acid as a substance that (a) _dissociates_ when dissolved in water and produces (b) _hydrogen_ ions. He defined a base as a substance that (c) _dissociate_ and releases (d) _____ ions when dissolved in water.

Section 9.2 The Brønsted Theory

According to the Brønsted theory an acid is any (a) _hydrogen_ containing substance that is capable of donating a (b) _proton_ to another substance. A base is any substance that accepts a (c) _____. The species that remains when a Brønsted acid donates a proton is called the (d) _____ base of the acid.

Section 9.3 Naming Acids

There are two classes of acids: those containing hydrogen and one other kind of atom called (a) _binary_ acids, and those containing hydrogen and a (b) _polyatomic_ ion. The name of the binary acid has a prefix (c) _hydro_ and ends with a (d) _-ic_ suffix. The name of the acids involving polyatomic ion is a modification of the name of the ion. If the ion name ends in -ate, the acid has a (e) _-ic_ suffix. If the ion name ends in -ite, the acid has a (f) _-ous_ suffix. The name of HBr is (g) _hydro bromic_. The name of HNO_2 is (h) _nitrous_..

Section 9.4 The Self Ionization of Water

In the pure state, water undergoes a self or (a) _auto_ ionization in which some water molecules function as Brønsted (b) _acids_ and some function as Brønsted (c) _bases_. The term *neutral* is used to describe any water solution that contains equal concentrations of (d) _hydronium_ and (e) _hydroxide_ ions. A solution is acidic when the (f) _hydronium_ ion concentration is higher, and basic or alkaline when the (g) _hydroxide_ ion concentration is higher.

Section 9.5 The pH Concept

Mathematically, pH is the (a) _negative logarithm_ of the (b) _molar_ concentration of the (c) _hydrogen_ ion in a solution. A solution with a pH higher than 7 is classified as (d) _basic_, and one lower than 7 is classified as (e) _acidic_.

Section 9.6 Properties of Acids

All acids taste (a) _sour_ and produce (b) _hydronium_ ions when dissolved in water. In addition, all acids react characteristically with solid (c) _oxides_, (d) _hydroxides_ (e) _carbonates_ and (f) _bicarbonates_. Acids react with certain metals to produce (g) _hydrogen_ gas. The tendency of metals to undergo such a reaction is given by an (h) _activity series_.

Section 9.7 Properties of Bases

The complete reaction of an (a) _acid_ with a (b) _base_ to produce a solution containing only (c) _water_ and a (d) _salt_ is called a (e) _neutralization_ reaction. Bases are often found in household cleaners because they react with (f) _fats_ and (g) _oils_ .

Section 9.8 Salts

At room temperature, salts are solid (a) _crystalline_ substances that contain the (b) _cation_ of a base and the (c) _anion_ of an acid. Salts that contain specific numbers of water molecules as a part of their crystalline structure are called (d) _hydrates_. The water in such compounds is called the (e) _water_ of _hydration_. An equivalent of a salt is the amount of salt that will produce one mole of (f) _positive charges_ when dissolved and dissociated.

Section 9.9 Strength of Acids and Bases

The strength of an acid or base is determined by the extent to which they (a) _dissociate_ when dissolved in water. Strong acids or bases dissociate almost (b) _completely_ in solution. Acids are classified as monoprotic, diprotic and triprotic on the basis of the number of (c) _hydrogens_ given up per molecule. In general, the anions produced by the dissociation of strong Brønsted acids behave as (d) _weak_ Brønsted bases.

Section 9.10 Analysis of Acids and Bases

A common procedure used to analyze acids and bases is called a (a) _titration_. When this procedure is used to analyze an acid, a (b) _base_ solution of known concentration is added to an acid solution until the (c) _equivalence_ point is reached. One way to detect when an acid and base have reacted completely, is to add an (d) _indicator_ which changes color as close as possible to the (e) _equivalence_ point of the titration.

Section 9.11 Titration Calculations

Titration calculations are done using the methods described earlier in Chapter (a) _seven_ of the text.

Section 9.12 Hydrolysis Reactions of Salts

In general, a hydrolysis reaction is any reaction with (a) _water_ . In the case of salts, hydrolysis reactions cause salt solutions to have a (b) _pH_ different from that of pure (c) _water_ . The hydrolysis of a salt containing the cation of a strong base and the anion of a weak acid produces a (d) _basic_ solution.

Section 9.13 Buffers

Buffers are solutions with the ability to resist changes in (a) _pH_ when (b) _acids_ or (c) _bases_ are added. The amount of H^+ or OH^- that a buffer system can absorb without changing (d) _pH_ significantly is called the (e) _buffer capacity_ .

SELF-TEST QUESTIONS

Multiple Choice

1. The Arrhenius definition of a base focuses on
   a) the acceptance of H^+
   b) the formation of covalent bonds
   c) the production of OH^-
   d) more than one response is correct

2. The pH of a 0.100 M solution of the weak acid HA is 4.91. Make appropriate assumptions, and evaluate K_a.
   a) 1.51×10^{-9}
   b) 1.23×10^{-4}
   c) 1.51×10^{-10}
   d) 6.61×10^8

3. A beaker contains 100 mL of a liquid with a pH of 7.0. When 0.5 mL of 0.2 M acid is added, the pH changes to 6.88. When 0.5 mL of 0.1 M base is added to another 100 mL sample of the liquid, the pH changes to 7.20. The liquid in the beaker is
   a) water   b) an acid solution   c) a base solution   d) a buffer solution

The following reaction refers to Questions 4 through 7: $HCl + NaOH \rightarrow NaCl + H_2O$

4. The above reaction between hydrochloric acid and sodium hydroxide is correctly classified as
   a) combustion   b) dehydration
   c) neutralization   d) more than one response is correct

5. The HCl solution is prepared to be 0.120 M. A 20.00 mL sample requires 18.50 mL of NaOH solution for complete reaction. What is the molarity of the NaOH solution?
   a) 0.130   b) 0.110   c) 0.0800   d) 0.200

6. How many moles of HCl would be contained in the 20.00 mL sample used in Question 5?
   a) 0.204   b) 2.40   c) 0.120   d) .0024

7. How many grams of HCl would be contained in the 20.00 mL sample used in Question 5?
   a) .0875   b) 4.38   c) 8.75   d) 36.5

8. A 25.00 mL sample of monoprotic acid is titrated with a standard 0.100 M base. Exactly 20.00 mL of base is required to titrate to the proper end point. What is the molarity of the acid?
   a) 0.0500   b) 0.100   c) 0.0800   d) 0.125

9. A certain solution has a pH of 1. This solution is best described as
   a) very basic   b) neutral   c) slightly acidic   d) very acidic

10. If the pH of an aqueous solution cannot be changed significantly by adding small amounts of strong acid or strong base, the solution contains
   a) an indicator   b) a buffer
   c) a protective colloid   d) a strong acid and a strong base

True-False

11. The terms *weak acid* and *dilute acid* can be used interchangeably.

12. Some H_3O^+ ions are present in pure water.

13. One of the products formed in the titration of an acid by a base is water.

14. A solution with a pH of 3 is correctly classified as acidic.

15. A solution with a pH of 6.00 has a concentration of 6.00 M H^+.

16. Three different sodium salts of phosphoric acid, H_3PO_4, are possible.

17. The anion produced by the first step in the dissociation of sulfurous acid, H_2SO_3, is SO_3^{2-}.

18. The second step in the dissociation of H_3PO_4 produces $H_2PO_4^-$.

19. In pure water, $[H_3O^+] = [OH^-]$.

20. In an acid-base titration, the point at which the acid and base have exactly reacted is called the equivalence point.

21. The pH is 7 at the equivalence point of all acid-base titrations.

Matching

Match the classifications given on the right to the species listed on the left. The species are involved in the following reversible reaction. Responses can be used more than once.

$$Na^+ + C_2H_3O_2^- + H_2O \rightleftharpoons Na^+ + HC_2H_3O_2 + OH^-$$

22. _____ Na^+

23. _____ $C_2H_3O_2^-$

24. _____ H_2O

25. _____ $HC_2H_3O_2$

a) behaves as a Brønsted acid

b) behaves as a Brønsted base

c) behaves as neither a Brønsted acid nor base

d) behaves as both a Brønsted acid and base

Choose the response that best completes each reaction below. The acid involved in each reaction is represented by HA.

26. _____ $2HA + ? \rightarrow H_2 + MgA_2$

27. _____ $HA + H_2O \rightarrow ? + A^-$

28. _____ $HA + NaOH \rightarrow H_2O + ?$

29. _____ $2HA + ? \rightarrow CO_2 + H_2O + 2NaA$

a) NaA

b) Na_2CO_3

c) Mg

d) H_3O^+

Choose a silver compound formula from the right to complete each of the reactions used to prepare silver salts. In each reaction, HA represents an acid.

30. _____ $2HA + ? \rightarrow 2AgA + H_2O$

31. _____ $2HA + ? \rightarrow 2AgA + CO_2 + H_2O$

32. _____ $HA + ? \rightarrow AgA + H_2O$

33. _____ $HA + ? \rightarrow AgA + CO_2 + H_2O$

a) AgOH

b) Ag_2O

c) $AgHCO_3$

d) Ag_2CO_3

Classify the systems described on the left into one of the pH ranges given as responses.

34. _____ the $[OH^-] = [H_3O^+]$ in pure H_2O

35. _____ oven cleaners are strongly basic

36. _____ the active ingredient in the stomach, digestive juice, is 0.01 M hydrochloric acid

37. _____ a carbonated soft drink has a tart taste

a) the pH is much lower than 7

b) the pH is much higher than 7

c) the pH is near 7

d) the pH is exactly 7

Choose a description from the right that best characterizes the solution made by dissolving in water each of the salts indicated on the left.

38. _____ Na_2SO_4

39. _____ Na_3PO_4

40. _____ NH_4Cl

a) it is acidic because hydrolysis occurs

b) it is basic because hydrolysis occurs

c) it is neutral because no hydrolysis occurs

d) more than one of the above is correct

SOLUTIONS

A. Answers to Programmed Review

9.1 a) dissociates   b) hydrogen (or H^+)   c) dissociates   d) hydroxide (or OH^-)

9.2 a) hydrogen   b) proton (H^+)   c) proton (H^+)   d) conjugate

9.3 a) binary   b) polyatomic   c) hydro-   d) -ic
     e) -ic   f) -ous   g) hydrobromic   h) nitrous

9.4 a) auto   b) acids   c) bases   d) hydronium (or H_3O^+)
     e) hydroxide (or OH^-)   f) hydronium (or H_3O^+)   g) hydroxide (or OH^-)

9.5 a) negative logarithm   b) molar   c) hydrogen (or H^+)   d) basic (or alkaline)
     e) acidic

9.6 a) sour   b) hydronium (or H_3O^+)   c) oxides   d) hydroxides
     e) carbonates   f) bicarbonates   g) hydrogen   h) activity series

9.7 a) acid   b) base   c) water   d) salt
     e) neutralization   f) fats   g) oils

9.8 a) crystalline   b) cation   c) anion   d) hydrates
     e) water of hydration   f) positive charges

9.9 a) dissociate   b) completely   c) hydrogens (or H^+)   d) weak

9.10 a) titration   b) base (or basic)   c) equivalence   d) indicator
      e) equivalence

9.11 a) seven

9.12 a) water   b) pH   c) water   d) basic (or alkaline)

9.13 a) pH   b) acids   c) bases   d) pH
      e) buffer capacity

B. Answers to Self-Test Questions

1. c	15. F	29. b
2. a	16. T	30. b
3. d	17. F	31. d
4. c	18. F	32. a
5. a	19. T	33. c
6. d	20. T	34. d
7. a	21. F	35. b
8. c	22. c	36. a
9. d	23. b	37. a
10. b	24. a	38. c
11. F	25. a	39. b
12. T	26. c	40. a
13. T	27. d	
14. T	28. a	

Chapter 10

Radioactivity and Nuclear Processes

CHAPTER OUTLINE

10.1 Radioactive Nuclei
10.2 Equations for Nuclear
        Reactions
10.3 Isotope Half-Life
10.4 Health Effects of Radiation

10.5 Measurement Units for Radiation
10.6 Medical Uses of Radioisotopes
10.7 Nonmedical Uses of Radioisotopes
10.8 Induced Nuclear Reactions
10.9 Nuclear Energy

LEARNING OBJECTIVES

When you have completed your study of this chapter, you should be able to:

1. Describe and characterize the common forms of radiation emitted during radioactive decay and other nuclear processes.
2. Write balanced equations for nuclear reactions.
3. Use the half-life concept to solve problems.
4. Describe the influence of radiation on health.
5. Describe the units used to measure radiation.
6. Describe the medical and nonmedical uses of radioisotopes.
7. Show an understanding of the concept of induced nuclear reactions.
8. Distinguish between and characterize nuclear fission and fusion reactions.

ANSWERS AND SOLUTIONS TO EVEN-NUMBERED PROBLEMS

Radioactive Nuclei (Section 10.1)

10.2   Group the common nuclear radiations (Table 10.1) into the following categories:
        a) those with a mass number of 0
        b) those with a positive charge
        c) those with a charge of 0

Solution:
a) The beta, gamma, and positron radiations have a mass number of 0.
b) The alpha and positron have a positive charge.
c) The neutron and gamma have 0 charge.

10.4   Characterize the following nuclear particles in terms of the fundamental particles—protons, neutrons, and electrons:
   a) a beta particle   b) an alpha particle   c) a positron

Solution:
a) A beta particle has the same characteristics as the electron. The beta particle comes from the nucleus. The electrons are found outside the nucleus.
b) An alpha particle is a combination of 2 protons and 2 neutrons.
c) A positron has the same mass as the electron, but it has a positive charge.

Equations for Nuclear Reactions (Section 10.2)

10.6   Summarize how the atomic number and mass number of daughter nuclei compare to the original nuclei after
   a) an alpha particle is emitted   b) a beta particle is emitted
   c) an electron is captured   d) a gamma ray is emitted
   e) a positron is emitted

Solution:
a) The atomic number decreases by 2; the mass number decreases by 4.
b) The atomic number increases by 1; the mass number is unchanged.
c) The atomic number decreases by 1; the mass number is unchanged.
d) Both the atomic number and mass number are unchanged.
e) The atomic number decreases by 1; the mass number is unchanged.

10.8   Write appropriate symbols for the following particles using the $^A_Z X$ symbolism:
   a) a nucleus of the element in period 5 group VB(5) with a mass number of 96
   b) a nucleus of element number 37 with a mass number of 80
   c) a nucleus of the calcium (Ca) isotope that contains 18 neutrons

Solution:
a) $^{96}_{41}Nb$ b) $^{80}_{37}Rb$ c) $^{38}_{20}Ca$

10.10   Complete the following equations, using appropriate notations and formulas:
   a) $^{204}_{82}Pb \rightarrow ? + ^4_2\alpha$ b) $^{84}_{35}Br \rightarrow ? + ^0_{-1}\beta$ c) $? + ^0_{-1}e^- \rightarrow ^{41}_{19}K$
   d) $^{149}_{62}Sm \rightarrow ^{145}_{60}Nd + ?$ e) $? \rightarrow ^{34}_{15}P + ^0_{-1}\beta$ f) $^{15}_8O \rightarrow ^0_{-1}\beta + ?$

Solution:
a) $^{204}_{82}Pb \rightarrow ^{200}_{80}Hg + ^4_2\alpha$ b) $^{84}_{35}Br \rightarrow ^{84}_{36}Kr + ^0_{-1}\beta$ c) $^{41}_{20}Ca + ^0_{-1}e^- \rightarrow ^{41}_{19}K$
d) $^{149}_{62}Sm \rightarrow ^{145}_{60}Nd + ^4_2\alpha$ e) $^{34}_{14}Si \rightarrow ^{34}_{15}P + ^0_{-1}\beta$ f) $^{15}_8O \rightarrow ^0_{-1}\beta + ^{15}_7N$

10.12   Write balanced equations to represent decay reactions of the following isotopes. The decay process or daughter isotopes is given in parentheses.
   a) $^{192}_{78}Pt$ (alpha emission)   b) $^{10}_4Be$ (beta emission)   c) $^{19}_8O$ (Daughter = $^{19}_9F$)

d) $^{238}_{92}U$ (alpha emission)   e) $^{108}_{50}Sn$ (electron capture)   f) $^{72}_{31}Ga$ (daughter = Ge-72)

Solution:

a) $^{192}_{78}Pt \rightarrow {}^{188}_{76}Os + {}^{4}_{2}\alpha$ b) $^{10}_{4}Be \rightarrow {}^{0}_{-1}\beta + {}^{10}_{5}B$ c) $^{19}_{8}O \rightarrow {}^{19}_{9}F + {}^{0}_{-1}\beta$

d) $^{238}_{92}U \rightarrow {}^{4}_{2}\alpha + {}^{234}_{90}Th$ e) $^{108}_{50}Sn + {}^{0}_{-1}e^- \rightarrow {}^{108}_{49}In$ f) $^{72}_{31}Ga \rightarrow {}^{72}_{32}Ge + {}^{0}_{-1}\beta$

Isotope Half-Life (Section 10.3)

10.14   Describe half-life in terms of something familiar such as a cake or cookies or your checking account.

Solution:
The half-life of the uncooked dough (after placing it in the oven) is time required to cook one half of the remaining dough. The half-life of the cookies in the cookie jar is the amount of time for 1/2 to be eaten. The half-life of a checking account is the time (usually very short) to spend half of the balance.

10.16   Technetium-99 has a half-life of 6 hours. This isotope is used diagnostically to perform brain scans. A patient is given a 6.0 nanogram dose. How many nanograms will be present in the patient 30 hours later?

Solution:
30 hrs/6 hrs/half-life = 5 half-life periods
fraction left = $1/2 \times 1/2 \times 1/2 \times 1/2 \times 1/2 = (1/2)^5 = 1/32$ of the original amount
$6.0 \times 1/32 = 0.19$ nanograms

10.18   An archaeologist unearths the remains of a wooden box, analyzes for the carbon-14 content, and finds that about 93.75% of the carbon-14 initially present has decayed. Estimate the age of the box. The half-life of the carbon-14 is 5600 years.

Solution:
Since 93.75% has reacted, then 6.25% is remaining or 1/16 is remaining.
$1/16 = (1/2)^n$; $n = 4$ half-lives
4 half-lives $\times$ 5600 yrs/half-life = 2.24×10^4 yrs old

10.20   A grain sample was found in a cave. The ratio of $^{14}_{6}C/^{12}_{6}C$ was about 1/8 the value in a fresh grain sample. How old was the grain in the cave?

Solution:
Since only 1/8 is left, $1/8 = (1/2)^n$; $n = 3$ half-lives.
3 half-lives $\times$ 5600 yrs/half-life = 1.68×10^4 yrs old
Note: Since the data stated "about 1/8," the answer should have 3 sig figs at most.

Health Effects of Radiation (Section 10.4)

10.22   A source of radiation has an intensity of 120 units at a distance of 15 feet. How far away from the source would you have to be to reduce the intensity to 20 units?

Solution:
Use Equation 10.2, where $I_x = 120$, $I_y = 20$, $d_x = 15$, and d_y is unknown.
$\frac{I_x}{I_y} = \frac{d_y^2}{d_x^2}$; $\frac{120}{20} = \frac{d_y^2}{15^2}$; $d_y^2 = 1350$; $d_y = \sqrt{1350}$; $d_y = 37$ feet

Measurement Units for Radiation (Section 10.5)

10.24   Explain the difference between physical and biological units of radiation.

Solution:

The physical units of radiation measure the number of radioactive decay events occurring in a specified time. (e.g., disintegrations/min). Biological units measure the damaging effect of the radiation in living cells. Since different radioactive isotopes undergo different types of decay, the same number of decay events giving off different forms of radiation (α, β, etc.) will cause different amounts of tissue damage.

10.26   An individual receives a short-term whole-body dose of 2.8 rads of beta radiation. How many roentgens of X-rays would represent the same health hazard?

Solution:

From Table 10.5, 1 R = 0.96 D, and 1 REM = 1 R

$2.8 \text{ D} \times \frac{1 \text{ R}}{0.96 \text{ D}} = 2.9 \text{ R}$

10.28   One Ci corresponds to 3.7×10^{10} nuclear disintegrations per second. How many disintegrations per second would take place in a sample containing 3.2 μCi of radioisotope?

Solution:

$3.2 \, \cancel{\mu\text{Ci}} \times \frac{1 \, \cancel{\text{Ci}}}{10^6 \, \cancel{\mu\text{Ci}}} \times \frac{3.7 \times 10^{10} \text{ dis/sec}}{1 \, \cancel{\text{Ci}}} = 1.2 \times 10^5 \text{ dis/sec}$

Medical Uses of Radioisotopes (Section 10.6)

10.30   Describe the importance of hot and cold spots in diagnostic work using tracers.

Solution:

If the tracer is accumulated in the organ (or diseased part), the radiation will be concentrated in that area. The organ will have more radiation coming from it than from surrounding areas (a hot spot). If the tracer is not assimilated by the organ (or diseased part), there will be less radiation coming from the organ than from surrounding areas (cold spot). Either way, the location of diseased tissue can be detected.

10.32   Chromium-51 is used medically to monitor kidney activity. Chromium-51 decays by electron capture. Write a balanced equation for the decay process and identify the daughter that is produced.

Solution:

$^{51}_{24}\text{Cr} + \, ^{0}_{-1}\text{e}^- \rightarrow \, ^{51}_{23}\text{V}$. The daughter is vanadium-51.

Nonmedical Uses of Radioisotope (Section 10.7)

10.34   A mixture of water (H_2O) and hydrogen peroxide (H_2O_2) will give off oxygen gas when solid manganese dioxide is added as a catalyst. Describe how you could use a tracer to determine if the oxygen comes from the water or the peroxide.

Solution:

Prepare water using a radioactive isotope of oxygen. Add H_2O_2 and collect the O_2 gas. If O_2 gas is radioactive then if the radioactivity showed up in the O_2 gas, the water is the source of the O_2. If the gas is not radioactive, the O_2, came from the H_2O_2.

10.36 Propose a method for measuring the volume of water in an irregular-shaped swimming pool. You have 1 gallon of water that contains a radioisotope with a long half-life, and a Geiger-Müller counter.

Solution:
Measure the radioactivity in the one gallon of radioactive water. Put the radioactive water into the swimming pool and allow for complete mixing. Refill the gallon container from the pool and measure the radioactivity in the diluted sample. The ratio of the radioactivity readings before and after is inversely proportional to the ratio of the 1 gallon to the volume of the pool; or

$$\frac{I_{initial}}{I_{final}} = \frac{V_{final}}{1 \text{ gallon}}; \text{ and } V_{final} = 1 \text{ gal} \times \frac{I_{initial}}{I_{final}}$$

Induced Nuclear Reactions (Section 10.8)

10.38 Write a balanced equation to represent the synthesis of silicon-27 that takes place when magnesium-24 reacts with an accelerated alpha particle. A neutron is also produced.

Solution:
$$^{24}_{12}Mg + ^{4}_{2}\alpha \rightarrow ^{27}_{14}Si + ^{1}_{0}n$$

10.40 Write a balanced equation to represent the (net breeding) reaction that occurs when uranium-238 reacts with a neutron to form plutonium-239 and two beta particles.

Solution:
$$^{238}_{92}U + ^{1}_{0}n \rightarrow ^{239}_{94}Pu + 2^{0}_{-1}\beta$$

10.42 Describe the role of a moderator in nuclear reactions involving neutrons.

Solution:
The moderator slows the neutrons, improving the chance for the neutron to be captured.

10.44 To make gallium-67 for diagnostic work, zinc-66 is bombarded with accelerated protons. Write a balanced equation to represent the process when a zinc-66 captures a single proton.

Solution:
$$^{66}_{30}Zn + ^{1}_{1}H \rightarrow ^{67}_{31}Ga$$

Nuclear Energy (Section 10.9)

10.46 Write one balanced equation that illustrates nuclear fusion.

Solution:
$$^{1}_{1}H + ^{1}_{1}H \rightarrow ^{2}_{1}H + ^{0}_{-1}\beta$$

10.48 Complete the following reactions which represent two additional ways uranium-235 can undergo nuclear fission.
a) $^{235}_{92}U + ^{1}_{0}n \rightarrow ^{160}_{62}Sm + ? + 4^{1}_{0}n$
b) $^{235}_{92}U + ^{1}_{0}n \rightarrow ^{87}_{35}Br + ? + 3^{1}_{0}n$

Solution:
a) $^{235}_{92}U + ^{1}_{0}n \rightarrow ^{160}_{62}Sm + ^{72}_{30}Zn + 4^{1}_{0}n$
b) $^{235}_{92}U + ^{1}_{0}n \rightarrow ^{87}_{35}Br + ^{146}_{57}La + 3^{1}_{0}n$

PROGRAMMED REVIEW

Section 10.1 Radioactive Nuclei

(a) _radioactive_ nuclei undergo spontaneous changes and emit energy. The emission of radiation by unstable nuclei is called (b) _radioactive decay_. The common types of emitted radiation are (c) _alpha_ particles, (d) _beta_ particles and (e) _gamma_ rays.

Section 10.2 Equations for Nuclear Reactions

In equations used to represent nuclear reactions, all particles are designated by a (a) _symbol_, a (b) _mass_ number and an (c) _atomic_ number. A nuclear equation is balanced when the sums of the (d) _mass_ numbers and (e) _atomic_ numbers are the same on both sides of the equation. A nucleus produced by radioactive decay is called a (f) _daughter nuclei_. When a nucleus emits a (g) _positron_, a nuclear proton is changed to a neutron.

Section 10.3 Isotope Half-life

A (a) _half_ life is the time required for (b) _one - half_ of the atoms in a sample to undergo radioactive decay. Radioisotopes found in nature have (c) _long_ half-lives, or are (d) _daughters_ of the decay of long-lived isotopes, or are continually produced by natural processes such as cosmic ray bombardment. After two half-lives have passed, the fraction of original atoms remaining is (e) _¼_.

Section 10.4 Health Effects of Radiation

The greatest danger of radiation to living organisms results from the ability of radiation to generate (a) _ions_ or (b) _free radicals_. The condition associated with short-term exposure to intense radiation is called (c) _radiation sickness_ Two protections against radiation are (d) _sheilding_ and (e) _distance_

Section 10.5 Measurement Units for Radiation

Two types of units used to describe quantities of radiation are (a) _physical_ units that indicate the activity of the source, and (b) _biological_ units that are related to the tissue damage caused by the radiation. Two physical units are the (c) _curie_ and the (d) _beckeral_. A biological unit used with x-rays and gamma rays is the (e) _roentgen_ Both the (f) _rad_ and (g) _gray_ describe the effects of radiation in terms of the energy transferred to tissue. A (h) _rem_ is a biological unit that accounts for differences in various types of radiation. Three devices that are used to detect radiation are (i) _film_ badges, (j) _scintillation_ counters and (k) _Geiger - Muller_ tubes.

Section 10.6 Medical Uses of Radioisotopes

In diagnostic applications, radioisotopes are used as (a) _tracers_. Radioisotopes used for this purpose ideally should have short (b) _half - lives_, but not too short since they must be prepared and administered conveniently. These isotopes also should produce (c) _daughters_ that are nontoxic and, ideally, (d) _stable_ toward further radioactive decay. The radioisotopes used diagnostically should preferably give off (e) _gamma_ radiation, and they should be absorbed by tissue to form (f) _cold hot spots_ or rejected by tissue to form (g) _cold spots_. Radioisotopes used therapeutically should ideally emit (h) _____ or (i) _____ radiation. Their (j) _____-_____ should be long enough to allow therapy to be accomplished. Their daughters should be non (k) _____ and give off little or no (l) _____. Therapeutic radioisotopes should be (m) _____ by the body in the target tissue.

Section 10.7 Nonmedical Uses of Radioisotopes

Radioisotopes are used as (a) _____ in some nonmedical applications such as following the path of a compound in a process, indicating the boundary between products in a (b) _____, and measuring the effectiveness of (c) _____. Some radioisotopes are used to determine the ages of artifacts or minerals in a process called (d) _____ _____.

Section 10.8 Induced Nuclear Reactions

(a) _____ and (b) _____ particles may be used to bombard nuclei. Of these two kinds of particles, only the (c) _____ can be accelerated.

Section 10.9 Nuclear Energy

Nuclear fusion is the (a) _____ of small nuclei to make (b) _____ nuclei. Nuclear (c) _____ is the splitting of (d) _____ nuclei to make smaller nuclei. Nuclear (e) _____ releases more energy per reaction than nuclear (f) _____. A nuclear reactor has nuclear (g) _____ occurring at a controlled (h) _____.

SELF-TEST QUESTIONS

Multiple Choice

1. The three common types of radiation emitted by naturally radioactive elements are
   a) electrons, protons and neutrons
   b) x-rays, gamma rays and protons
   c) alpha rays, beta rays and neutrons
   d) alpha rays, beta rays and gamma rays

2. Which of the following types of radiation is composed of particles which carry a +2 charge?
   a) alpha
   b) beta
   c) gamma
   d) neutrons

3. Which of the following types of radiation is not composed of particles?
   a) alpha
   b) beta
   c) gamma
   d) neutrons

4. After four half-lives have elapsed, the amount of a radioactive sample which has not decayed is
   a) 40% of the original amount
   b) 1/4 of the original amount
   c) 1/8 of the original amount
   d) 1/16 of the original amount

5. If 1/8 of an isotope sample is present after 22 days, what is the half-life of the isotope?
   a) 10 days
   b) 5 days
   c) 2 3/4 days
   d) 7 1/3 days

6. By doubling the distance between yourself and a source of radiation how is the intensity of the radiation getting to you changed?
   a) it is 1/2 as great
   b) it is 1/3 as great
   c) it is 1/4 as great
   d) it is 1/8 as great

7. Which of the following would be the most convenient unit to use when determining the total dose of radiation received by an individual who was exposed to several different types of radiation?
   a) Roentgen
   b) rad
   c) gray
   d) rem

8. Which type of radiation has the lowest penetration?
   a) α
   b) β
   c) δ
   d) X

9. The reaction, $^{24}_{11}Na \rightarrow {}^{24}_{12}Mg + {}^{0}_{-1}\beta$, is an example of
   a) alpha decay
   b) beta decay
   c) positron decay
   d) fission

10. The reaction, $^{235}_{92}U + {}^{1}_{0}n \rightarrow {}^{103}_{41}Nb + {}^{131}_{51}Sb + 2{}^{1}_{0}n$, is an example of
    a) alpha decay
    b) beta decay
    c) positron decay
    d) fission

True-False

11. Radioactive tracers are useful in both medical and nonmedical applications.

12. Radioactive isotopes are not taken into the body during medical uses of radioisotopes.

13. A Curie is a physical measurement of the quantity of radiation.

14. The rem is a biological radiation measurement unit.

Matching

For each of the nuclear equations on the left choose the correct identity of X from the choices on the right.

15. _____ $^{13}_{7}N \rightarrow ^{13}_{6}C + X$ a) beta, $^{0}_{-1}\beta^{+}$

16. _____ $^{27}_{13}Al + ^{2}_{1}H \rightarrow ^{25}_{12}Mg + X$ b) neutron, $^{1}_{0}n$

17. _____ $^{9}_{4}Be + ^{4}_{2}He \rightarrow ^{12}_{6}C + X$ c) positron, $^{0}_{1}\beta$

          d) alpha, $^{4}_{2}\alpha$

SOLUTIONS

A. Answers to Programmed Review

10.1 a) radioactive   b) radioactive decay   c) alpha   d) beta
      e) gamma

10.2 a) symbol   b) mass   c) atomic   d) mass
      e) atomic   f) daughter nuclei   g) positron

10.3 a) half   b) one-half   c) long   d) daughters
      e) 1/4

10.4 a) ions   b) free radicals   c) radiation sickness   d) shielding
      e) distance

10.5 a) physical   b) biological   c) Curie   d) Becquerel
      e) Roentgen   f) rad   g) gray   h) rem
      i) film   j) scintillation   k) Geiger-Müller

10.6 a) tracers   b) half-lives   c) daughters   d) stable
      e) gamma   f) hot spots   g) cold spots   h) alpha
      i) beta   j) half-lives   k) toxic   l) radiation
      m) concentrated

10.7 a) tracers   b) pipeline   c) lubricants   d) radioactive dating

10.8 a) charged
      b) neutral or uncharged
      c) charged

10.9 a) combining   b) larger   c) fission   d) large
      e) fusion   f) fission   g) fission   h) rate

B. Answers to Self-Test Questions

1. d	7. d	13. F
2. a	8. a	14. T
3. c	9. b	15. c
4. d	10. d	16. d
5. d	11. T	17. b
6. c	12. F	

Chapter 11

Organic Compounds: Alkanes

CHAPTER OUTLINE

11.1 Carbon, The Element of Organic Compounds

11.2 Organic and Inorganic Compounds Compared

11.3 Bonding Characteristics and Isomerism

11.4 Functional Groups: The Organization of Organic Chemistry

11.5 Alkane Structures

11.6 Conformations of Alkanes

11.7 Alkane Nomenclature

11.8 Cycloalkanes

11.9 The Shape of Cycloalkanes

11.10 Physical Properties of Alkanes

11.11 Alkane Reactions

LEARNING OBJECTIVES

When you have completed your study of this chapter, you should be able to:

1. Show that you understand the general importance of organic chemical compounds.
2. Recognize the molecular formulas of organic and inorganic compounds.
3. Explain some general differences between inorganic and organic compounds.
4. Use structural formulas to identify compounds that are isomers of each other.
5. Write condensed or expanded structural formulas for compounds.
6. Classify alkanes as normal, branched, or cycloalkanes.
7. Use structural formulas to determine whether compounds are structural isomers.
8. Assign IUPAC names and draw structural formulas for alkanes and cycloalkanes.
9. Name and draw structural formulas for geometric isomers of cycloalkanes.
10. Describe the key physical properties of alkanes.
11. Write alkane combustion reactions.

ANSWERS AND SOLUTIONS TO EVEN-NUMBERED PROBLEMS

Carbon, The Element of Organic Compounds (Section 11.1)

11.2   Why were the compounds of carbon originally called organic compounds?

Solution:

The term, *organic compounds*, originally referred to those compounds found in living, "organic" matter.

11.4   What is the unique structural feature shared by all organic compounds?

Solution:
They all contain carbon atoms.

Organic and Inorganic Compounds Compared (Section 11.2)

11.6   What kind of bond between atoms is most prevalent among organic compounds?

Solution:
Covalent bonds are the most prevalent. Most organic compounds have nonpolar bonds because they are between the same kind of atoms (carbon to carbon).

11.8   Indicate for each of the following characteristics whether it more likely describes an inorganic or organic compound. Give one reason for your answer.
a) This compound is a liquid that readily burns.
b) A white solid upon heating is found to melt at 735°C.
c) A liquid added to water floats on the surface and does not dissolve.
d) This compound exists as a gas at room temperature and ignites easily.
e) A solid substance melts at 65°C.

Solution:
a) Organic. Most organic compounds are flammable, but relatively few inorganic compounds are flammable.
b) Inorganic. Inorganic compounds tend to have higher melting points.
c) Organic. Organic compounds tend not to dissolve in water and many are less dense than water.
d) Organic. Flammable compounds are usually organic.
e) Organic. Organic solids tend to have low melting points, due to their covalent bonding.

Bonding Characteristics and Isomerism (Section 11.3)

11.10   Give two reasons for the existence of the tremendous number of organic compounds.

Solution:
(1) Carbon atoms can bond to other carbon atoms; and
(2) carbon atoms may have different arrangements (isomerism).

11.12   Describe what atomic orbitals overlap to produce a carbon-hydrogen bond in CH_4.

Solution:
The s orbital of the hydrogen overlaps with the sp^3 hybridized orbital of the carbon.

11.14   Compare the shapes of unhybridized p and hybridized sp^3 orbitals.

Solution:
Each of the three p orbitals has two equal lobes and is mutually perpendicular (at right angles) to the other two. Each of the sp^3 orbitals has two unequal lobes and the set of four is arranged tetrahedrally. Each sp^3 orbital forms an angle of about 109° with the other three sp^3 orbitals.

11.16   Complete the following structures by adding hydrogen atoms where needed.

a) $C—C=C$ b) $C—\overset{\overset{\displaystyle O}{\|}}{C}—C$ c) $C—\overset{\overset{\displaystyle O}{\|}}{C}—O$ d) $C—N—C$

Solution:

a) $H—\overset{\overset{\displaystyle H}{|}}{\underset{\underset{\displaystyle H}{|}}{C}}—\overset{}{\underset{\underset{\displaystyle H}{|}}{C}}=\overset{}{\underset{\underset{\displaystyle H}{|}}{C}}—H$

b) $H—\overset{\overset{\displaystyle H}{|}}{\underset{\underset{\displaystyle H}{|}}{C}}—\overset{\overset{\displaystyle O}{\|}}{C}—\overset{\overset{\displaystyle H}{|}}{\underset{\underset{\displaystyle H}{|}}{C}}—H$

c) $H—\overset{\overset{\displaystyle H}{|}}{\underset{\underset{\displaystyle H}{|}}{C}}—\overset{\overset{\displaystyle O}{\|}}{\underset{\underset{\displaystyle H}{|}}{C}}—O$

d) $H—\overset{\overset{\displaystyle H}{|}}{\underset{\underset{\displaystyle H}{|}}{C}}—N—\overset{\overset{\displaystyle H}{|}}{\underset{\underset{\displaystyle H}{|}}{C}}—H$

11.18   Which of the following pairs of compounds are structural isomers?

a) $CH_3—CH=CH—CH_3$ and $CH_3—CH_2—CH_2—CH_3$

b) $CH_3—CH_2—CH_2—CH_2—CH_3$ and $CH_3—\overset{\overset{\displaystyle CH_3}{|}}{\underset{\underset{\displaystyle CH_3}{|}}{C}}—CH_3$

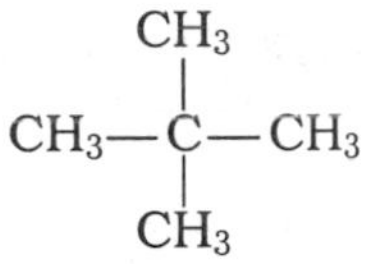

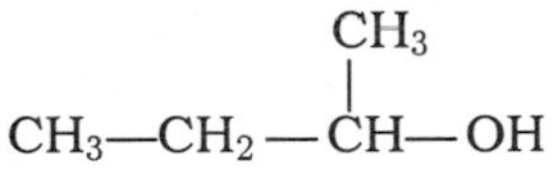

c) $CH_3—CH_2—\overset{\overset{\displaystyle CH_3}{|}}{CH}—OH$ and $CH_3—\overset{\overset{\displaystyle O}{\|}}{C}—CH_2—CH_3$

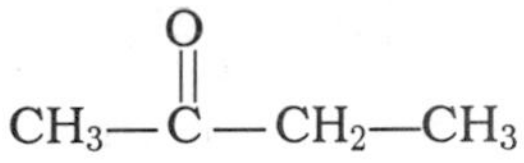

d) 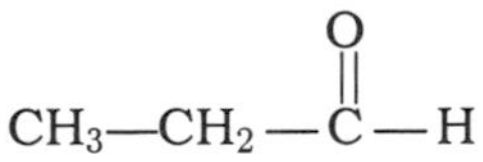 $CH_3—CH_2—\overset{\overset{\displaystyle O}{\|}}{C}—H$ and $CH_3—\overset{\overset{\displaystyle O}{\|}}{C}—CH_3$

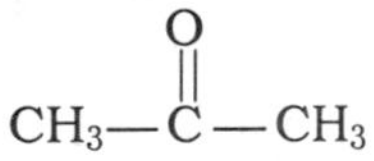

e) $CH_3—CH_2—CH_2—NH_2$ and $CH_3—CH_2—NH—CH_3$

Solution:
a) Not isomers. They are C_4H_8 and C_4H_{10} respectively.
b) Isomers. Both are C_5H_{12}.
c) Not isomers. Both are $C_4H_{10}O$ and C_4H_8O respectively.
d) Isomers. Both are C_3H_6O.
e) Isomers. Both are C_3H_9N respectively.

11.20 On the basis of the number of covalent bonds possible for each atom, determine which of the following structural formulas are correct. Explain what is wrong with the incorrect structures.

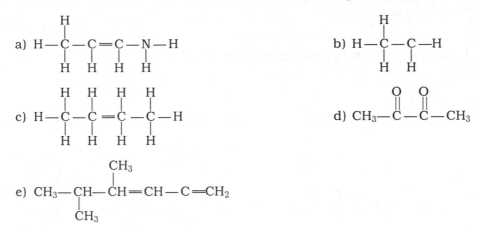

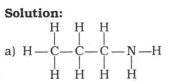

Solution:

a) Correct.

b) Incorrect. The carbon on the right only has 3 bonds.

c) Incorrect. The two interior carbons have 5 bonds each.

d) Correct.

e) Incorrect. The carbon in the center of the chain has 5 bonds. The second carbon from the right has only 3 bonds.

Functional Groups: The Organization of Organic Chemistry (Section 11.4)

11.22 Write the condensed structural formula for the following compounds:

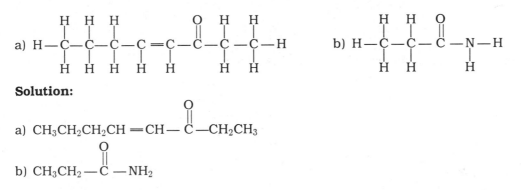

Solution:

a)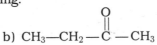
$CH_3CH_2CH_2CH = CH - \overset{\displaystyle O}{\overset{\|}{C}} - CH_2CH_3$

b)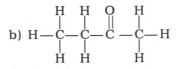
$CH_3CH_2 - \overset{\displaystyle O}{\overset{\|}{C}} - NH_2$

11.24 Write an expanded structural formula for the following:

a) $CH_3 - CH_2 - CH_2 - NH_2$

b) $CH_3 - CH_2 - \overset{\displaystyle O}{\overset{\|}{C}} - CH_3$

Solution:

a)
```
    H  H  H
    |  |  |
H — C — C — C — N — H
    |  |  |  |
    H  H  H  H
```

b)
```
    H  H  O  H
    |  |  ‖  |
H — C — C — C — C — H
    |  |     |
    H  H     H
```

Alkane Structures (Section 11.5)

11.26 The name of the normal alkane containing 9 carbon atoms is nonane. Where are the molecular and condensed structural formulas for nonane?

   Solution:
   The formula for an alkane is C_nH_{2n+2}; the molecular formula is C_9H_{20}.
   The condensed formula for nonane is $CH_3(CH_2)_7CH_3$.

Alkane Nomenclature (Section 11.7)

11.28 For each of the following carbon skeletons, give the number of carbon atoms in the longest continuous chain:

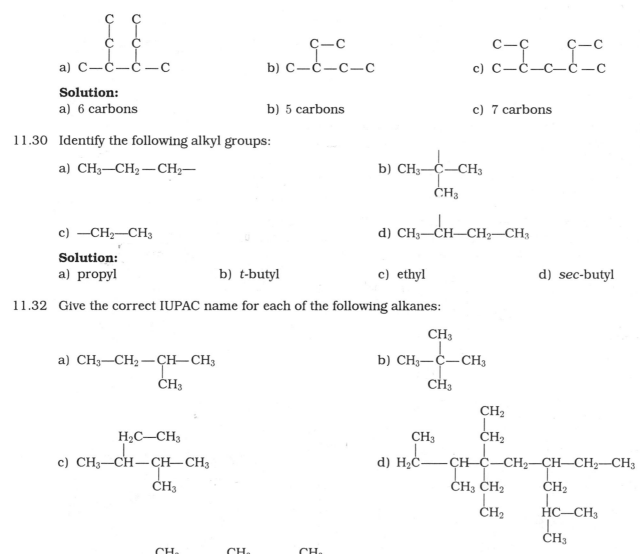

   Solution:
   a) 6 carbons   b) 5 carbons   c) 7 carbons

11.30 Identify the following alkyl groups:

   a) CH_3—CH_2—CH_2—

   b) CH_3—$\overset{\displaystyle |}{\underset{\displaystyle CH_3}{C}}$—$CH_3$

   c) —CH_2—CH_3

   d) CH_3—$\overset{\displaystyle |}{CH}$—$CH_2$—$CH_3$

   Solution:
   a) propyl   b) *t*-butyl   c) ethyl   d) *sec*-butyl

11.32 Give the correct IUPAC name for each of the following alkanes:

   a) CH_3—CH_2—$\overset{\displaystyle |}{\underset{\displaystyle CH_3}{CH}}$—$CH_3$

   b) CH_3—$\overset{\displaystyle CH_3}{\underset{\displaystyle CH_3}{\overset{|}{\underset{|}{C}}}}$—$CH_3$

   c) CH_3—$\overset{\displaystyle H_2C—CH_3}{\underset{\displaystyle CH_3}{\overset{|}{\underset{|}{CH}}}}$—$CH$—$CH_3$

   d) H_2C——CH—$\overset{\displaystyle CH_2}{\overset{|}{\underset{|}{\underset{\displaystyle CH_2}{C}}}}$—$CH_2$—$CH$—$CH_2$—$CH_3$

   e) CH_3—CH_2—$\overset{\displaystyle CH_3}{\overset{|}{CH}}$—$CH$—$\overset{\displaystyle CH_3}{\overset{|}{CH}}$—$CH_2$—$\overset{\displaystyle CH_3}{\overset{|}{CH}}$—$CH_3$

Solution:

a) 2-methylbutane

b) 2,2-dimethylpropane

c) Hint: Check longest chain. 2,3-dimethylpentane

d) Hint: Check longest chain. 4,6,6-triethyl-2,7-dimethylnonane

e) Hint: Check longest chain. 5-sec-butyl-2,4-dimethylnonane

11.34 Draw a condensed structural formula for each of the following compounds:

a) 2,3-dimethylbutane

b) 3-isopropylheptane

c) 5-t-butyl-2,3-dimethyloctane

d) 4-ethyl-4-methyloctane

Solution:

a) 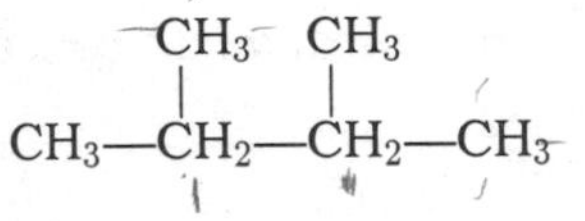
$$CH_3-CH_2-CH_2-CH_3$$
with CH_3 and CH_3 substituents

b) 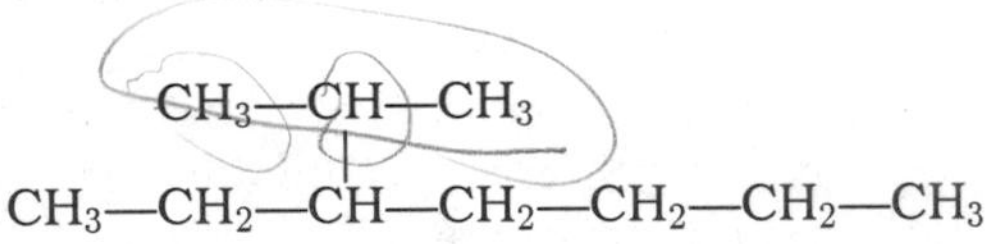
$$CH_3-CH_2-CH-CH_2-CH_2-CH_2-CH_3$$
with $CH_3-CH-CH_3$ substituent

c) 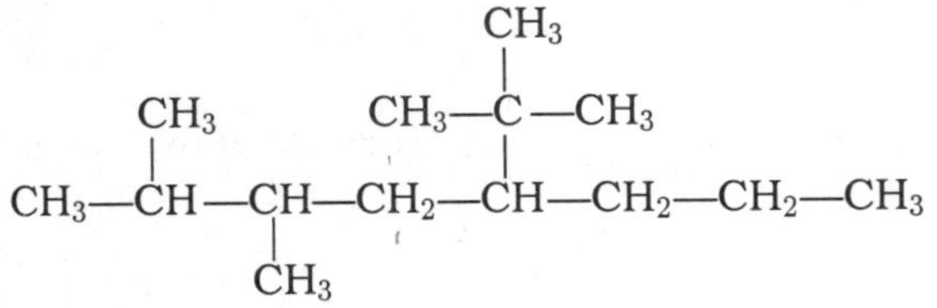
$$CH_3-CH-CH-CH_2-CH-CH_2-CH_2-CH_3$$
with CH_3, CH_3 substituents and CH_3-C-CH_3 with CH_3

d) 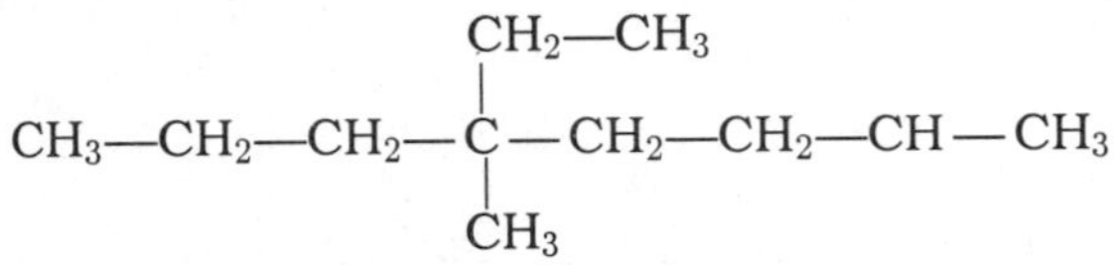
$$CH_3-CH_2-CH_2-C-CH_2-CH_2-CH-CH_3$$
with CH_2-CH_3 and CH_3 substituents

11.36 Draw the condensed structural formula for each of the three structural isomers of C_5H_{12} and give the correct IUPAC names.

Solution:

(1) $CH_3-CH_2-CH_2-CH_2-CH_3$ pentane

(2) 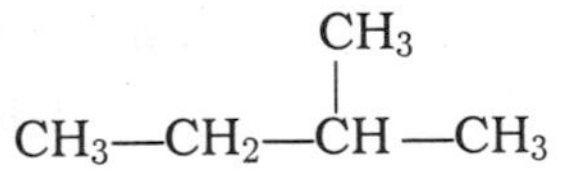 $CH_3-CH_2-CH-CH_3$ 2-methylbutane
with CH_3 substituent

(3) CH_3-C-CH_3 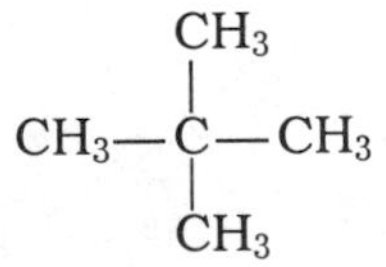   2,2-dimethylpropane
with CH_3 top and CH_3 bottom substituents

11.38 Draw structural formulas for the compounds and give correct IUPAC names for the five structural isomers of C_6H_{16}.

Solution:

(1) $CH_3-CH_2-CH_2-CH_2-CH_2-CH_3-$ hexane

(2) 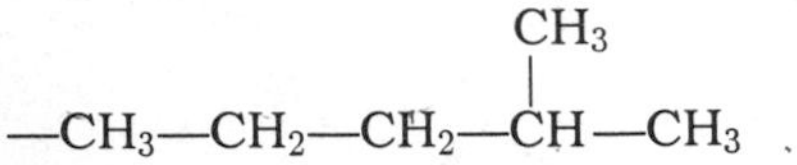 $-CH_3-CH_2-CH_2-CH-CH_3$ 2-methylpentane
with CH_3 substituent

(3)   —CH₃—CH₂—$\overset{\displaystyle \underset{|}{CH_3}}{CH}$—CH₂—CH₃   3-methylpentane

(4)   —CH₃—CH₂—$\overset{\displaystyle \underset{|}{CH_3}}{\underset{\displaystyle \underset{|}{CH_3}}{C}}$—CH₃   2,2-dimethylbutane

(5)   —CH₃—$\overset{\displaystyle \underset{|}{CH_3}}{CH}$—$\overset{\displaystyle \underset{|}{CH_3}}{CH}$—CH₃   2,3-dimethylbutane

11.40 The following names are incorrect according to IUPAC rules. Draw the structural formulas and tell why each name is incorrect. Write the correct name for each compound.
a) 1,2-dimethylpropane
b) 3,4-dimethylpentane
c) 2-ethyl-4-methylpentane
d) 2-bromo-3-ethylbutane

Solution:
a) The longest chain is four carbons long. (A branch cannot be on carbon #1. That would extend the chain.)

CH₃—CH₂—$\overset{\displaystyle \underset{|}{CH_3}}{CH}$—CH₃   2-methylbutane

b) The chain is numbered from the wrong end.

CH₃—CH₂—$\overset{\displaystyle \underset{|}{CH_3}}{CH}$—$\overset{\displaystyle \underset{|}{CH_3}}{CH}$—CH₃   2,3-dimethylpentane

c) The longest chain is six carbons long. The ethyl on carbon #2 extends the chain.

CH₃—$\overset{\displaystyle \underset{|}{CH_3}}{CH}$—CH₂—$\overset{\displaystyle \underset{|}{\overset{\displaystyle CH_2}{\overset{|}{CH_3}}}}{CH}$—CH₃   2,4-dimethylhexane

d) The longest chain is five carbons long. The ethyl on carbon #3 of butane extends the chain.

CH₃—$\overset{\displaystyle \underset{|}{\overset{\displaystyle CH_2}{\overset{|}{CH_3}}}}{CH}$—$\overset{\displaystyle \underset{|}{Br}}{CH}$—CH₃   2-bromo-3-methylpentane

Cycloalkanes (Section 11.8)
11.42 Write the correct IUPAC name for each of the following:
a)

b)
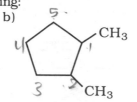

C_3H_6

c)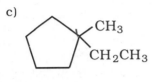

CH₃

CH₂CH₃

d) 3 ⬜ 4   CH₂ CH₂ OH₂

2

CH₃ / CHCH₃   C₆

CH₃

Solution:

a) cyclopropane

c) 1-ethyl-1-methylcyclopentane

b) 1,2-dimethylcyclopentane

d) 1-isopropyl-2-methylcyclobutane

11.44 Draw the structural formulas corresponding to each of the following IUPAC names:

a) 1,2-isopropylcyclopentane

b) 1,1-dimethylcyclobutane

c) 1-isobutyl-3-isopropylcyclohexane

Solution:

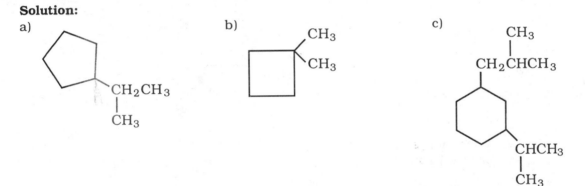

a)

b)   CH₃

CH₃

CH₂CH₃

CH₃

c)   CH₃

CH₂CHCH₃

CHCH₃

CH₃

11.46 Which of the following pairs of cycloalkanes represent structural isomers?

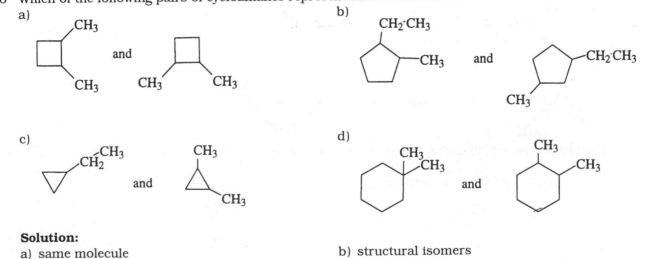

a)   CH₃

and

CH₃   CH₃   CH₃

b)   CH₂CH₃

and   CH₂CH₃

CH₃   CH₃

c)   CH₃

CH₂   CH₃

and

CH₃   CH₃

d)   CH₃   CH₃

CH₃   CH₃

and   CH₃   CH₃

Solution:

a) same molecule

c) structural isomers

b) structural isomers

d) structural isomers

The Shape of Cycloalkanes (Section 11.9)

11.48 Why does cyclohexane assume a chair form rather than a planar hexagon?

Solution:
In a planar hexagon, the angles are all 120°. The carbon-carbon bond angle would prefer to be about 109°. By assuming a chair shape, the bond angles can assume the tetrahedral shape.

11.50 Which of the following cycloalkanes could show geometric isomerism? For each that could, draw structural formulas and name both the *cis* and *trans* isomers.

a)
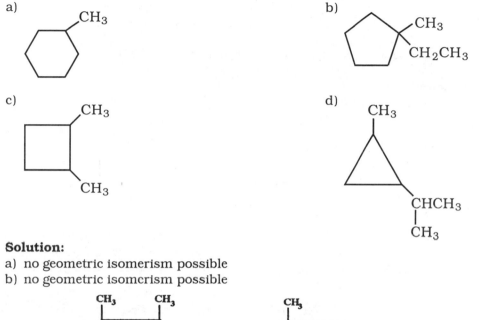
b)

c)   d)

Solution:
a) no geometric isomerism possible
b) no geometric isomerism possible

c)
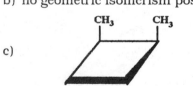
cis-1-2-dimethylcyclobutane   *trans*-1-2-dimethylcyclobutane

d)
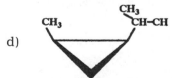
cis-1-isopropyl-2-methylcyclopropane *trans*-1-isopropyl-2-methylcyclopropane

11.52 Using the prefix *cis*- or *trans*- name each of the following:

a)   b)

c)

H$_2$C–CH$_2$-CH$_3$

CH$_3$

d)

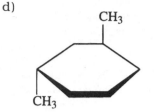

CH$_3$

CH$_3$

Solution:
a) *trans*-1-ethyl-2-methylcyclopropane
b) *cis*-1-bromo-2-chlorocyclopentane
c) *trans*-1-methyl-2-propylcyclobutane
d) *trans*-1,3-dimethylcyclohexane

Physical Properties of Alkanes (Section 11.10)

11.54 The compound decane is a straight-chain alkane. Predict the following:
   a) Is decane a solid, liquid, or gas at room temperature?
   b) Is it soluble in water?
   c) Is it soluble in hexane?
   d) Is it more or less dense than water?

Solution:
a) liquid   b) not soluble in water
c) soluble in hexane   d) less dense than water

11.56 Suppose you have a sample of 2-methylhexane and a sample of 2-methylheptane. Which sample would you expect to have a higher melting point? Boiling point?

Solution:
2-methylheptane, the larger, nonpolar molecule, has the larger interparticle forces and hence a higher melting point and a higher boiling point.

Alkane Reactions (Section 11.11)

11.58 Write a balanced equation to represent the complete combustion of each of the following:
   a) propane   b) CH$_3$—CH—CH$_3$ c)
                                              |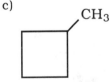
                                             CH$_3$ CH$_3$

Solution:
a) CH_3—CH_2—CH_3 + $5O_2 \rightarrow 3CO_2 + 4H_2O$

b) $2CH_3$—CH—CH_3 + $13O_2 \rightarrow 8CO_2 + 10H_2O$
        |
       CH_3

c) $2C_5H_{10} + 15O_2 \rightarrow 10CO_2 + 10H_2O$

11.60   Write a balanced equation for the *incomplete* combustion of hexane, assuming the formation of carbon monoxide and water as the only products.

Solution:

$$2C_6H_{14} + 13\,O_2 \rightarrow 12CO + 14H_2O$$

Chemistry Around Us and Key Chemicals

11.62   What is the goal of combinatorial chemistry?

Solution:

The goal of combinatorial chemistry is to synthesize a large variety of compounds and then test for desired properties rather than trying to synthesize a high yield of a compound and then test.

11.64   What petroleum products are sometimes found in skin moisturizers?

Solution:

The petroleum products in skin moisturizers include petroleum jelly and mineral oil to serve as a barrier protecting from water loss.

11.66   Give another reason why petroleum is so valuable to our society besides serving as a source of fuels.

Solution:

Petroleum is the starting material for most organic chemicals, including plastics, detergents, dyes, and other materials.

11.68   What is the antidote for CO poisoning?

Solution:

Increase exposure to oxygen. Administer oxygen gas, if available, or get plenty of fresh air.

PROGRAMMED REVIEW

Section 11.1 Carbon: The Element of Organic Compounds

Wöhler's synthesis of the compound (a) _____ led to the downfall of the vital force theory. The element (b) _____ is present in all organic compounds. Elements and compounds not studied in organic chemistry are considered to be a part of (c) _____.

Section 11.2 Organic and Inorganic Compounds Compared

The number of known organic compounds is more than (a) _____. The bonding forces normally present within organic molecules are (b) _____. The solubility of organic compounds in water is often (c) _____. (d) _____ compounds are usually nonflammable.

Section 11.3 Bonding Characteristics and Isomerism

Four (a) _____ hybrid orbitals of carbon form bonds with hydrogen atoms in CH_4. Compounds with the same molecular formula but with the atoms bonded in different patterns are called (b) _____ _____. The number of covalent bonds normally surrounding a carbon atom is (c) _____. The number of covalent bonds normally surrounding an oxygen atom is (d) _____.

Section 11.4 Functional Groups: The Organization of Organic Chemistry

Organic compounds are arranged into classes on the basis of structural features called (a) _____ _____. Structural formulas that show all covalent bonds are referred to as (b) _____. Structural formulas that show only certain covalent bonds are referred to as (c) _____.

Section 11.5 Alkane Structures

Organic compounds containing only the elements carbon and hydrogen are called (a) _____.
Saturated hydrocarbons are also known as (b)_____. An alkane with six carbon atoms has
(c) _____ hydrogen atoms. Alkanes are classified as normal or branched. The compound
$CH_3CH_2CH_2CH_2CH_3$ is a (d) _____ alkane.

Section 11.6 Conformations of Alkanes

Different orientations produced in a molecule by the rotation about single bonds are called
(a) _____.

Section 11.7 Alkane Nomenclature

The IUPAC ending for the name of an alkane compound is (a) _____. The root word hept- is
used in naming an alkane with (b) _____ carbon atoms in the longest chain. The alkyl group

CH_3—$\underset{\underset{CH_3}{|}}{CH}$— is called (c) _____. The alkyl group CH_3—CH_2—$\underset{\underset{CH_3}{|}}{CH}$— is called (d) _____.

The longest carbon chain in CH_2—$\underset{\underset{\underset{\underset{CH_3}{|}}{CH_2}}{|}}{CH}$—$CH_2$—$CH_2$—$CH_3$ has (e) _____ carbon atoms. The

IUPAC name for CH_3—$\underset{\underset{CH_3}{|}}{CH}$—$CH_2$—$\underset{\underset{\underset{\underset{CH_3}{|}}{CH_2}}{|}}{CH}$—$CH_2$ is (f) _____.

Section 11.8 Cycloalkanes

The number of hydrogen atoms present in a molecule of [cyclopentane with CH$_3$] is (a) _____ A cycloalkane

with 14 hydrogen atoms has (b) _____ carbon atoms. The cycloalkane [cyclopentane with CH$_3$ and CH$_3$] has methyl

groups located at positions (c) _____. The IUPAC name for [cyclobutane with CH$_3$ and CH$_2$–CH$_2$–CH$_3$] is (d) _____

carbon atoms.

Section 11.9 The Shape of Cycloalkanes

The most stable geometric arrangement of four atoms attached to a carbon atom is (a) _____.
Another name for *cis-trans* isomers is (b) _____ isomers. The prefix (c) _____ is used to
denote an isomer in which is two groups are attached to the same side of a cycloalkane ring. The

IUPAC name for [cyclopropane with CH$_3$ and CH$_3$] is (d) _____.

Section 11.10 Physical Properties of Alkanes

A compound in an homologous series differs from the next member in the series by a (a) _____ unit. Hydrophobic molecules are (b) _____ in water. Alkanes are (c) _____ dense than water. Liquid alkanes generally contain from (d) _____ to 20 carbon atoms.

Section 11.11 Alkane Reactions

Alkanes are the (a) _____ reactive of all organic compounds. The products of complete combustion of an alkane are (b) _____ and H_2O. The products of incomplete combustion of an alkane may be (c) _____ or C and H_2O.

SELF-TEST QUESTIONS

Multiple Choice

1. The maximum number of covalent bonds which carbon can form is
    a) 1   b) 2   c) 3   d) 4

2. Which of the following is considered an organic compound?
    a) CH_4 b) NaOH   c) Na_2CO_3 d) KCN

3. How many hydrogen atoms are needed to complete the following structure:

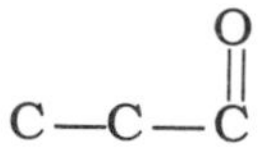

    $$C-C-\overset{\displaystyle \overset{O}{\|}}{C}$$

    a) 2   b) 4   c) 6   d) 8

4. A C—H bond in CH_4 is formed by the overlap of what orbitals?
    a) sp^3 and 1s   b) 1s and 1s   c) p and 1s   d) sp and 1s

5. Which of the following are the same compound?

    I. $CH_3-CH_2-CH_2-CH_3$

    II. 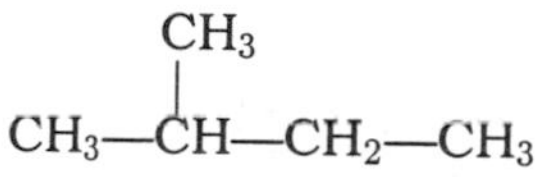 $CH_3-\overset{\displaystyle \overset{CH_3}{|}}{CH}-CH_2-CH_3$

    III. 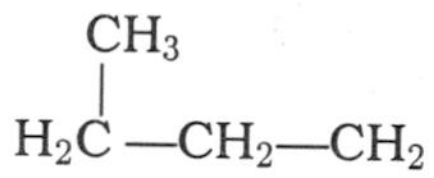 $H_2\overset{\displaystyle \overset{CH_3}{|}}{C}-CH_2-CH_2$

    IV. 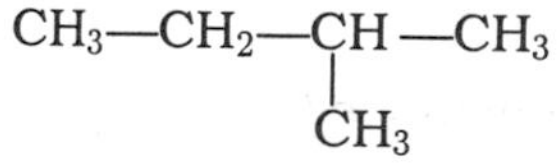 $CH_3-CH_2-\overset{\displaystyle \underset{CH_3}{|}}{CH}-CH_3$

    V. 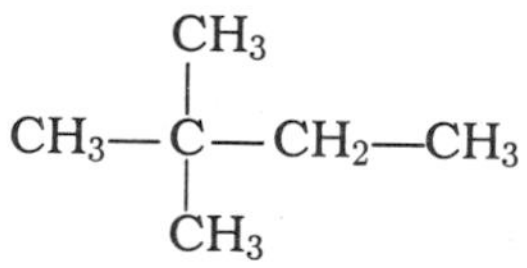 $CH_3-\overset{\displaystyle \overset{CH_3}{|}}{\underset{\displaystyle \underset{CH_3}{|}}{C}}-CH_2-CH_3$

    a) I and II   b) I and III   c) II and III   d) IV and V

6. The molecular formula for [cyclobutane with a CH_3 group] is

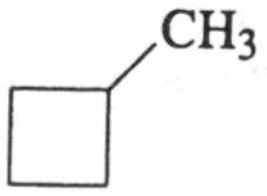

    a) C_5H_{12} b) C_5H_{10} c) C_5H_9 d) C_4H_9

7. Which of the following compounds is a structural isomer of $CH_3-CH_2-\underset{\underset{\displaystyle CH_3}{|}}{CH}-CH_3$?

a) $CH_3-CH_2-CH_2-CH_3$

b) $CH_3-\underset{\underset{\displaystyle CH_3}{|}}{\overset{\overset{\displaystyle CH_3}{|}}{C}}-CH_3$

c) $CH_3-CH_2-\underset{\underset{}{}}{\overset{\overset{\displaystyle CH_3}{|}}{CH}}-CH_3$

d) $CH_3-\overset{\overset{\displaystyle CH_3}{|}}{CH}-\underset{\underset{\displaystyle CH_3}{|}}{CH}-CH_3$

8. What is a structural isomer of ◁—CH_3 ?

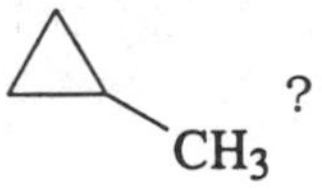

a)  ☐

b)  △—CH_3 (with CH_3 on top)

c) $CH_3-CH_2-CH_2-CH_3$

d) $CH_3-\overset{\overset{\displaystyle CH_3}{|}}{CH}-CH_3$

9. The unbranched alkane that contains eight carbon atoms is called
    a) hexane   b) heptane   c) octane   d) nonane

10. How many structural isomers have the formula C_4H_{10}?
    a) 2   b) 3   c) 4   d) 5

11. The number of carbon atoms in the longest chain of $CH_3-\underset{\underset{\displaystyle CH_2CH_3}{\diagdown}}{\overset{\overset{\displaystyle CH_3}{|}}{CH}}-CH_2$ is

    a) 3   b) 4   c) 5   d) 6

12. The correct IUPAC name for $CH_3-\underset{\underset{\underset{\underset{\displaystyle CH_3}{|}}{CH_2}}{|}}{CH}-CH_2-\underset{\underset{\displaystyle CH_3}{|}}{CH}-CH_3$ is

    a) 2-ethyl-4-methylpentane   b) 2-methyl-4-ethylpentane
    c) 3,5-dimethylhexane   d) 2,4-dimethylhexane

13. The correct IUPAC name for CH_3—⬡—CH_3 is

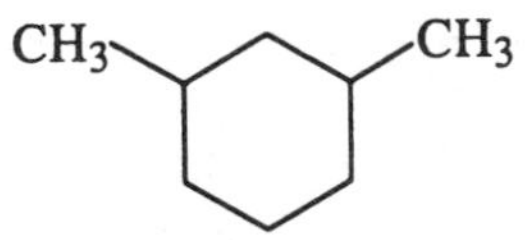

    a) 1,3-dimethylhexane   b) 1,3-methylcyclohexane
    c) 1,5-dimethylcyclohexane   d) 1,3-dimethylcyclohexane

14. The bromine in 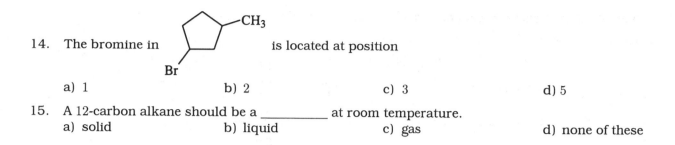 is located at position

   a) 1   b) 2   c) 3   d) 5

15. A 12-carbon alkane should be a _____ at room temperature.
   a) solid   b) liquid   c) gas   d) none of these

Matching

Match an alkyl group name to each structure on the left.

16. $CH_3—CH_2—$ a) propyl

17. $CH_3—CH_2—CH_2—$ b) *sec*-butyl

18.    c) isobutyl

                                           d) ethyl

Match the structures on the left to the descriptions on the right.

19.   a) a *cis* compound

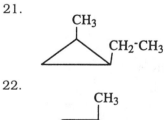   b) a *trans* compound

                                           c) neither *cis* nor *trans*

20.

21.

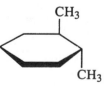

22.

True-False

23. Most organic compounds are very soluble in water.

24. Covalent bonds are more prevalent in inorganic compounds than in organic compounds.

25. Solutions of inorganic compounds are better electrical conductors than solutions of organic compounds.

26. There are more known inorganic compounds than organic compounds.

27. A few compounds of carbon, such as CO_2, are classified as inorganic.

28. Structural isomers always have the same molecular formula.

29. Structural isomers always have the same functional group.

30. A molecule may have more than one functional group.

31. A condensed structural formula may show some bonds.

32. An expanded structural formula may not show all the bonds.

33. Pentane and cyclopentane are isomers of each other.

34. Alkanes have lower boiling points than other organic compounds.

35. The main component of natural gas is butane.

36. Complete combustion of pentane produces H_2O and CO_2.

37. Alkanes are polar molecules.

SOLUTIONS

A. Answers to Programmed Review

11.1 a) urea   b) carbon   c) inorganic chemistry

11.2 a) 6,000,000   b) covalent   c) low   d) inorganic

11.3 a) sp^3 b) structural isomers c) four   d) two

11.4 a) functional groups b) expanded   c) condensed

11.5 a) hydrocarbons   b) alkanes   c) 14   d) normal

11.6 a) conformations

11.7 a) -ane   b) 7   c) isopropyl   d) *sec*-butyl
      e) 6   f) 2,4-dimethylhexane

11.8 a) 12   b) 7   c) 1 and 3
      d) 1-methyl-2-propylcyclobutane

11.9 a) tetrahedral   b) geometric   c) *cis*
      d) *trans*-1,2-dimethylcyclopropane

11.10 a) CH_2 b) insoluble   c) less   d) 5

11.11 a) least   b) CO_2 c) CO

B. Answers to Self-Test Questions

1.	d	14.	a	27.	T
2.	a	15.	b	28.	T
3.	c	16.	d	29.	F
4.	a	17.	a	30.	T
5.	b	18.	c	31.	T
6.	b	19.	c	32.	F
7.	b	20.	b	33.	F
8.	a	21.	a	34.	T
9.	c	22.	b	35.	F
10.	a	23.	F	36.	T
11.	c	24.	F	37.	F
12.	d	25.	T		
13.	d	26.	F		

Chapter 12

Unsaturated Hydrocarbons

CHAPTER OUTLINE

12.1 Nomenclature of Alkenes
12.2 Geometry of Alkenes
12.3 Properties of Alkenes
12.4 Addition Polymers
12.5 Alkynes

12.6 Aromatic Compounds and the Benzene Structure
12.7 Nomenclature of Benzene Derivatives
12.8 Properties and Uses of Aromatic Compounds

LEARNING OBJECTIVES

When you have completed your study of this chapter, you should be able to:
1. Classify unsaturated hydrocarbons as alkenes, alkynes, or aromatics.
2. Write the IUPAC names of alkenes and alkynes from their molecular structures.
3. Predict the existence of geometric (*cis- trans-*) isomers from formulas of compounds.
4. Write names and structural formulas for geometric isomers.
5. Write equations for addition reactions of alkenes and use Markovnikov's rule to predict the major products of certain reactions.
6. Write equations for addition polymerization and list uses for addition polymers.
7. Name and draw structural formulas for aromatic compounds.

ANSWERS AND SOLUTIONS TO EVEN-NUMBERED PROBLEMS

Nomenclature of Alkenes (Section 12.1) and Alkynes (Section 12.5)

12.2   Give the IUPAC name for the following compounds:

a) $CH_3-CH=CH-CH_3$

b) $CH_3-CH_2-\underset{\underset{\displaystyle CH_2-CH_3}{|}}{C}=CH-CH_3$

c) $CH_3-\overset{1}{C}\equiv\overset{2}{C}-\overset{3}{\underset{\underset{CH_3}{|}}{\overset{\overset{CH_3}{|}}{C}}}-\overset{5}{CH_2}-\overset{6}{CH_3}$

d)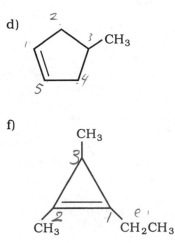

e) $\overset{7}{CH_3}-\overset{6}{\underset{\underset{CH_3}{|}}{\overset{\overset{Br}{|}}{CH}}}-\overset{5}{CH_2}-\overset{4}{C}\equiv\overset{3}{C}-\overset{2}{CH}-\overset{1}{CH_3}$

f)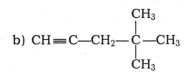

g) $CH_3-\overset{\overset{CH_3}{|}}{CH}-CH=CH-CH_2-CH=CH_2$

Solution:

a) 2-butene

b) 3-ethyl-2-pentene

c) 4,4-dimethyl-2-hexyne

d) 4-methylcyclopentene

e) 6-bromo-2-methyl-3-heptyne

f) 1-ethyl-2,3-dimethylcyclopropene

g) 6-methyl-1,4-heptadiene

12.4   Draw structural formula for the following compounds:

a) 4-methyl-2-hexene

b) 4,4-dimethyl-1-pentyne

c) 1,3-butadiene

d) 1-ethyl-3-methylcyclopentene

e) 1,6-dimethylcyclohexene

Solution:

a) $CH_3-CH=CH-\overset{\overset{}{\underset{\underset{CH_3}{|}}{CH}}}-CH_2-CH_3$

b) $CH\equiv C-CH_2-\overset{\overset{CH_3}{|}}{\underset{\underset{CH_3}{|}}{C}}-CH_3$

c) $CH_2=CH-CH=CH_2$

d)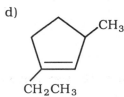

e)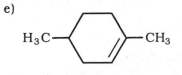

12.6   A compound has the molecular formula C_5H_8. Draw a structural formula for a compound with this formula that would be classified as (a) an alkyne, (b) a diene, and (c) a cyclic alkene. Give the IUPAC name for each compound.

Solution:

a) $H—C\equiv C—CH_2CH_2CH_3$ 1-pentyne

b) $CH_2\!=\!CHCH\!=\!CHCH_3$ 1,3-pentadiene

c)

cyclopentene

12.8 α-farnesene is a constituent of the natural wax found on apples. Given that a 12 carbon chain is named as a dodecane, what is the IUPAC name of α-farnesene?

$$CH_3—\underset{\underset{\displaystyle CH_3}{|}}{C}\!=\!CH—CH_2—CH_2—\underset{\underset{\displaystyle CH_3}{|}}{C}\!=\!CH—CH_2—CH\!=\!\underset{\underset{\displaystyle CH_3}{|}}{C}—CH\!=\!CH_2$$

Solution:
3,7,11-trimethyl-1,3,6,10-dodecatetraene

12.10   What is wrong with each of the following names? Give the structure and correct name for each compound.
a) 2-methyl-4-hexene   b) 3,5-heptadiene   c) 4-methylcyclobutene

Solution:

a) The name given assigns the wrong carbon as the number 1 carbon. The carbon with the double bond is closest to the terminal carbon. The methyl group is not near the number 1 carbon, but at the toward the other end of the molecule.

$$CH_3—\underset{\underset{\displaystyle CH_3}{|}}{C}HCH_2CH\!=\!CHCH_3 \quad \text{5-methyl-2-hexene}$$

b) The name is correct.

$$CH_3CH_2CH\!=\!CHCH\!=\!CHCH_3$$

c) The problem with this name is that the methyl group is on a carbon associated with the double bond. That carbon becomes the number 1 carbon, not the other member of the double bond.

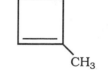

1-methylcyclobutene

Geometry of Alkenes (Section 12.2)

12.12   What type of orbitals overlap to form a pi bond in an alkene? What symbol is used to represent a pi bond? How many electrons are shared in a pi bond?

Solution:
The unhybridized p orbital on each carbon overlaps to form a π (greek letter, pi) bond containing two shared electrons.

12.14 Explain the difference between geometric and structural isomers of alkenes.

Solution:
Structural isomers differ in the order of linkage of the atoms. Geometrical isomers have the same order but have a different spatial arrangement of the atoms in space.

12.16 Which of the following alkenes can exist as *cis-trans* isomers? Draw structural formulas and name the *cis* and *trans* isomers.

a) $CH_3—CH_2—CH_2—CH=CH—CH_2—CH_2—CH_3$

b) 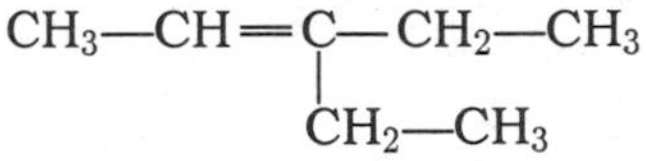 $CH_3—CH=C—CH_2—CH_3$
$\qquad\qquad\quad |$
$\qquad\qquad CH_2—CH_3$

c) $Br—CH_2—CH=CH—CH_2—Br$

Solution:

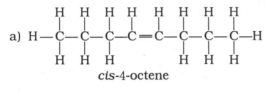

a)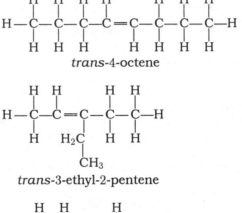

cis-4-octene   *trans*-4-octene

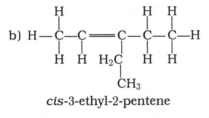

b)

cis-3-ethyl-2-pentene   *trans*-3-ethyl-2-pentene

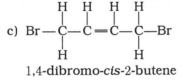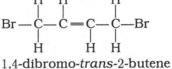

c)

1,4-dibromo-*cis*-2-butene   1,4-dibromo-*trans*-2-butene

12.18 Draw structural formulas for the following:
a) *trans*-2-pentene   b) *cis*-2-hexene

Solution:

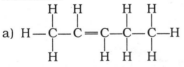

a)   b)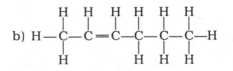

Properties of Alkenes (Section 12.3)

12.20 In what ways are the physical properties of alkenes similar to those of alkanes?

Solution:
Both alkanes and alkenes are nonpolar compounds. They are not soluble in water, but are soluble in nonpolar solvents. They have a density less than water.

12.22 State Markovnikov's rule and write a reaction that illustrates its application.

Solution:
When H–X adds to an alkene, the principle product has the H of the H–X adding to the carbon already having a greater number of H atoms.

$$HCl + H_2C{=}CH{-}CH_3 \longrightarrow CH_3{-}\underset{\underset{Cl}{|}}{C}H{-}CH_3$$

12.24 Complete the following reactions. Where more than one product is possible, show only the one expected according to Markovnikov's rule.

a) $CH_2{=}CH{-}CH_2{-}CH_3 + H_2 \xrightarrow{\text{Pt}}$

b) $CH_2{=}CH{-}CH_2{-}CH_3 + Br_2 \longrightarrow$

c)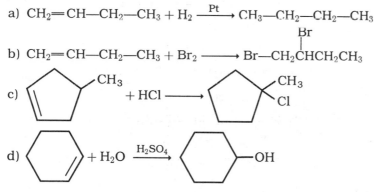

    $+ HCl \longrightarrow$

d)

    $+ H_2O \xrightarrow{\text{H}_2\text{SO}_4}$

Solution:

a) $CH_2{=}CH{-}CH_2{-}CH_3 + H_2 \xrightarrow{\text{Pt}} CH_3{-}CH_2{-}CH_2{-}CH_3$

b) $CH_2{=}CH{-}CH_2{-}CH_3 + Br_2 \longrightarrow Br{-}CH_2\underset{\underset{Br}{|}}{C}HCH_2CH_3$

c) [cyclopentene with CH₃] $+ HCl \longrightarrow$ [cyclopentane with CH₃ and Cl]

d) [cyclohexene] $+ H_2O \xrightarrow{\text{H}_2\text{SO}_4}$ [cyclohexane—OH]

12.26 What reagents would you use to prepare each of the following from cyclohexene?

a) [cyclohexane with two Cl on adjacent carbons]

b) [cyclohexane]

c) [cyclohexane with Cl]

d) [cyclohexane with OH]

Solution:

a)

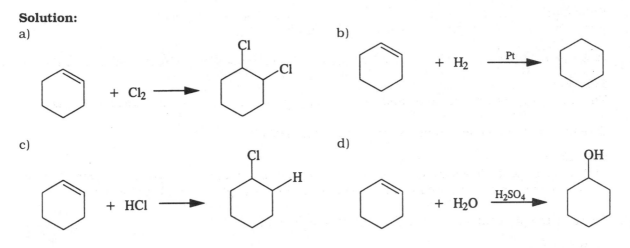

b)

12.28   Cyclohexane and 2-hexene both have the molecular formula C_6H_{12}. Describe a simple chemical test that would distinguish one from the other.

Solution:
Add Br_2 to both compounds. The cyclohexane would not have any reaction with Br_2 and would become orange, but the 2-hexene would react. The reaction can be detected by the loss of the colored Br_2.

Addition Polymers (Section 12.4)

12.30   Explain what is meant by each of the following terms: *monomer, polymer, addition polymer,* and *copolymer.*

Solution:
monomer: The starting material in polymerization, where many units of this monomer are combined to make large molecules made up of these repeating monomer units.
polymer: A large molecule made up of many repeating monomer units. addition
polymer: Polymers made by numerous addition reactions involving alkene monomers.
copolymer: An addition polymer where two different alkene monomers are used.

12.32   Identify a structural feature characteristic of all monomers listed in Table 12.3.

Solution:
They all have an alkene group or a double bond between carbon atoms.

12.34   Much of today's plumbing in newly built homes is made from a plastic called poly (vinyl chloride) or PVC. Using Table 12.3, write and equation for the formation of poly (vinyl chloride).

Solution:

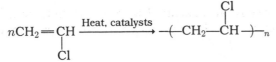

Alkynes (Section 12.5)

12.36   What type of hybridized orbitals are present on carbon atoms bonded by a triple bond? How many of these hybrid orbitals are on each carbon atom?

Solution:
Each carbon has two hybrid *sp* orbitals and 2 unhybridized *p* orbitals.

12.38 Describe the geometry in an alkyne of the carbon-carbon triple bond and the two attached atoms.

Solution:
The triple bonded carbon atoms and the two other atoms attached to them are all co-linear.

12.40 Give the common name and major uses of the simplest alkyne.

Solution:
The simplest possible alkyne has two carbons. Its common name is acetylene. Acetylene is used as a fuel especially for welding, and as the raw material for many plastics and synthetic fibers.

12.42 Describe the physical and chemical properties of alkynes.

Solution:
The physical and chemical properties of alkynes are very similar to the alkenes. Specifically, both are nonpolar, insoluble in water, and soluble in nonpolar solvents. They burn and can undergo addition reactions, but the addition reaction can occur twice in alkynes.

Aromatic Compounds and the Benzene Structure (Section 12.6)

12.44 What type of orbitals overlap to form the pi bonding in a benzene ring?

Solution:
The unhybridized *p* orbital on each carbon atom overlaps with those on the adjacent carbon atom to form the π bonds.

12.46 Define the terms *aromatic* and *aliphatic*.

Solution:
An organic compound containing a benzene ring structure is aromatic. Aliphatic compounds do not contain a benzene ring structure.

12.48 A disubstituted cycloalkane such as (a) exhibits *cis-trans* isomerism, where as a disubstituted benzene (b) does not. Explain.

a)

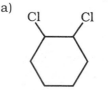

b)

Cl   Cl

Solution:
The ring structure of cycloalkanes does not allow for free rotation around carbon-carbon bonds; therefore, the position of the substituents is fixed relative to each other, and *cis-trans* isomerism can occur.

In order for *cis-trans* isomerism to occur across the double bond, there must be different groups at each carbon. In the case of (b) the attachments to the carbons are the same; therefore, no isomerism.

Nomenclature of Benzene Derivatives (Section 12.7)

12.50   Give the IUPAC name for each of the following hydrocarbons as a derivative of benzene:

a)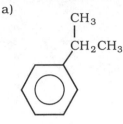

b)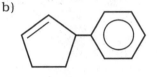

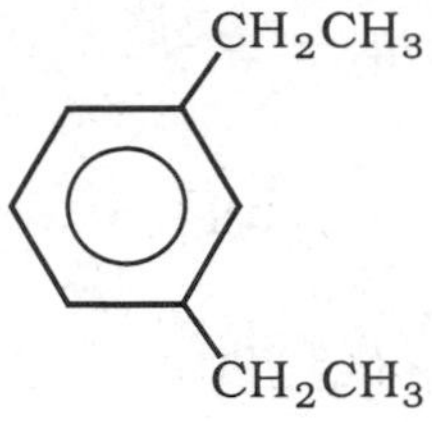

Solution:
a) isopropylbenzene

b) 1,3-diethylbenzene

12.52   Give the IUPAC name for the following as hydrocarbons with the benzene ring as a substituent:

a)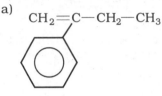

b)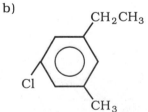

Solution:
a) 2-phenyl-1-butene

b) 3-phenylcylopentene

12.54   Name the following compounds, using prefixed abbreviations for ortho, meta, and para and assigning IUPAC-acceptable common names:

a)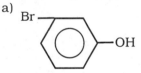

b)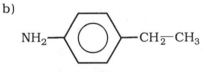

Solution:
a) *m*-bromophenol

b) *p*-ethylaniline

12.56   Name the following by numbering the benzene ring. IUPAC-acceptable common names may be used where appropriate.

a)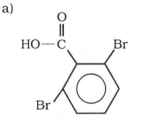

b)

Solution:
Because there are three substitutions, the ring number method is used; benzoic acid and toluene are recognized.

a) 2,5-dibromobenzoic acid

b) 3-chloro-5-ethyltoluene

12.58 Draw structural formulas for the following:

a) *p*-propylphenol   b) *o*-bromobenzoic acid   c) 3-methyl-2-phenylbutane

Solution:

a)

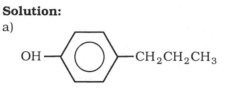

b)

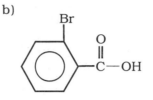

c)

$$CH_3{-}CH{-}\overset{\displaystyle CH_3}{\overset{|}{CH}}{-}CH_3$$

Properties aad Uses of Aromatic Compounds (Section 12.8)

12.60 Describe the chief physical properties of aromatic hydrocarbons.

Solution:

Aromatic hydrocarbons have similar properties to the alkanes and alkenes. Specifically, they have a density less than water. They are nonpolar compounds not soluble in water, but are soluble in nonpolar solvents.

12.62 Compare the chemical behavior of benzene and cyclohexene.

Solution:

Cyclohexene will readily undergo addition reactions; benzene resists addition reactions, favoring substitution reactions. Both will burn in air.

12.64 For each of the following uses, list an appropriate aromatic compound:
a) Used in the production of Formica.
b) A starting material for polystyrene.
c) Used to manufacture drugs.
d) A starting material for Bakelite.

Solution:

a) phenol   b) styrene   c) aniline   d) phenol

Chemistry Around Us, Key Chemicals, and How Reactions Occur

12.66 Spinach, broccoli, cantaloupe, and pumpkins are dietary sources for what important nutrient?

Solution:

Beta-carotene is found the these foods and is also found in foods that are orange.

12.68 Why do terpene molecules contain a multiple of five carbon atoms?

Solution:
The terpenes are composed of two or more isoprene units connected to produce molecules that include molecules with 40 or more carbons.

12.70   What is a carbocation? How is it formed in the hydration of an alkene?

Solution:
A carbocation is an ion of the form $^+CH_3$ when a H^+ ion bonds to a carbon of a carbon-carbon double bond.

12.72   What is the active ingredient in patches, gum, and sprays designed to help break the smoking habit?

Solution:
Nicotine is included in patches, gum, and sprays to help reduce the withdrawal symptoms that accompany dropping the smoking habit. Since the nicotine is being delivered constantly, the use of tobacco products with these nicotine-containing aides can lead to a nicotine overdose.

12.74   What is a polycyclic aromatic compound? Under what conditions do they form?

Solution:
A polycyclic aromatic compound contains "fused" benzene rings, where the rings share a common side (two carbon atoms). They are formed when aromatic compounds are heated to a high temperature with limited air present.

PROGRAMMED REVIEW

Section 12.1 Nomenclature of Alkenes
A compound containing a carbon-carbon double bond has an IUPAC name with the ending (a) _____. The compound CH_3—CH=CH—$\underset{\underset{CH_3}{|}}{CH}$—$CH_3$ has the IUPAC name (b) _____. The IUPAC name for a compound containing two carbon-carbon double bonds has the ending (c) _____. The IUPAC name for 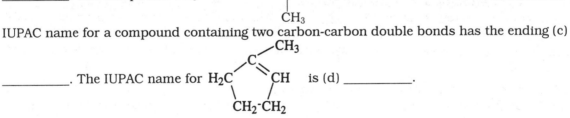 is (d) _____.

Section 12.2 Geometry of Alkenes
Three (a) _____ hybrid orbitals are present about each carbon atom in CH_2=CH_2. The geometry of a carbon-carbon double bond and the four attached atoms is such that all six atoms lie in the same (b) _____. Geometric isomers may be designated as *cis* or *trans* compounds. The IUPAC name for

$$\underset{CH_3}{\overset{H}{\diagdown}}C=C\overset{\diagup CH_3}{\underset{\diagdown H}{}}$$

is (c) _____ isomer. To complete the structure

$$\underset{CH_3}{\overset{Br}{\diagdown}}C=C\overset{\diagup}{\underset{\diagdown H}{}}$$

so that a *cis* compound is formed, (d) _____ must be attached.

Section 12.3 Properties of Alkenes

The physical properties of the alkenes are similar to those of the (a) _____. The characteristic reactions of alkenes are referred to as (b) _____ reactions. The product of the reaction $CH_3CH_2CH{=}CH_2 + Br_2 \rightarrow$ is (c) _____. Another name for an alkyl halide is (d) _____. An alkane can be produced from an alkene by a reaction called (e) _____. According to

Markovnikov's Rule, the (f) _____ carbon atom of
$$\underset{1\quad 2\quad 3\quad 4}{CH_3{-}\overset{\displaystyle \overset{CH_3}{|}}{C}{=}CH{-}CH_3}$$
would gain a hydrogen atom

in the reaction with HCl. The chemical reaction in which a molecule of water adds to a carbon-carbon double bond is called (g) _____.

Section 12.4 Addition Polymers

The chemical reaction in which hundreds of alkene molecules react to form a large molecule is called (a) _____. The starting materials for this same process are referred to as (b) _____. A fluorine-containing polymer is (c) _____. A polymer produced by using two different starting materials is referred to as a (d) _____.

Section 12.5 Alkynes

The characteristic structural feature of alkynes is the carbon-carbon (a) _____ bond. Two (b) _____ hybrid orbitals are present about each carbon atom in H—C≡C—H. IUPAC names for alkynes end with the letters (c) _____. The simplest alkyne is well-known by the common name (d) _____. Alkynes undergo the same kinds of reactions as the (e) _____.

Section 12.6 Aromatic Compounds and the Benzene Structure

Aromatic compounds are those which contain a (a) _____ ring. Non-aromatic compounds are referred to as (b) _____ compounds. The circle within the benzene structure contains (c) _____ electrons. Each carbon atom of benzene contains three (d) _____ hybrid orbitals. The six carbons and six hydrogens of benzene are arranged in a (e) _____ geometry.

Section 12.7 Nomenclature of Benzene Derivatives

Another name for methylbenzene is (a) _____. Aminobenzene is more simply named as (b) _____. In naming a benzene derivative with groups attached at positions 1 and 3, the prefix (c) _____ may be used. The aromatic group C_6H_{5-}, when named as a branch, is called a (d) _____ group.

Section 12.8 Properties and Uses of Aromatic Compounds

Aromatic compounds are nonpolar and (a) _____ in water. The typical reaction of aromatic compounds is a (b) _____ reaction. Benzene and (c) _____ are aromatic compounds which are useful as laboratory solvents. The vitamin (d) _____ is an aromatic compound. The plastic bakelite is prepared commercially from the aromatic compound (e) _____.

SELF-TEST QUESTIONS

Multiple Choice

1. What is the correct IUPAC name for $CH_3{-}CH{=}CH{-}CH_2{-}CH_2{-}\underset{\underset{\displaystyle Cl}{|}}{CH_2}$?

   a) 1-chloro-2-hexene   b) 1-chloro-4-hexene
   c) 6-chloro-2-hexene   d) 1-chlorohexene

2. What is the correct IUPAC name for 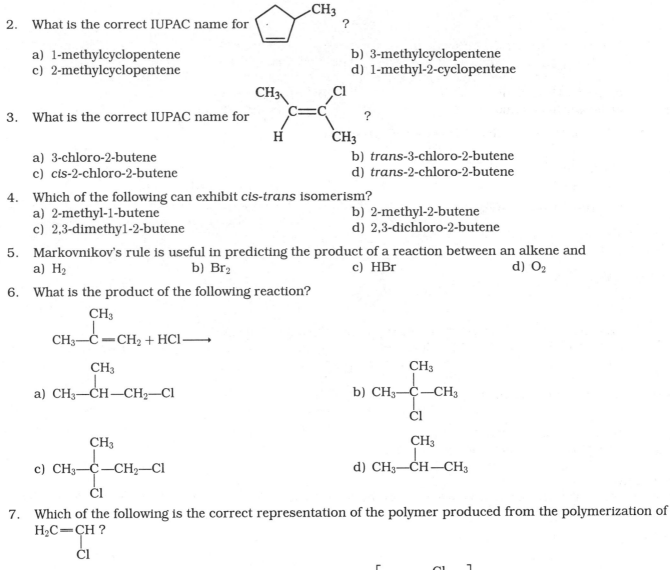 ?

   a) 1-methylcyclopentene
   c) 2-methylcyclopentene

   b) 3-methylcyclopentene
   d) 1-methyl-2-cyclopentene

3. What is the correct IUPAC name for

   $$CH_3\!\!\diagdown C\!\!=\!\!C\diagup Cl$$
   $$H\diagup \qquad \diagdown CH_3$$

   ?

   a) 3-chloro-2-butene
   c) *cis*-2-chloro-2-butene

   b) *trans*-3-chloro-2-butene
   d) *trans*-2-chloro-2-butene

4. Which of the following can exhibit *cis-trans* isomerism?
   a) 2-methyl-1-butene
   c) 2,3-dimethyl-2-butene

   b) 2-methyl-2-butene
   d) 2,3-dichloro-2-butene

5. Markovnikov's rule is useful in predicting the product of a reaction between an alkene and
   a) H_2 b) Br_2 c) HBr   d) O_2

6. What is the product of the following reaction?

   $$CH_3\!-\!\overset{\overset{\displaystyle CH_3}{|}}{C}\!=\!CH_2 + HCl \longrightarrow$$

   a) $CH_3\!-\!\overset{\overset{\displaystyle CH_3}{|}}{CH}\!-\!CH_2\!-\!Cl$

   b) $CH_3\!-\!\overset{\overset{\displaystyle CH_3}{|}}{\underset{\underset{\displaystyle Cl}{|}}{C}}\!-\!CH_3$

   c) $CH_3\!-\!\overset{\overset{\displaystyle CH_3}{|}}{\underset{\underset{\displaystyle Cl}{|}}{C}}\!-\!CH_2\!-\!Cl$

   d) $CH_3\!-\!\overset{\overset{\displaystyle CH_3}{|}}{CH}\!-\!CH_3$

7. Which of the following is the correct representation of the polymer produced from the polymerization of
   $H_2C\!=\!\underset{\underset{\displaystyle Cl}{|}}{CH}$?

   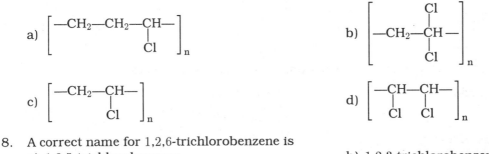

   a) $\left[\!-CH_2\!-\!CH_2\!-\!\overset{}{\underset{\underset{\displaystyle Cl}{|}}{CH}}\!-\!\right]_n$

   b) $\left[\!-CH_2\!-\!\overset{\overset{\displaystyle Cl}{|}}{\underset{\underset{\displaystyle Cl}{|}}{CH}}\!-\!\right]_n$

   c) $\left[\!-CH_2\!-\!\overset{}{\underset{\underset{\displaystyle Cl}{|}}{CH}}\!-\!\right]_n$

   d) $\left[\!-\overset{}{\underset{\underset{\displaystyle Cl}{|}}{CH}}\!-\!\overset{}{\underset{\underset{\displaystyle Cl}{|}}{CH}}\!-\!\right]_n$

8. A correct name for 1,2,6-trichlorobenzene is
   a) 1,2,5-trichlorobenzene
   c) 1,3,5-trichlorobenzene

   b) 1,2,3-trichlorobenzene
   d) 1,2,4-trichlorobenzene

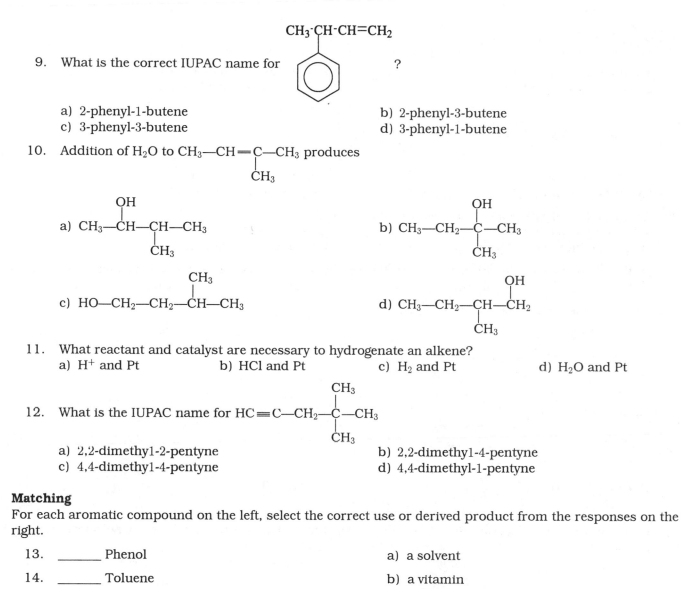

9. What is the correct IUPAC name for [structure: CH$_3$-CH-CH=CH$_2$ with benzene ring] ?

    a) 2-phenyl-1-butene
    c) 3-phenyl-3-butene

    b) 2-phenyl-3-butene
    d) 3-phenyl-1-butene

10. Addition of H$_2$O to CH$_3$—CH=C—CH$_3$ produces
    (with CH$_3$ substituent below)

    a) CH$_3$—CH—CH—CH$_3$
       (with OH above first CH, CH$_3$ below)

    b) CH$_3$—CH$_2$—C—CH$_3$
       (with OH above C, CH$_3$ below)

    c) HO—CH$_2$—CH$_2$—CH—CH$_3$
       (with CH$_3$ above)

    d) CH$_3$—CH$_2$—CH—CH$_2$
       (with OH above last CH$_2$, CH$_3$ below)

11. What reactant and catalyst are necessary to hydrogenate an alkene?
    a) H$^+$ and Pt   b) HCl and Pt   c) H$_2$ and Pt   d) H$_2$O and Pt

12. What is the IUPAC name for HC≡C—CH$_2$—C—CH$_3$
    (with CH$_3$ above and CH$_3$ below)

    a) 2,2-dimethyl-2-pentyne
    c) 4,4-dimethyl-4-pentyne

    b) 2,2-dimethyl-4-pentyne
    d) 4,4-dimethyl-1-pentyne

Matching
For each aromatic compound on the left, select the correct use or derived product from the responses on the right.

13. _____ Phenol   a) a solvent

14. _____ Toluene   b) a vitamin

15. _____ Aniline   c) formica

16. _____ Riboflavin   d) dyes

For each description on the left, select a correct polymer from the responses on the right.

17. _____ Used for insulation   a) saran wrap

18. _____ A copolymer   b) plexiglass

19. _____ Used in airplane windows c) polystyrene

                                      d) PVC

True-False

20. Carbon-carbon double bonds do not rotate as freely as carbon-carbon single bonds.

21. The compound $BrCH\!=\!CHCH_3$ can exhibit *cis-trans* isomerism.

22. Alkenes are polar substances.

23. All aromatic compounds are cyclic.

24. Two hydrogens may be bonded to the same benzene carbon atom.

25. Benzene is a completely planar molecule.

26. An *sp* hybrid orbital is obtained by hybridizing an *s* orbital and two *p* orbitals.

27. A copolymer is formed from two different monomers.

28. All the atoms in ethene lie in the same plane.

29. Acetylene is the common name of the simplest alkene.

SOLUTIONS

A. Answers to Programmed Review

12.1 a) -ene	b) 4-methyl-2-pentene	c) diene	d) 1-methylcyclopentene
12.2 a) sp^2	b) plane	c) *trans*-2-butene	d) Br

12.3 a) alkanes   b) addition   c) $CH_3\!-\!CH_2\!-\!\overset{\overset{\displaystyle Br}{\displaystyle |}}{CH}\!-\!CH_2\!-\!Br$
      d) haloalkane   e) hydrogenation   f) third   g) addition, or hydration

12.4 a) polymerization	b) monomers	c) teflon	d) copolymer
12.5 a) triple e) alkenes	b) *sp*	c) -yne	d) acetylene
12.6 a) benzene e) planar	b) aliphatic	c) six	d) sp^2
12.7 a) toluene	b) aniline	c) meta	d) phenyl
12.8 a) insoluble e) phenol	b) substitution	c) toluene	d) riboflavin

B. Answers to Self-Test Questions

1. c	11. c	21. T
2. b	12. d	22. F
3. d	13. c	23. T
4. d	14. a	24. F
5. c	15. d	25. T
6. b	16. b	26. T
7. c	17. c	27. T
8. b	18. a	28. T
9. d	19. b	29. F
10. b	20. T	

Chapter 13

Alcohols, Phenols, and Ethers

CHAPTER OUTLINE

LEARNING OBJECTIVES

When you have completed your study of this chapter, you should be able to:

1. Name and draw structural formulas for alcohols, phenols, and ethers.
2. Classify alcohols as primary, secondary, or tertiary on the basis of their structural formulas.
3. Discuss how hydrogen bonding influences the physical properties of alcohols and ethers.
4. Write equations for alcohol dehydration and oxidation reactions.
5. Recognize uses for specific alcohols, phenols, and ethers.
6. Write equations for a thiol reaction with heavy-metal ions and production of disulfides.
7. Identify functional groups in polyfunctional compounds.

ANSWERS AND SOLUTIONS TO EVEN-NUMBERED PROBLEMS

Nomenclature of Alcohols and Phenols (Section 13.1)

13.2   Assign IUPAC names to the following alcohols:

a) CH_3—CH_2—CH_2—OH

b)

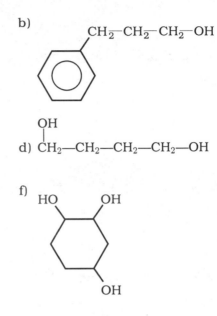

c)

$$CH_3-\overset{\overset{\displaystyle CH_3}{|}}{CH}-\underset{\underset{\displaystyle Cl}{|}}{CH}-CH_2-OH$$

d) $\overset{\overset{\displaystyle OH}{|}}{CH_2}$—$CH_2$—$CH_2$—$CH_2$—OH

e)

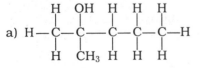

f)

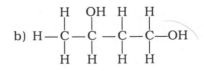

Solution:

a) 1-propanol

b) 3-phenol-1-propanol

c) 2-chloro-3-methyl-1-butanol

d) 1,4-butanediol

e) 2-chloro-5-ethylcyclopentanol

f) 1,2,4-cyclohexanetriol

13.4   Give each of the structures in Exercise 13.3 an IUPAC name. (The structures are shown below.)

a) CH_3—OH

b) $CH_3-\overset{\overset{\displaystyle OH}{|}}{CH}-CH_3$

c) CH_3—CH_2—OH

d) $\overset{\overset{\displaystyle OH}{|}}{H_2C}$ —— $\overset{\overset{\displaystyle OH}{|}}{CH_2}$

e) $\overset{\overset{\displaystyle OH}{|}}{H_2C}$ —— $\overset{\overset{\displaystyle OH}{|}}{CH}$ —— $\overset{\overset{\displaystyle OH}{|}}{CH_2}$

Solution:

a) methanol

b) 2-propanol

c) ethanol

d) 1,2-ethanediol

e) 1,2,3-propanetriol

13.6   Draw structural formulas for each of the following:

a) 2-methyl-2-pentanol

b) 1,3-butanediol

c) 1-ethylcyclopentanol

Solution:

a) $$H-\overset{\overset{\displaystyle H}{|}}{\underset{\underset{\displaystyle H}{|}}{C}}-\overset{\overset{\displaystyle OH}{|}}{\underset{\underset{\displaystyle CH_3}{|}}{C}}-\overset{\overset{\displaystyle H}{|}}{\underset{\underset{\displaystyle H}{|}}{C}}-\overset{\overset{\displaystyle H}{|}}{\underset{\underset{\displaystyle H}{|}}{C}}-\overset{\overset{\displaystyle H}{|}}{\underset{\underset{\displaystyle H}{|}}{C}}-H$$

b) $$H-\overset{\overset{\displaystyle H}{|}}{\underset{\underset{\displaystyle H}{|}}{C}}-\overset{\overset{\displaystyle OH}{|}}{\underset{\underset{\displaystyle H}{|}}{C}}-\overset{\overset{\displaystyle H}{|}}{\underset{\underset{\displaystyle H}{|}}{C}}-\overset{\overset{\displaystyle H}{|}}{\underset{\underset{\displaystyle H}{|}}{C}}-OH$$

c)

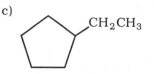

13.8 Name each of the following as a derivative of phenol:

a)

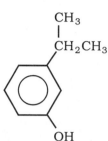

b)

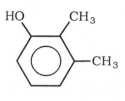

Solution:

a) 2-isopropylphenol

b) 2,3-dimethylphenol

13.10 Draw structural formulas for each of the following:

a) *m*-methylphenol

b) 2,3-dicholorophenol

Solution:

a)

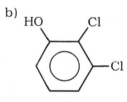

b)

HO   Cl

Cl

Classification of Alcohols (Section 13.2)

13.12 Classify the following alcohols as primary, secondary, or tertiary:

a)
$$CH_3—CH_2—\overset{\overset{\displaystyle CH_3}{|}}{\underset{\underset{\displaystyle CH_3}{|}}{C}}—OH$$

b) $CH_3—CH_2—CH_2—CH_2—OH$

c)
$$CH_3—CH_2—\underset{\underset{\displaystyle CH_3}{|}}{CH}—OH$$

Solution:

a) tertiary   b) primary   c) secondary

13.14 Draw structural formulas for the four aliphatic alcohols with the molecular formula $C_4H_{10}O$. Name each compound using the IUPAC system and classify it as a *primary, secondary,* or *tertiary* alcohol.

Solution:

(1) $CH_3—CH_2—CH_2—CH_2—OH$ 1-butanol   primary

OH
|
(2) CH₃—CH₂—CH—CH₃   2-butanol   secondary

CH₃
|
(3) CH₃—CH—CH₂—OH   2-methyl-1-propanol   primary

OH
|
(4) CH₃—C—CH₃   2-methyl-2-propanol   tertiary
|
CH₃

Physical Properties of Alcohols (Section 13.3)

13.16 Arrange the compounds of each group in order of increasing boiling point.

    a) ethanol, 1-propanol, methanol   b) butane, ethylene glycol, 1-propanol

Solution:

    a) methanol, ethanol, 1-propanol. All have hydrogen bonding, so the largest molecule has the highest boiling point.

    b) butane, 1-propanol, ethylene glycol. Butane has no hydrogen bonding, and ethylene glycol has two groups capable of hydrogen bonding. Butane has the lowest boiling point, and ethylene glycol has the highest.

13.18 Draw structural formulas for the following molecules and use a dotted line to show the formation of hydrogen bonds:

    a) one molecule of 1-butanol and one molecule of ethanol

    b) cyclohexanol and water

Solution:

a)

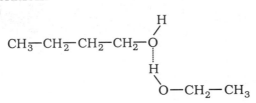

b)

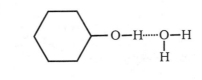

Reactions of Alcohols (Section 13.4)

13.20 Draw the structures of the chief product formed when the following alcohols are dehydrated to alkenes:

a)

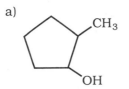

OH
|
b) CH₃—CH—CH—CH₃
              |
              CH₂—CH₃

Solution:

a)

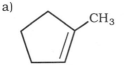

b) CH₃—CH=CH—CH₃
                 |
                CH₂—CH₃

13.22 Draw the structures of the ethers that can be produced from the following alcohols:

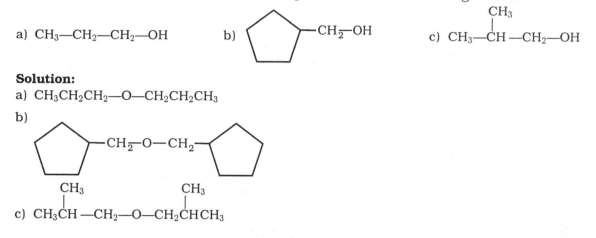

a) CH₃—CH₂—CH₂—OH

b)

c) CH₃—CH—CH₂—OH

Solution:
a) CH₃CH₂CH₂—O—CH₂CH₂CH₃

b)

c)

13.24 Give the structure of an alcohol that could be used to prepare each of the following compounds

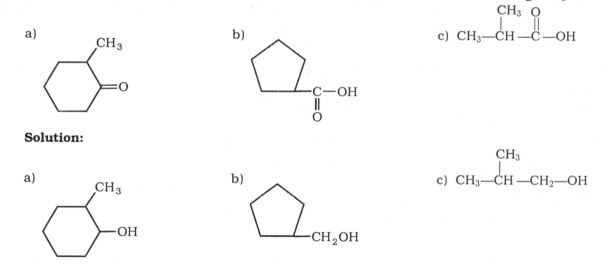

a)

b)

c) CH₃—CH—C—OH

Solution:

a)

b)

c) CH₃—CH—CH₂—OH

13.26 What products would result from the following processes? Write an equation for each reaction.
a) 2-methyl-2-butanol is subjected to controlled oxidation.
b) 1-propanol is heated to 140°C in the presence of sulfuric acid.
c) 3-pentanol is subjected to controlled oxidation.
d) 3-pentanol is heated to 180°C in the presence of sulfuric acid.
e) 1-hexanol is subjected to an excess of oxidizing agent.

Solution:

a)
$$\text{CH}_3\text{—CH}_2\text{—}\underset{\underset{\text{CH}_3}{|}}{\overset{\overset{\text{OH}}{|}}{\text{C}}}\text{—CH}_3 + (\text{O}) \longrightarrow \text{no reaction}$$

b) 2CH₃—CH₂—CH₂—OH $\xrightarrow[140°C]{\text{H}_2\text{SO}_4}$ CH₃—CH₂—CH₂—O—CH₂—CH₂—CH₃ + H₂O

c)
$$\underset{\underset{\displaystyle OH}{|}}{CH_3-CH_2-CH-CH_2-CH_3} + (O) \longrightarrow CH_3-CH_2-\overset{\displaystyle O}{\overset{\|}{C}}-CH_2-CH_3 + H_2O$$

d)
$$\underset{\underset{\displaystyle OH}{|}}{CH_3-CH_2-CH-CH_2-CH_3} \xrightarrow[180°C]{H_2SO_4} CH_3-CH=CH-CH_2-CH_3 + H_2O$$

e)
$$CH_3-CH_2-CH_2-CH_2-CH_2-CH_2-OH + 2(O) \longrightarrow CH_3-CH_2-CH_2-CH_2-CH_2-\overset{\displaystyle O}{\overset{\|}{C}}-OH + H_2O$$

13.28 Each of the following conversions requires more than one step, and some reactions studied in previous chapters may be needed. Show the reagents you would use and draw structural formulas for intermediate compounds formed in each conversion.

a)
$$CH_3-CH_2-CH=CH_2 \longrightarrow CH_3-CH_2-\overset{\displaystyle O}{\overset{\|}{C}}-CH_3$$

b)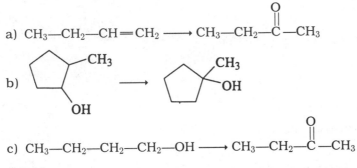

c)
$$CH_3-CH_2-CH_2-CH_2-OH \longrightarrow CH_3-CH_2-\overset{\displaystyle O}{\overset{\|}{C}}-CH_3$$

Solution:
a) (1) Addition of H_2O to alkene using H_2SO_4 as a catalyst →
$$CH_3-CH_2-CH=CH_2 + H_2O \xrightarrow{H_2SO_4} \underset{\underset{\displaystyle OH}{|}}{CH_3-CH_2-CH-CH_3}$$
(2) Oxidation of the secondary alcohol to give ketone
$$\underset{\underset{\displaystyle OH}{|}}{CH_3-CH_2-CH-CH_3} + (O) \longrightarrow CH_3-CH_2-\overset{\displaystyle O}{\overset{\|}{C}}-CH_3 + H_2O$$

b) (1) Elimination of H_2O using H_2SO_4 at temperature 180°C →

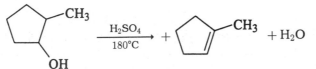

(2) Addition of water using H_2SO_4 as a catalyst →

c) (1) Elimination of H_2O using H_2SO_4 at temperature 180°C →
$$CH_3-CH_2-CH_2-CH_2-OH \xrightarrow[180°C]{H_2SO_4} CH_3-CH_2-CH=CH_2 + H_2O$$
(2) Addition of water using H_2SO_4 as a catalyst →
$$CH_3-CH_2-CH=CH_2 + H_2O \xrightarrow{H_2SO_4} \underset{\underset{\displaystyle OH}{|}}{CH_3-CH_2-CH-CH_3}$$

(3) Oxidation of the secondary alcohol to give ketone

$$CH_3-CH_2-\overset{\overset{\displaystyle OH}{|}}{C}H-CH_3 + (O) \longrightarrow CH_3-CH_2-\overset{\overset{\displaystyle O}{\|}}{C}-CH_3 + H_2O$$

13.30 The three-carbon diol used in antifreeze is

$$CH_3-\overset{\overset{\displaystyle OH}{|}}{C}H-\overset{\overset{\displaystyle OH}{|}}{C}H_2 \cdot$$

It is nontoxic and is used as a moisturizing agent in foods. Oxidation of this substance within the liver produces pyruvic acid, which can be used by tbe body to supply energy. Give the structure of pyruvic acid.

Solution:
Oxidation of the two alcohol groups gives an acid and ketone

$$H_3C-\overset{\overset{\displaystyle O}{\|}}{C}-\overset{\overset{\displaystyle O}{\|}}{C}-OH$$

Important Alcohols (Section 13.5)

13.32 Why is methanol such an important industrial chemical?

Solution:
Methanol is used as a fuel and in the preparation of formaldehyde, a starting material for certain plastics.

13.34 Name an alcohol used in each of the following ways:
a) A moistening agent in many cosmetics
b) The solvent in solutions called tinctures
c) Automobile antifreeze
d) Rubbing alcohol
e) A flavoring in cough drops
f) Present in gasohol

Solution:
a) glycerol
c) ethylene glycol or 1,2-ethanediol
e) menthol

b) ethyl alcohol or ethanol
d) isopropyl alcohol or 2-propanol
f) ethanol

Characteristics and Uses of Phenols (Section 13.6)

13.36 Name a phenol used in each of the following ways:
a) A disinfectant used for cleaning walls
b) An antiseptic found in some mouthwashes
c) An antioxidant used to prevent rancidity in foods

Solution:
a) *o*-phenylphenol or 2-benzyl-4-chlorophenol
b) 4-hexylresorcinol or phenol
c) BHA or BHT

Ethers (Section 13.7)

13.38 Assign a common name to each of the following ethers:

a) $CH_3\!-\!O\!-\!CH_2\!-\!CH_3$

b)

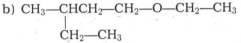

c) $CH_3\!-\!CH_2\!-\!\overset{\overset{\displaystyle CH_3}{\displaystyle |}}{CH}\!-\!O\!-\!\overset{\overset{\displaystyle CH_3}{\displaystyle |}}{CH}\!-\!CH_2\!-\!CH_3$

Solution:

a) ethyl methyl ether   b) phenyl propyl ether   c) di-sec-butyl ether

13.40 Assign the IUPAC name to each of the following ethers. Name the smaller alkyl group as the alkoxy group.

a) $CH_3\!-\!O\!-\!CH_2\!-\!CH_2\!-\!CH_3$

b) $CH_3\!-\!\underset{\underset{\displaystyle CH_2\!-\!CH_3}{\displaystyle |}}{CH}CH_2\!-\!CH_2\!-\!O\!-\!CH_2\!-\!CH_3$

c)

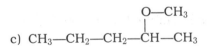

d)

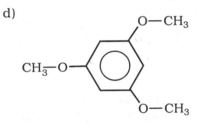

Solution:

a) methoxypropane   b) 1-ethoxy-3-methylpentane
c) 1-cyclopropoxypropane or propoxycyclopropane d) 1,3,5-trimethoxybenzene

13.42 Draw structural formulas for the following:

a) methyl isopropyl ether   b) phenyl propyl ether
c) 2-methoxypentane   d) 1,2-diethoxycyclopentane
e) 2-phenoxy-2-butene

Solution:

a) $CH_3\!-\!O\!-\!\overset{\overset{\displaystyle CH_3}{\displaystyle |}}{CH}\!-\!CH_3$

b)

c) $CH_3\!-\!CH_2\!-\!CH_2\!-\!\overset{\overset{\displaystyle O\!-\!CH_3}{\displaystyle |}}{CH}\!-\!CH_3$

d)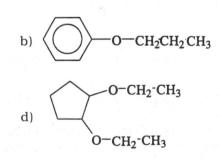

e)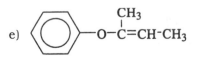

Properties of Ethers (Section 13.8)

13.44 What is the chief chemical property of ethers?

Solution:

Ethers are very inert toward chemical change, except for their combustibility in air.

13.46 Arrange the following compounds in order of decreasing solubility in water. Explain the basis for your decisions.

$CH_3—CH_2—CH_2—OH$ $\quad$ $CH_3—CH_2—CH_2—CH_3$ $\quad$ $CH_3—CH_2—O—CH_3$

Solution:

Based on the decreasing ability to form hydrogen bonds with water, the alcohol is more soluble than the ether, which is more soluble than the alkane.

$CH_3—CH_2—CH_2—OH$ $\quad$ $CH_3—CH_2—O—CH_3$ $\quad$ $CH_3—CH_2—CH_2—CH_3$

13.48 Draw structural formulas and use a dotted line to show hydrogen bonding between a molecule of diethyl ether and water.

Solution:

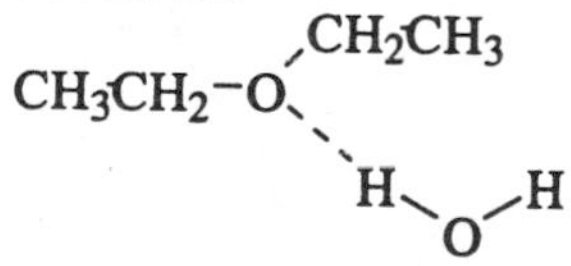

Thiols (Section 13.9)

13.50 Complete the following equations:

a) $2\,CH_3—CH_2—CH_2—SH + Hg^{2+} \longrightarrow$

b) $2\,CH_3—CH_2—CH_2—SH + (O) \longrightarrow$

c) 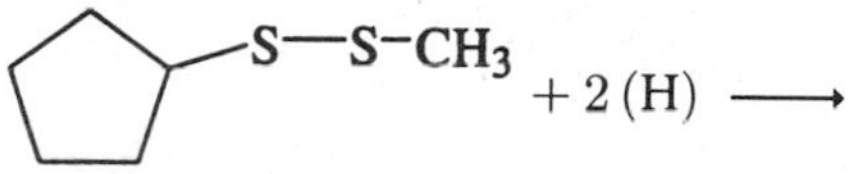

Solution:

a) $CH_3—CH_2—CH_2—SH + Hg^{2+} \longrightarrow CH_3—CH_2—CH_2—S—Hg—S—CH_2—CH_2—CH_3 + 2H^+$

b) $2\,CH_3—CH_2—CH_2—SH + (O) \longrightarrow CH_3—CH_2—CH_2—S—S—CH_2—CH_2—CH_3 + H_2O$

c) 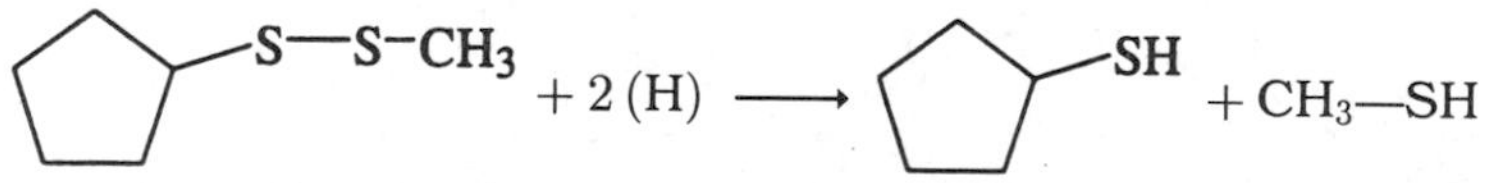

13.52 Alcohols and thiols can both be oxidized in a controlled way. What are the differences in the products?

Solution:

Oxidation of alcohols give aldehydes (or acids), or ketones. Oxidation of thiols give disulfides.

Chemistry Around Us, Key Chemicals, and How Reactions Occur

13.54 Write a mechanism for the dehydration reaction:

$$CH_3—\underset{}{\overset{\displaystyle OH}{\overset{|}{C}H}}—CH_3 \xrightarrow[180°C]{H_2SO_4} CH_3—CH{=}CH_2 + H_2O$$

Solution:

(1) A proton from H_2SO_4 attacks the –OH group.

$$CH_3-\underset{\underset{OH}{|}}{CH}-CH_3 + H_2SO_4 \rightleftharpoons \left[CH_3-\underset{\underset{CH_3}{|}}{CH}-\overset{H}{\underset{H}{O}}+ \right] + HSO_4^-$$

(2) The positive change moves to the adjacent carbon atom and H_2O is eliminated.

$$\left[CH_3-\underset{\underset{CH_3}{|}}{CH}-\overset{H}{\underset{H}{O}}+ \right] \rightleftharpoons CH_3-\underset{\underset{CH_3}{|}}{\overset{\overset{H}{|}}{C}} + \ + H_2O$$

(3) The H^+ leaves and the electron pair forms a double bond.

$$CH_3-\underset{\underset{CH_3}{|}}{\overset{\overset{H}{|}}{C}} + \ + HSO_4^- \rightleftharpoons CH_3-CH=CH_2 + H_2SO_4$$

13.56 What oil is the culprit in the poison ivy plant?

Solution:
The sap of the plant contains urushiol, which causes a rash, blisters, and itching.

13.58 What is believed to be the cellular role of vitamin E?

Solution:
It is an antioxidant. Vitamin E can be oxidized preferentially rather than many other cell components, such as lipids and cell membranes.

PROGRAMMED REVIEW

Section 13.1 Nomenclature of Alcohols and Phenols
IUPAC names for alcohols end with the letters (a) _____ . If two –OH groups are present in a molecule, the IUPAC name ends with the letters (b) _____. The IUPAC name for $CH_3CH_2CH_2$—OH is (c) _____. The structure of 2-chlorophenol is (d)_____.

Section 13.2 Classification of Alcohols
The hydroxyl-bearing carbon in a secondary alcohol is attached to (a) _____ carbon atoms. 1-butanol is classified as a (b) _____ alcohol. Cyclopentanol is classified as a (c) _____ alcohol. If a hydroxyl-bearing carbon is attached to three carbon atoms, the compound is classified as a (d) _____ alcohol.

Section 13.3 Physical Properties of Alcohols
Alcohol molecules and water interact to form (a) _____ bonds. Alcohols with (b) _____ or fewer carbon atoms are completely soluble in water. Alcohols have higher boiling points than alkanes of similar molecular weight because of the formation of (c) _____ bonds.

Section 13.4 Reactions of Alcohols

A reaction in which water is chemically removed from an alcohol is termed (a) _____. Reaction of an alcohol with sulfuric acid at 180°C produces an (b) _____. Oxidation of a secondary alcohol produces a (c) _____ Primary alcohols undergo oxidation to (d) _____ which can be further oxidized to (e) _____.

Section 13.5 Important Alcohols

A major industrial source of formaldehyde is the alcohol (a) _____. Solutions which are referred to as tinctures contain the alcohol (b) _____ as the solvent. The products of fermentation of sugars are ethanol and (c) _____. The IUPAC name for rubbing alcohol is (d) _____.

Section 13.6 Characteristics and Uses of Phenols

A phenol used in mouthwashes and throat lozenges is (a) _____. A substance which prevents another substance from being oxidized is called an (b) _____. A common additive to vegetable oils to prevent rancidity is (c) _____. The active ingredient in Lysol is (d) _____.

Section 13.7 Ethers

The common name for $CH_3-O-CH_2CH_2CH_2$ is (a)_____. The IUPAC name for $CH_3-O-CH_2CH_2CH_3$ is (b) _____. A ring containing an atom other than carbon is referred to as (c) _____.

Section 13.8 Properties of Ethers

Ethers are (a) _____ soluble in water than hydrocarbons of comparable molecular weight. In terms of reactivity, ethers are quite (b) _____. The boiling points of ethers are similar to those of the (c) _____.

Section 13.9 Thiols

The functional group of a thiol is (a) _____. Compounds containing the group —S—S— are referred to as (b) _____. An example of an ion which can react with thiols is (c) _____. Disulfides can be converted to thiols by the use of (d) _____.

SELF-TEST QUESTIONS

Multiple Choice

1. The structure of 1-propoxypropane is

    a) $CH_3-CH_2-O-CH_2-CH_2-CH_3$

    b) $CH_3-CH_2-CH_2-O-CH_2- CH_2-CH_3$

    c) $CH_3-CH_2-O-\overset{\overset{\displaystyle CH_3}{|}}{C}H-CH_3$

    d) $CH_3-CH_2-CH_2-O-\overset{\overset{\displaystyle CH_3}{|}}{\underset{\underset{\displaystyle CH_3}{|}}{C}}H$

2. What is the correct IUPAC name for $CH_3-CH_2-CH_2-\overset{\overset{\displaystyle CH_3}{|}}{\underset{\underset{\underset{\displaystyle CH_3}{|}}{CH_2}}{C}}-OH$

    a) 1-ethyl-1-methylbutanol

    b) 3-methyl-3-hexanol

    c) 2-ethyl-2-pentanol

    d) 3-methylheptanol

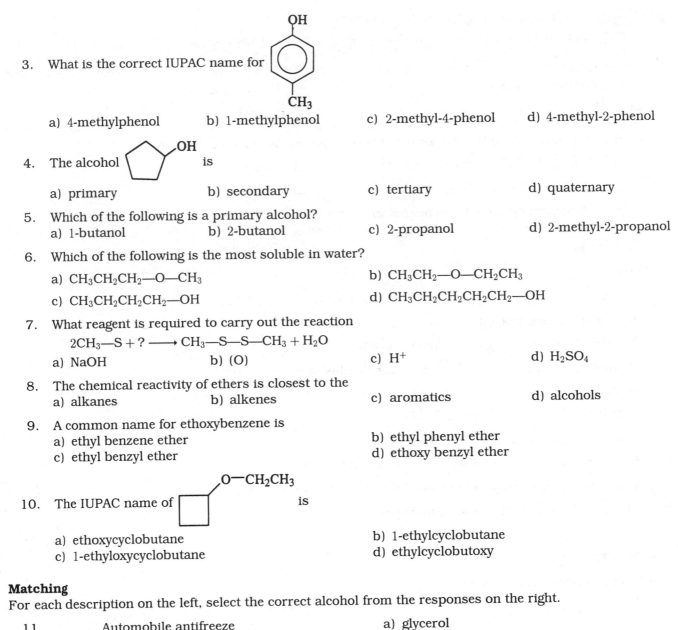

3. What is the correct IUPAC name for

    a) 4-methylphenol   b) 1-methylphenol   c) 2-methyl-4-phenol   d) 4-methyl-2-phenol

4. The alcohol   is

    a) primary   b) secondary   c) tertiary   d) quaternary

5. Which of the following is a primary alcohol?
    a) 1-butanol   b) 2-butanol   c) 2-propanol   d) 2-methyl-2-propanol

6. Which of the following is the most soluble in water?

    a) $CH_3CH_2CH_2-O-CH_3$ b) $CH_3CH_2-O-CH_2CH_3$

    c) $CH_3CH_2CH_2CH_2-OH$ d) $CH_3CH_2CH_2CH_2CH_2-OH$

7. What reagent is required to carry out the reaction
    $$2CH_3-S + ? \longrightarrow CH_3-S-S-CH_3 + H_2O$$
    a) NaOH   b) (O)   c) H^+ d) H_2SO_4

8. The chemical reactivity of ethers is closest to the
    a) alkanes   b) alkenes   c) aromatics   d) alcohols

9. A common name for ethoxybenzene is
    a) ethyl benzene ether   b) ethyl phenyl ether
    c) ethyl benzyl ether   d) ethoxy benzyl ether

10. The IUPAC name of   is

    a) ethoxycyclobutane   b) 1-ethylcyclobutane
    c) 1-ethyloxycyclobutane   d) ethylcyclobutoxy

Matching
For each description on the left, select the correct alcohol from the responses on the right.

11. _____ Automobile antifreeze   a) glycerol

12. _____ A moistening agent in cosmetics   b) ethylene glycol

13. _____ Rubbing alcohol   c) isopropyl alcohol

14. _____ Present in alcoholic beverages   d) ethyl alcohol

Select the correct phenol for each description on the left.

15. _____ An antioxidant used in food packaging

16. _____ Used in throat lozenges

17. _____ Present in Lysol

a) o-phenylphenol

b) BHA

c) 4-hexylresorcinol

d) eugenol

For each reaction on the left, select the correct product from the responses on the right.

18. _____ An alcohol is dehydrated at 140°C

19. _____ An alcohol is dehydrated at 180°C

20. _____ A secondary alcohol is subjected to oxidation

a) a ketone

b) an ether

c) an alkene

d) an aldehyde

True-False

21. Methyl alcohol is a product of the fermentation of sugars and starch.

22. Vitamin E is a natural antioxidant.

23. Tertiary alcohols undergo oxidation to produce ketones.

24. Hydrogen bonding accounts for the water solubility of certain alcohols.

25. Alcohols have a higher boiling point than ethers of similar molecular weight.

26. Oxidation of a thiol produces a disulfide.

27. Two Hg^{2+} ions are required for the reaction with one thiol.

28. Ethers can form hydrogen bonds with water molecules.

29. Thiols are responsible for the pleasant fragrance of many flowers.

30. Ethylene glycol is used as a moisturizer in foods.

SOLUTIONS

A. Answers to Programmed Review

13.1   a) ol   b) diol   c) 1-propanol   d)

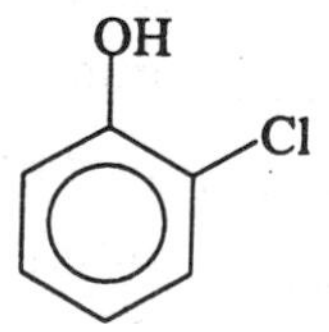

13.2   a) two   b) primary   c) secondary   d) tertiary

13.3   a) hydrogen   b) three   c) hydrogen

13.4   a) dehydration b) alkene   c) ketone   d) aldehydes
       e) carboxylic acids

13.5 a) methanol   b) ethanol   c) CO_2 d) 2-propanol

13.6 a) hexylresorcinol b) antioxidant   c) vitamin E   d) *o*-phenylphenol

13.7 a) methyl propyl ether b) 1-methoxypropane c) heterocyclic

13.8 a) more   b) unreactive   c) hydrocarbons

13.9 a) SH   b) disulfides   c) Hg^{2+} d) reducing agent (H)

B. Answers to Self-Test Questions

1. b	11. b	21. F
2. b	12. a	22. T
3. a	13. c	23. F
4. b	14. d	24. T
5. a	15. b	25. T
6. c	16. c	26. T
7. b	17. a	27. F
8. a	18. b	28. T
9. b	19. c	29. F
10. a	20. a	30. F

Chapter 14

CHAPTER OUTLINE

14.1 Naming Aldehydes and Ketones
14.2 Physical Properties
14.3 Chemical Properties
14.4 Important Aldehydes and Ketones

LEARNING OBJECTIVES

When you have completed your study of this chapter, you should be able to:
1. Recognize the carbonyl group in compounds and classify the compounds as aldehydes or ketones.
2. Assign IUPAC names to aldehydes and ketones.
3. Compare the physical properties of aldehydes and ketones to those of compounds in other classes.
4. Write key reactions for aldehydes and ketones.
5. Give specific uses for aldehydes and ketones.

ANSWERS AND SOLUTIONS TO EVEN-NUMBERED PROBLEMS

Naming Aldehydes and Ketones (Section 14.1)

14.2   Identify each of the following compounds as an aldehyde, ketone, or neither:

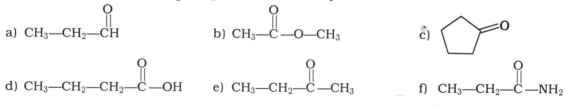

Solution:

a) aldehyde

c) ketone

e) ketone

b) neither, the carbonyl is attached to an O atom as well as a C atom.

d) neither, the carbonyl is attached to an O atom as well as a C atom.

f) neither, the carbonyl is attached to an N atom as well as a C atom.

14.4   Assign IUPAC names to the following aldehydes and ketones:

a) CH₃—CH₂—C̈—H
       ‖
       O

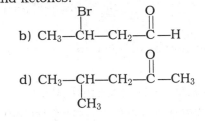

b) CH₃—CH—CH₂—C—H
            | ‖
            Br   O

c)

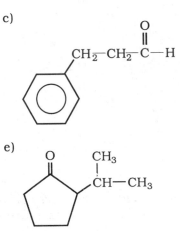

d) CH₃—CH—CH₂—C—CH₃
            | ‖
            CH₃   O

e)

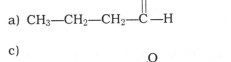

Solution:

a) propanal

d) 4-methyl-2-pentanone

b) 3-phenylpropanal

e) 2-isopropylcyclopentanone

c) 3-phenypropanal

14.6   Draw structural formulas for each of the following compounds:

a) butanal

c) 2-ethyl-4-propylcyclohexanone

b) 2-methyl-3-pentanone

d) 2-chloro-4-phenylpentanal

Solution:

a) CH₃—CH₂—CH₂—C—H
                           ‖
                           O

b) CH₃—CH—C—CH₂—CH₃
            | ‖
            CH₃ O

c)

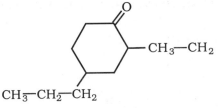

d)

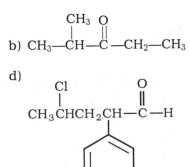

14.8   What is wrong with each of the following names? Give the structure and correct name for each compound.

a) 3-ethyl-2-methylbutanal

b) 2-methyl-4-butanone

c) 4,5-dibromocyclopentanone

Solution:

a) An ethyl on carbon #2 of a 4 carbon chain extends the chain. A 5 carbon aldehyde is present.

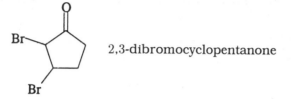

2,3-dimethylpentanal

b) A ketone cannot be on the terminal carbon. It must be an aldehyde.

$$CH_3—\underset{\underset{CH_3}{|}}{CH}—CH_2—\overset{\overset{O}{||}}{CH}$$ 3-methylbutanal

c) Ring was numbered in the wrong direction.

2,3-dibromocyclopentanone

Physical Properties (Section 14.2)

14.10 Most of the remaining water in washed laboratory glassware can be removed by rinsing the glassware with acetone (propanone). Explain.how this process works (acetone is much more volatile than water).

Solution:

Since water and acetone are miscible, the water dissolves in the acetone. When the rinse solution is dumped out, most of the water goes with the acetone. The remaining acetone readily evaporates. Note: The evaporation of the acetone often cools the glass enough to get condensation of water vapor making the surface wet once again.

14.12 Explain why propane boils at $-42°C$, whereas ethanal, which has the same molecular weight, boils at $20°C$.

Solution:

Ethanal is a polar molecule and propane is nonpolar. The ethanal has stronger intermolecular forces, giving it a higher boiling point.

14.14 Use a dotted line to show hydrogen bonding between molecules in each of the following pairs:

a)   b)

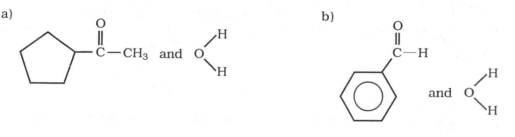

Solution:

a)

b)

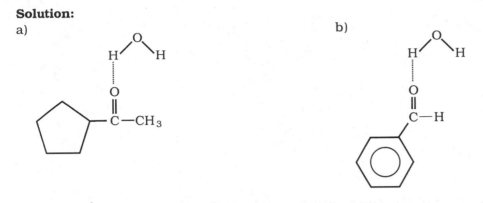

14.16 The compounds menthone and menthol are fragrant substances present in an oil produced by mint plants:

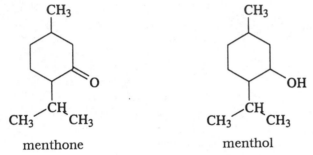

menthone   menthol

When pure, one of these pleasant-smelling compounds is a liquid at room temperature; the other is a solid. Identify the solid and the liquid and explain your reasoning.

Solution:
Because the alcohol can hydrogen-bond and the ketone cannot, the alcohol will have the higher melting point. Menthol is the solid at room temperature (the melting point is above room temperature) and menthone is the liquid.

Chemical Properties (Section 14.3)

14.18 Write an equation for the formation of the following compounds from the appropriate alcohol:

a)

b) $CH_3—\overset{O}{\overset{\|}{C}}—\underset{Br}{\overset{|}{CH}}—CH_3$

c)

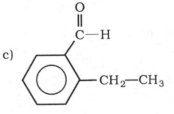

Solution:

a)

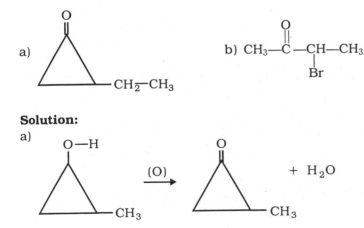

b)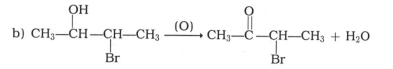

c)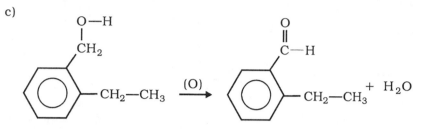

14.20 Identify the following structures as hemiacetals, hemiketals or neither:

c)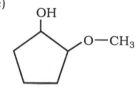

d)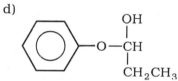

Solution:
a) hemiketal   b) hemiacetal   c) neither   d) hemiacetal

14.22 Label each of the following as acetals, ketals, or neither:

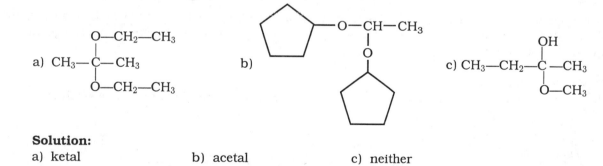

Solution:
a) ketal   b) acetal   c) neither

14.24 Label each of the following structures as a hemiacetal, hemiketal, acetal, ketal, or none of these:

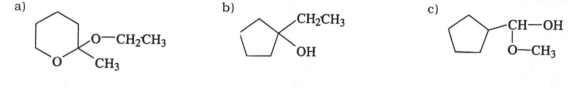

Solution:
a) ketal   b) none (an alcohol)   c) hemiacetal

14.26 Hemiacetals are sometimes referred to as potential aldehydes. Explain.

Solution:
They are unstable and readily break apart to make an aldehyde and alcohol (cyclic hemiacetals excepted).

14.28 What observation characterizes a positive Tollen's test?

Solution:
the formation of a silver "mirror" on the clean glass surface and/or a dark precipitate

14.30 Which of the following compounds will react with Tollens' reagent (Ag^+, NH_3, and H_2O)? For those that do react draw a structural formula for the organic product.

a)

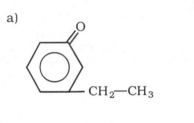

b)

c) $CH_3-O-\overset{\overset{\displaystyle OH}{|}}{C}H-CH_3$

d) $CH_3-CH_2-\overset{\overset{\displaystyle O}{||}}{C}-H$ e) $CH_3-\overset{\overset{\displaystyle O}{||}}{C}-CH_3$

Solution:
a) no reaction
b)

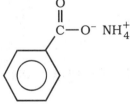

c) no reaction

d) $CH_3-CH_2-\overset{\overset{\displaystyle O}{||}}{C}-O^-\ NH_4^+$
e) no reaction

14.32 Not all aldehydes give a positive Benedict's test. Which of the following aldehydes do?

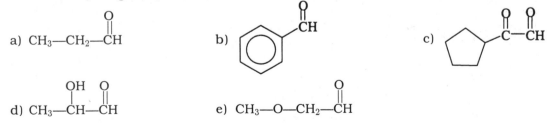

a) $CH_3-CH_2-\overset{\overset{\displaystyle O}{||}}{C}H$ b)   c)

d) $CH_3-\overset{\overset{\displaystyle OH}{|}}{C}H-\overset{\overset{\displaystyle O}{||}}{C}H$ e) $CH_3-O-CH_2-\overset{\overset{\displaystyle O}{||}}{C}H$

Solution:

a) no

b) no

c) yes

d) yes

e) no

14.34 Glucose, the sugar present within the blood, gives a positive Benedict's test. Circle the structural features that enable glucose to react.

OH OH OH OH OH O
| | | | | ||
H₂C—CH—CH—CH—CH—CH

Solution:

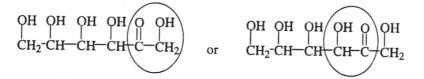

14.36 Fructose, present with glucose in honey, reacts with Benedict's reagent. Circle the structural features that enable fructose to react.

OH   OH   OH   OH   O   OH
| | | | || |
CH₂—CH—CH—CH—C —CH₂

Solution:

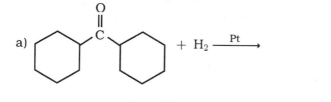 or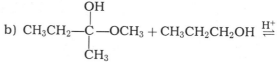

14.38 Complete the following equations. If no reaction occurs, write "no reaction."

a) 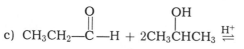 + H₂ $\xrightarrow{Pt}$

b) CH₃CH₂—$\overset{\overset{\displaystyle OH}{|}}{\underset{\underset{\displaystyle CH_3}{|}}{C}}$—OCH₃ + CH₃CH₂CH₂OH $\overset{H^+}{\rightleftharpoons}$

c) CH₃CH₂—$\overset{\overset{\displaystyle O}{||}}{C}$—H + 2CH₃$\overset{\overset{\displaystyle OH}{|}}{C}$HCH₃ $\overset{H^+}{\rightleftharpoons}$

d) 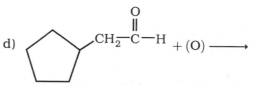 + (O) $\longrightarrow$

e) CH₃CH₂$\overset{\overset{\displaystyle CH_3}{|}}{C}$H—$\overset{\overset{\displaystyle O}{||}}{C}$—H + H₂ $\xrightarrow{Pt}$

Solution:

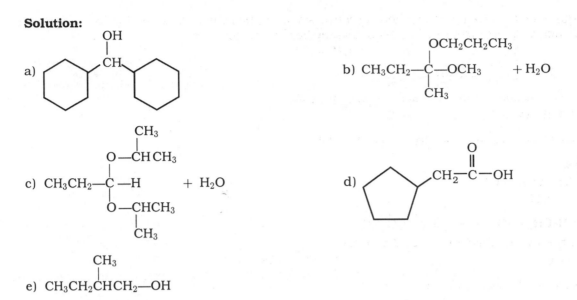

a)

b) $CH_3CH_2-\overset{\displaystyle OCH_2CH_2CH_3}{\underset{\displaystyle CH_3}{\overset{\displaystyle |}{\underset{\displaystyle |}{C}}}}-OCH_3 \quad + H_2O$

c) $CH_3CH_2-\overset{\displaystyle O-\overset{\displaystyle CH_3}{\overset{|}{C}HCH_3}}{\underset{\displaystyle O-CHCH_3}{\overset{|}{\underset{|}{C}}}}-H \quad + H_2O$

d)

e) $CH_3CH_2\overset{\displaystyle CH_3}{\overset{|}{C}H}CH_2-OH$

14.40 Describe the products that result when hydrogen (H_2) is added to alkenes, alkynes, aldehydes, and ketones. You might review Chapter 12 briefly.

Solution:
When H_2 is added, alkenes give alkanes; alkynes give alkenes and then alkanes; aldehyes give primary alcohols; and ketones give secondary alcohols.

14.42 The following comounds are cyclic acetals or ketals. Write structural formulas for the hydrolysis products.

a)

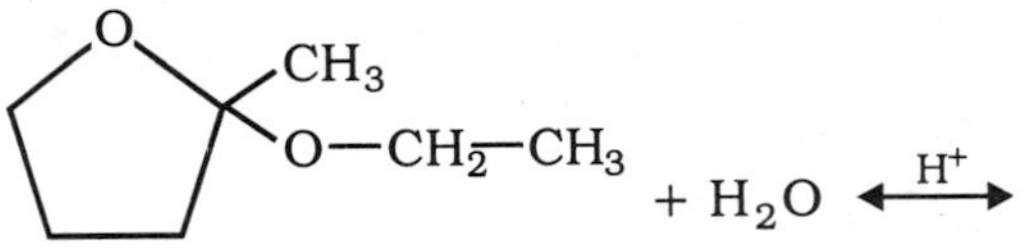

b)

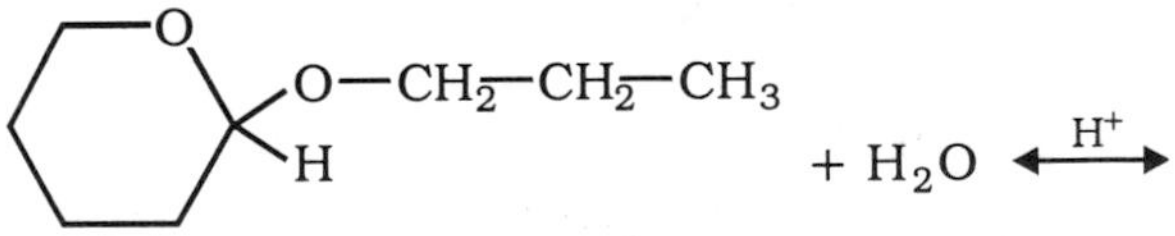

Solution:

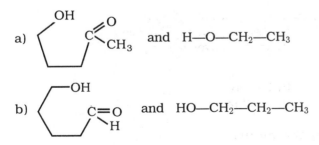

a)   and H—O—CH₂—CH₃

b)   and HO—CH₂—CH₂—CH₃

14.44 Write equations to show how the following conversions can be achieved. More than one reaction is required and reactions from earlier chapters may be necessary.

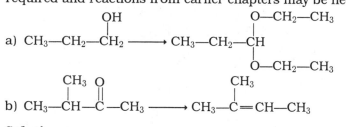

a) CH_3—CH_2—$\overset{\displaystyle OH}{\underset{\displaystyle |}{CH_2}}$ ⟶ CH_3—CH_2—$\overset{\displaystyle O{-}CH_2{-}CH_3}{\underset{\displaystyle O{-}CH_2{-}CH_3}{CH}}$

b) CH_3—$\overset{\displaystyle CH_3}{\underset{\displaystyle |}{CH}}$—$\overset{\displaystyle O}{\underset{\displaystyle \|}{C}}$—$CH_3$ ⟶ CH_3—$\overset{\displaystyle CH_3}{\underset{\displaystyle |}{C}}$=$CH$—$CH_3$

Solution:

a) Oxidize the alcohol:

$CH_3CH_2\overset{\displaystyle OH}{\underset{\displaystyle |}{CH_2}} + (O)$ ⟶ $CH_3CH_2\overset{\displaystyle O}{\underset{\displaystyle \|}{C}}$—H

Then react with ethanol in an acid solution:

$CH_3CH_2\overset{\displaystyle O}{\underset{\displaystyle \|}{C}}$—H + $2CH_3CH_2OH$ $\overset{H^+}{\rightleftharpoons}$ CH_3CH_2—$\overset{\displaystyle O{-}CH_2{-}CH_3}{\underset{\displaystyle O{-}CH_2{-}CH_3}{CH}}$ $+ H_2O$

b) Reduce the ketone to an alcohol:

CH_3—$\overset{\displaystyle CH_3}{\underset{\displaystyle |}{CH}}$—$\overset{\displaystyle O}{\underset{\displaystyle \|}{C}}$—$CH_3 + H_2$ $\overset{Pt}{\longrightarrow}$ CH_3—$\overset{\displaystyle CH_3}{\underset{\displaystyle |}{CH}}$—$\overset{\displaystyle OH}{\underset{\displaystyle |}{CH}}$—$CH_3$

Then eliminate the water from the alcohol:

CH_3—$\overset{\displaystyle CH_3}{\underset{\displaystyle |}{CH}}$—$\overset{\displaystyle OH}{\underset{\displaystyle |}{CH}}$—$CH_3$ $\overset{H_2SO_4}{\underset{180°C}{\longrightarrow}}$ CH_3—$\overset{\displaystyle CH_3}{\underset{\displaystyle |}{C}}$=$CH$—$CH_3$ + H_2O

a) Oxidize the alcohol:

CH_3—CH_2—$\overset{\displaystyle OH}{\underset{\displaystyle |}{CH_2}}$ + H—O—CH_2—CH_3 ⟶ CH_3—CH_2—$\overset{\displaystyle O{-}CH_2{-}CH_3}{\underset{\displaystyle |}{CH_2}}$ + H_2O

Then react with more ethanol in an acid solution:

CH_3—CH_2—$\overset{\displaystyle OH}{\underset{\displaystyle O{-}CH_2{-}CH_3}{CH_2}}$ + H—O—CH_2—CH_3 $\overset{H^+}{\longrightarrow}$ CH_3—CH_2—$\overset{\displaystyle O{-}CH_2{-}CH_3}{\underset{\displaystyle O{-}CH_2{-}CH_3}{CH}}$

Important Aldehydes and Ketones (Section 14.4)

14.46 Use Table 14.3 and name an aldehyde or ketone used in the following ways:

a) peppermint flavoring   b) flavoring for margarines
c) cinnamon flavoring   d) vanilla flavoring

Solution:

a) menthone   b) biacetyl
c) cinnamaldehyde   d) vanillin

Chemistry Around Us, How Reactions Occur, and Key Chemicals:

14.48 What is the active ingredient in some self-tanning lotions?

Solution:
Dihydroxyacetone. DHA is a colorless skin dye that reacts with the dead skin cells to produce a brown. color.

14.50   Explain why norethynodrel is effective as a centraceptive.

Solution:
Norethynodrel is effective because it mimics the natural hormone progesterone.

14.52   What flavor is characteristic of the vanilloids:

Solution:
The vallinoids have a hot and spicy flavor. One of the best known vanilloid is capsaicin, which is the spicy compound found in various types of red peppers.

PROGRAMMED REVIEW

Section 14.1 Naming Aldehydes and Ketones

An aldehyde differs from a ketone in that an aldehyde has a (a) _____ atom is attached to the carbonyl carbon. The IUPAC ending for naming an aldehyde is (b) _____. The aldehyde carbon is always located at position (c)_____. The characteristic IUPAC ending for naming a ketone is (d) _____.

Section 14.2 Physical Properties

Aldehydes and ketones have boiling points (a) _____ those of alcohols with similar molecular weights. The lack of a hydrogen atom on the oxygen prevents the formation of (b) _____ bonds in aldehydes and ketones. Aldehydes and ketones have boiling points (c) _____ those of alkanes with similar molecular weights. In terms of water solubility, low-molecular-weight aldehydes and ketones are (d) _____.

Section 14.3 Chemical Properties

Oxidation of an aldehyde produces a (a) _____ _____. Ketones do not react when treated with (b) _____ agents. A positive reaction with Tollens' reagent produces metallic (c) _____. Addition of hydrogen to an aldehyde or ketone requires a catalyst like (d) _____. The reaction of two molecules of alcohol with an aldehyde produces an (e) _____.

Section 14.4 Important Aldehydes and Ketones

An aldehyde used in preserving biological specimens is (a) _____. The ketone produced in the greatest quantity is (b) _____. Vanillin, the compound present in vanilla flavoring, belongs to the (c) _____ functional class.

SELF-TEST QUESTIONS

Multiple Choice

1.   What is the correct IUPAC name for $CH_3-CH_2-CH_2-CH_2-\overset{\overset{\textstyle O}{\textstyle \|}}{C}H$?
     a) pentanal   b) 1-pentanol   c) 5-pentanal   d) 1-pentanone

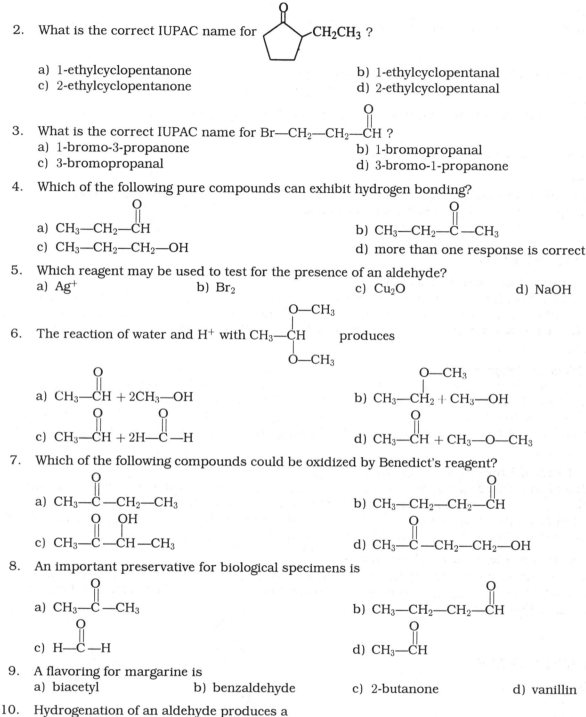

2. What is the correct IUPAC name for [structure] CH_2CH_3 ?

a) 1-ethylcyclopentanone
b) 1-ethylcyclopentanal
c) 2-ethylcyclopentanone
d) 2-ethylcyclopentanal

3. What is the correct IUPAC name for Br—CH_2—CH_2—CH (O) ?
a) 1-bromo-3-propanone
b) 1-bromopropanal
c) 3-bromopropanal
d) 3-bromo-1-propanone

4. Which of the following pure compounds can exhibit hydrogen bonding?

a) CH_3—CH_2—CH (O)
b) CH_3—CH_2—C (O) —CH_3
c) CH_3—CH_2—CH_2—OH
d) more than one response is correct

5. Which reagent may be used to test for the presence of an aldehyde?
a) Ag^+
b) Br_2
c) Cu_2O
d) NaOH

6. The reaction of water and H^+ with CH_3—CH (O—CH_3)(O—CH_3) produces

a) CH_3—CH (O) + 2CH_3—OH
b) CH_3—CH_2 (O—CH_3) + CH_3—OH
c) CH_3—CH (O) + 2H—C (O) —H
d) CH_3—CH (O) + CH_3—O—CH_3

7. Which of the following compounds could be oxidized by Benedict's reagent?

a) CH_3—C (O) —CH_2—CH_3
b) CH_3—CH_2—CH_2—CH (O)
c) CH_3—C (O) —CH (OH) —CH_3
d) CH_3—C (O) —CH_2—CH_2—OH

8. An important preservative for biological specimens is

a) CH_3—C (O) —CH_3
b) CH_3—CH_2—CH_2—CH (O)
c) H—C (O) —H
d) CH_3—CH (O)

9. A flavoring for margarine is
a) biacetyl
b) benzaldehyde
c) 2-butanone
d) vanillin

10. Hydrogenation of an aldehyde produces a
a) acetal
b) primary alcohol
c) secondary alcohol
d) ketone

Matching

For each structure on the left, select the correct class of compound from the responses on the right.

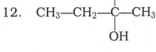

11.   CH₃—CH₂—CH—O—CH₃

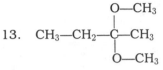

12.   CH₃—CH₂—C—CH₃

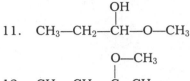

13.   CH₃—CH₂—C—CH₃

14.

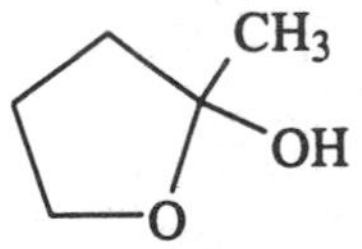

a) acetal

b) ketal

c) hemiacetal

d) hemiketal

Select the correct product for each of the reactions on the left.

15.   aldehyde + (O) ⟶

16.   ketone + hydrogen ⟶

17.   aldehyde + alcohol ⟶

18.   ketone + alcohol ⇌

a) hemiacetal

b) hemiketal

c) carboxylic acid

d) alcohol

True-False

19.   Both aldehydes and ketones contain a carbonyl group.

20.   Pure ketones can hydrogen bond.

21.   Pure aldehydes can hydrogen bond.

22.   Both aldehydes and ketones can form hydrogen bonds with water.

23.   Acetone is an important organic solvent.

24.   Camphor is used in foods as peppermint flavoring.

25.   Cinnamon flavoring contains cinnamaldehyde.

SOLUTIONS

A. Answers to Programmed Review

14.1   a) hydrogen   b) -al   c) one   d) -one

14.2   a) below   b) hydrogen   c) above   d) soluble

14.3 a) carboxylic acid   b) oxidizing   c) Ag   d) Pt
      e) acetal

14.4 a) formaldehyde   b) acetone   c) aldehyde

B. Answers to Self-Test Questions

1. a   10. b   19. T
2. c   11. c   20. F
3. c   12. d   21. F
4. c   13. b   22. T
5. a   14. d   23. T
6. a   15. c   24. F
7. c   16. d   25. T
8. c   17. a
9. a   18. b

Chapter 15

Carboxylic Acids and Esters

CHAPTER OUTLINE

15.1 Nomenclature of Carboxylic Acids
15.2 Physical Properties of Carboxylic Acids
15.3 The Acidity of Carboxylic Acids
15.4 Salts of Carboxylic Acids

15.5 Carboxylic Esters
15.6 Nomenclature of Esters
15.7 Reactions of Esters
15.8 Esters of Inorganic Acids

LEARNING OBJECTIVES

When you have completed your study of this chapter, you should be able to:
1. Assign IUPAC names and draw structural formulas for carboxylic acids.
2. Explain how hydrogen bonding affects the physical properties of carboxylic acids.
3. Assign common and IUPAC names to carboxylic acid salts and esters.
4. Recognize and write key reactions of carboxylic acids and esters.
5. Describe uses for carboxylic acids, carboxylate salts, and esters.
6. Write reactions for the formation of phosphate esters.

ANSWERS AND SOLUTIONS TO EVEN-NUMBERED PROBLEMS

Nomenclature of Carboxylic Acids (Section 15.1)

15.2   What structural features are characteristic of fatty acids? Why are fatty acids given that name?

Solution:
Fatty acids are carboxylic acids with 10 or more carbon atoms. They got their names because they were first made by saponification of fats.

15.4   What carboxylic acid is present in sour milk and sauerkraut?

Solution:
Lactic acid

15.6   Write the correct IUPAC name for each of the following:

a) 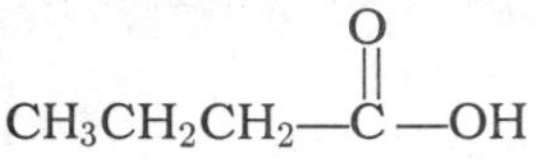 $CH_3CH_2CH_2-\overset{\overset{O}{\|}}{C}-OH$

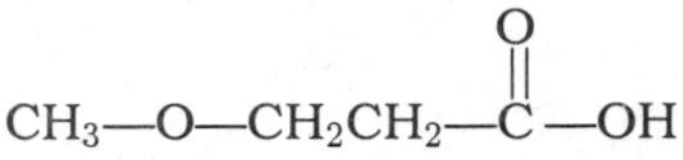

b) $CH_3-O-CH_2CH_2-\overset{\overset{O}{\|}}{C}-OH$

c) $CH_3\overset{\overset{CH_3}{|}}{C}H\,CHCH_2-\overset{\overset{O}{\|}}{C}-OH$ with Br on the CH below

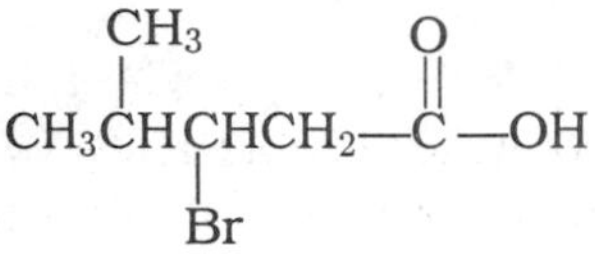

d)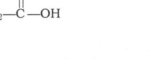

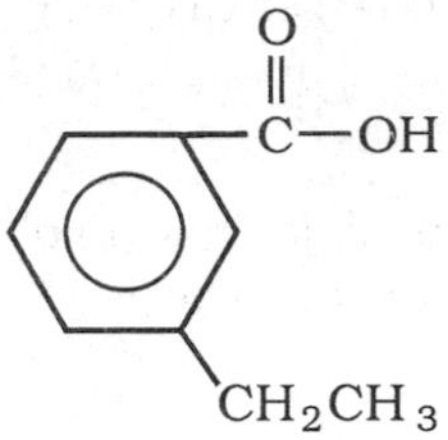

e) 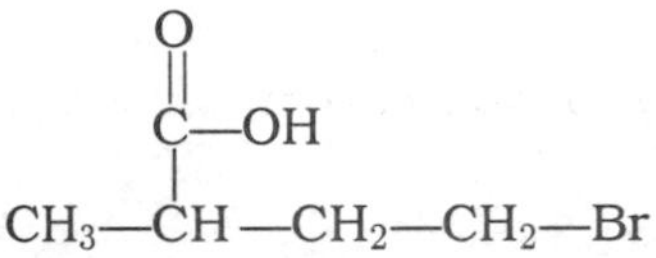 $\underset{\underset{CH_3-\overset{|}{C}H-CH_2-CH_2-Br}{}}{\overset{\overset{\overset{O}{\|}}{C}-OH}{}}$

Solution:
a) butanoic acid
c) 3-bromo-4-methylpentanoic acid
e) 4-bromo-2-methylbutanoic acid

b) 3-methoxyproprionic acid
d) 3-ethylbenzoic acid or *m*-ethylbenzoic acid

15.8   Write a structural formula for each of the following:
a) hexanoic acid
c) *o*-ethylbenzoic acid

b) 4-bromo-3-methylpentanoic acid

Solution:

a)  $CH_3CH_2CH_2CH_2CH_2-\overset{\overset{O}{\|}}{C}-OH$

b) $CH_3\overset{\overset{Br}{|}}{C}H-\overset{\overset{CH_3}{|}}{C}H\,CH_2-\overset{\overset{O}{\|}}{C}-OH$

c)

(benzene ring with) $-\overset{\overset{O}{\|}}{C}-OH$ and $-CH_2CH_3$

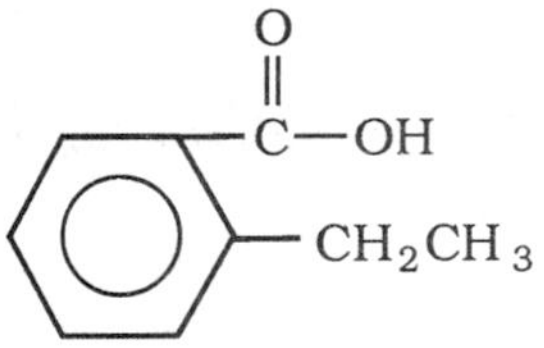

Physical Properties of Carboxylic Acids (Section 15.2)

15.10   Which compound in each of the following pairs would you expect to have the higher boiling point? Explain your answer.
a) acetic acid or propanol
c) acetic acid or butyric acid

b) propanoic acid or butanone

Solution:
a) Acetic acid is the higher boiling compound. It has more hydrogen bonding. The liquid consists of dimers.
b) Propanoic acid is the higher boiling compound. It has more hydrogen bonding. The liquid consists of dimers.
c) Butyric acid is the higher boiling compound. It is the larger of the two acids. It has the higher molecule weight.

15.12 Draw the structure of the dimer formed when two molecules of propanoic acid hydrogen bond with each other.

Solution:

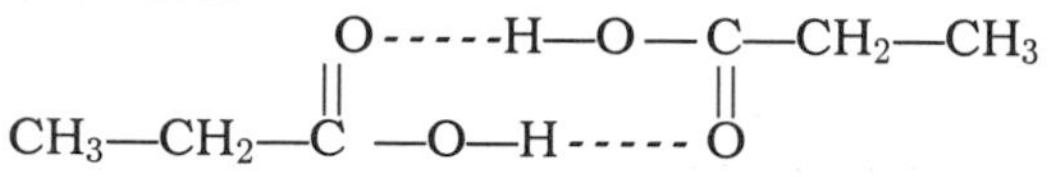

$$\begin{array}{ccc} & O\text{-----}H\text{—}O\text{—}C\text{—}CH_2\text{—}CH_3 \\ & \| & \| \\ CH_3\text{—}CH_2\text{—}C\text{ —}O\text{—}H\text{-----}O & & \end{array}$$

15.14 Caproic acid, a six-carbon acid, has a solubility in water of 1 g/100 mL of water (Table 15.2). Which part of the structure of caproic acid is responsible for its solubility in water, and which part prevents greater solubility?

Solution:

The hydrophilic carboxyl group makes it soluble. The aliphatic, hydrophobic carbon-carbon chain limits the solubility.

15.16 List the following compounds in order of increasing water solubility:

a) 1-butanol   b) pentane   c) butyric acid   d) butanal

Solution:

Pentane, butanal, l-butanol, butyric acid. Water solubility increases as the polarity increases, and as the amount of hydrogen bonding increases. The pentane is nonpolar. The butanal is polar with no hydrogen bonding. The 1-butanol is polar and has hydrogen bonding. The butyric acid is polar and can form multiple hydrogen bonds.

The Acidity of Carboxylic Acids (Section 15.3)

15.18 As we discuss the cellular importance of lactic acid in a later chapter, we will refer to this compound as lactate. Explain why.

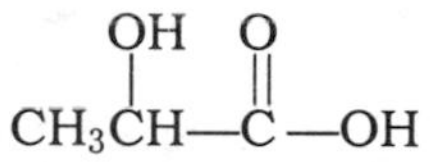

$$\begin{array}{cc} OH & O \\ | & \| \\ CH_3CH\text{—}C\text{—}OH \end{array}$$

Solution:

At the pH found in living cells, the acidic hydrogen on the carboxyl group is removed, making the lactate anion the predominant form.

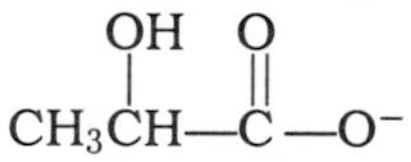

$$\begin{array}{cc} OH & O \\ | & \| \\ CH_3CH\text{—}C\text{—}O^- \end{array} \quad \text{lactate ion}$$

15.20 Complete each of the following reactions:

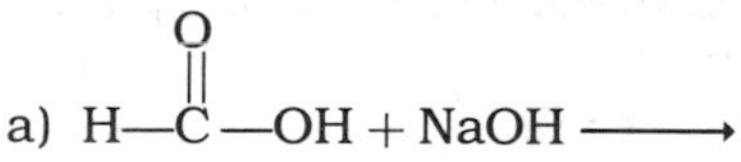

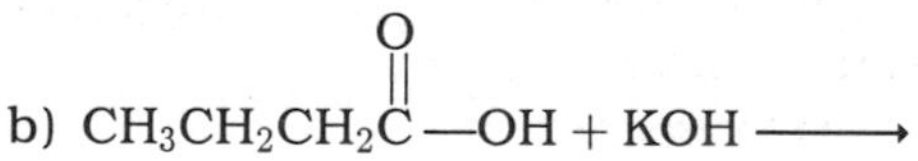

$$\text{a) } H\overset{\overset{\displaystyle O}{\|}}{C}\text{—}OH + NaOH \longrightarrow \qquad\qquad \text{b) } CH_3CH_2CH_2\overset{\overset{\displaystyle O}{\|}}{C}\text{—}OH + KOH \longrightarrow$$

Solution:

$$\text{a) } H\overset{\overset{\displaystyle O}{\|}}{C}\text{—}OH + NaOH \longrightarrow H\overset{\overset{\displaystyle O}{\|}}{C}\text{—}O\text{—}Na^+ + H_2O$$

$$\text{b) } CH_3CH_2CH_2\overset{\overset{\displaystyle O}{\|}}{C}\text{—}OH + KOH \longrightarrow CH_3CH_2CH_2\text{—}C\text{—}O\text{—}K^+ + H_2O$$

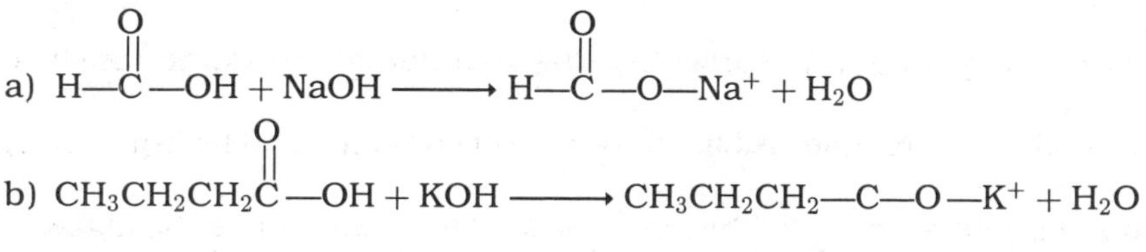

15.22   Write a balanced equation for the reaction of propanoic acid with each of the following:
a) KOH   b) H$_2$O   c) NaOH

Solution:

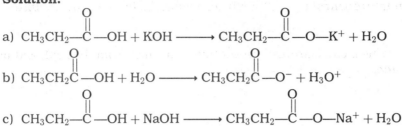

a) CH_3CH_2—$\overset{\overset{\displaystyle O}{\|}}{C}$—OH + KOH $\longrightarrow$ CH_3CH_2—$\overset{\overset{\displaystyle O}{\|}}{C}$—O—K$^+$ + H$_2$O

b) $CH_3CH_2\overset{\overset{\displaystyle O}{\|}}{C}$—OH + H$_2$O $\longrightarrow$ $CH_3CH_2\overset{\overset{\displaystyle O}{\|}}{C}$—O$^-$ + H$_3$O$^+$

c) CH_3CH_2—$\overset{\overset{\displaystyle O}{\|}}{C}$—OH + NaOH $\longrightarrow$ CH_3CH_2—$\overset{\overset{\displaystyle O}{\|}}{C}$—O—Na$^+$ + H$_2$O

Salts of Carboxylic Acids (Section 15.4)

15.24   Write the IUPAC name for each of the following:

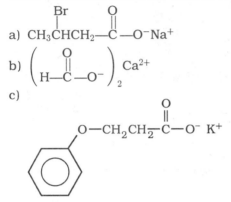

a) $CH_3\overset{\overset{\displaystyle Br}{|}}{C}HCH_2$—$\overset{\overset{\displaystyle O}{\|}}{C}$—O$^-Na^+$

b) $\left(H-\overset{\overset{\displaystyle O}{\|}}{C}-O^- \right)_2$ Ca^{2+}

c)

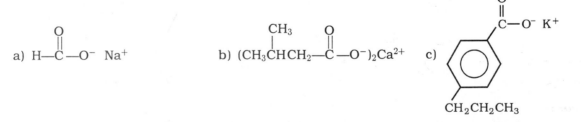

O—CH_2CH_2—$\overset{\overset{\displaystyle O}{\|}}{C}$—O$^-$ K$^+$

Solution:
a) sodium 3-bromobutanoate   b) calcium formate   c) potassium 3-phenoxypropanoate

15.26   Draw structural formulas for the following:
a) sodium methanoate   b) calcium 3-methylbutanoate   c) potassium p-propylbenzoate

Solution:

a) H—$\overset{\overset{\displaystyle O}{\|}}{C}$—O$^-$ Na$^+$

b) $(CH_3\overset{\overset{\displaystyle CH_3}{|}}{C}HCH_2$—$\overset{\overset{\displaystyle O}{\|}}{C}$—O$^-)_2$Ca^{2+}

c)

$\overset{\overset{\displaystyle O}{\|}}{C}$—O$^-$ K$^+$

CH$_2$CH$_2$CH$_3$

15.28   Give the name of a carboxylic acid or carboxylate salt used in each of the following ways:
a) As a soap
b) As a general food preservative used to pickle vegetables
c) As a preservative used in soft drinks
d) As a treatment of athelete's foot
e) As a mold inhibitor used in bread
f) As a food additive noted for its pH buffering ability

Solution:

a) sodium stearate or sodium palmitate

b) acetic acid

c) sodium benzoate

d) zinc 10-undecylenate

e) calcium propanoate and sodium propanoate

f) a citric acid and sodium citrate mixture

Carboxylic Esters (Section 15.5)

15.30 For each ester in Exercise 15.29 draw a circle around the portion that came from the acid, and use an arrow to point out the *ester linkage.*

Solution:

a) not an ester

b)

c) not an ester

d)

e) not an ester

f)

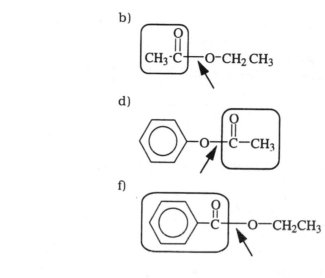

15.32 Complete the following reactions:

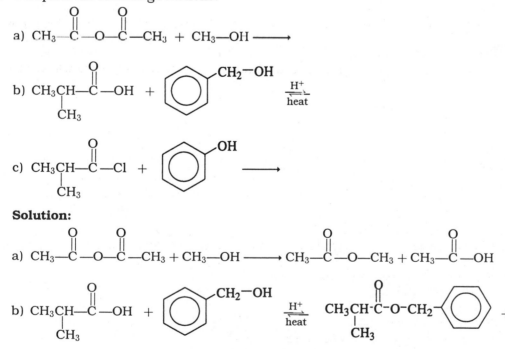

c) 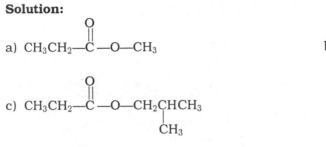 + HCl

15.34   Give the structure of the ester that forms when propanoic acid is reacted with following:
a) methyl alcohol   b) phenol   c) 2-methyl-1-propanol

Solution:

a) $CH_3CH_2-\overset{O}{\underset{}{C}}-O-CH_3$

b)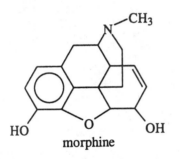

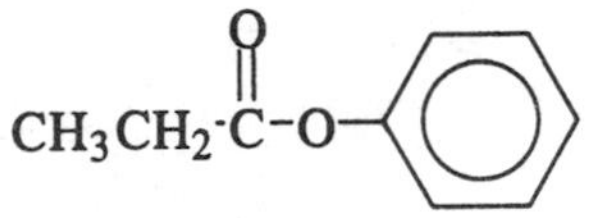

c)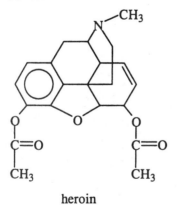

15.36   Heroin is formed by reacting morphine with two mol of $CH_3-\overset{O}{\underset{}{C}}-Cl$ to form the diester. Show the structure of heroin.

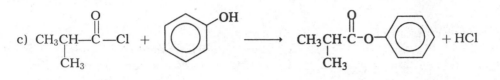

morphine

Solution:

heroin

Nomenclature of Esters (Section 15.6)

15.38 Assign common names to the following esters. Refer to Table 15.1 for the common names of the acids.

a) H—C(=O)—O—CH$_2$CH$_3$

b) CH$_3$(CH$_2$)$_4$—C(=O)—O—CHCH$_3$ with CH$_3$ below

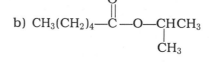

Solution:

a) ethyl formate

b) isopropyl caproate

15.40 Give the IU name to each of the following:

a) CH$_3$CH$_2$CH$_2$—C(=O)—O—CH$_2$CH$_3$

b)

Solution:

a) ethyl butanoate

b) isopropyl *m*-methylbenzoate

15.42 Assign common names to the simple esters produced by a reaction between methanol and the following:

a) propionic acid   b) butyric acid   c) lactic acid

Solution:

a) methyl propionate   b) methyl butyrate   c) methyl lactate

15.44 Draw structural formulas for the following:

a) methyl ethanoate   b) propyl 2-bromobenzoate   c) ethyl 3,4-dimethylpentanoate

Solution:

a) CH$_3$—C(=O)—O—CH$_3$

b)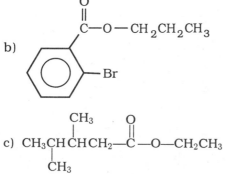

c) CH$_3$CHCHCH$_2$—C(=O)—O—CH$_2$CH$_3$ with CH$_3$ groups

Reactions of Esters (Section 15.7)

15.46   Write the equations for the hydrolysis and saponification of phenyl butanoate and name the component parts.

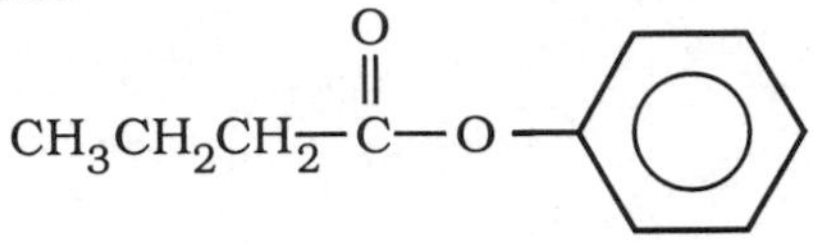

$$CH_3CH_2CH_2{-}\overset{\overset{\displaystyle O}{\|}}{C}{-}O{-}\bigcirc$$

Solution:

Hydrolysis:

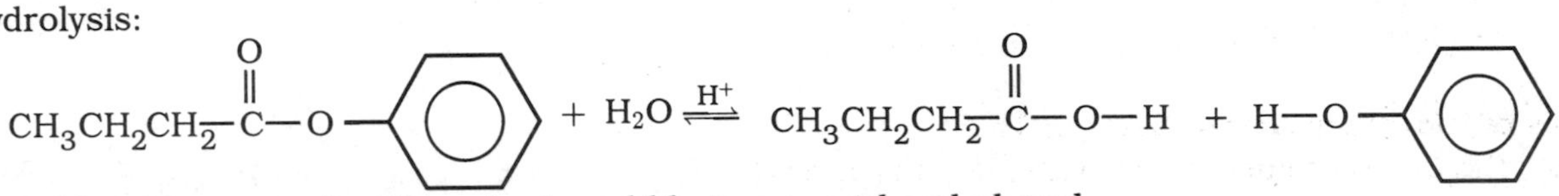

$$CH_3CH_2CH_2{-}\overset{\overset{\displaystyle O}{\|}}{C}{-}O{-}\bigcirc \ +\ H_2O \overset{H^+}{\rightleftharpoons}\ CH_3CH_2CH_2{-}\overset{\overset{\displaystyle O}{\|}}{C}{-}O{-}H\ +\ H{-}O{-}\bigcirc$$

Phenyl butanoate reacts with water to yield butanoic acid and phenol.

Saponification:

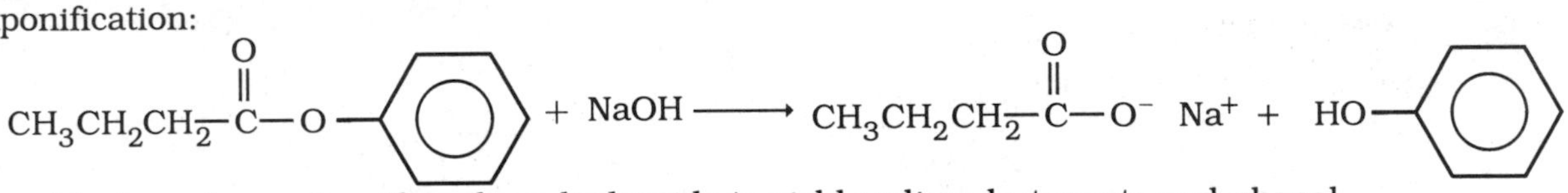

$$CH_3CH_2CH_2{-}\overset{\overset{\displaystyle O}{\|}}{C}{-}O{-}\bigcirc \ +\ NaOH \longrightarrow\ CH_3CH_2CH_2{-}\overset{\overset{\displaystyle O}{\|}}{C}{-}O^-\ Na^+\ +\ HO{-}\bigcirc$$

Phenyl butanoate reacts with sodium hydroxide to yield sodium butanoate and phenol.

15.48   Complete the following reactions:

a) $CH_3CH_2CH{-}\overset{\overset{\displaystyle O}{\|}}{C}{-}O{-}CH_2CH_3 + NaOH \longrightarrow$
   $\qquad\quad\ \underset{CH_3}{|}$

b)

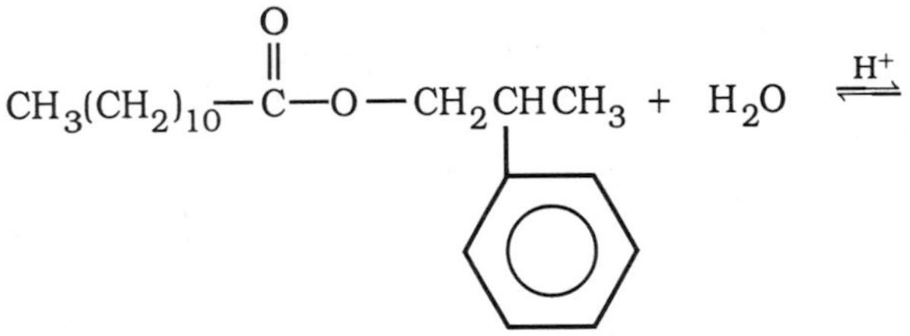

$$CH_3(CH_2)_{10}{-}\overset{\overset{\displaystyle O}{\|}}{C}{-}O{-}CH_2CHCH_3\ +\ H_2O\ \overset{H^+}{\rightleftharpoons}$$

Solution:

a) $CH_3CH_2CH{-}\overset{\overset{\displaystyle O}{\|}}{C}{-}O^-Na^+ + H{-}O{-}CH_2CH_3$
   $\qquad\quad\ \underset{CH_3}{|}$

b)

$$CH_3(CH_2)_{10}{-}\overset{\overset{\displaystyle O}{\|}}{C}{-}O{-}H\ +\ HO{-}CH_2CHCH_3$$

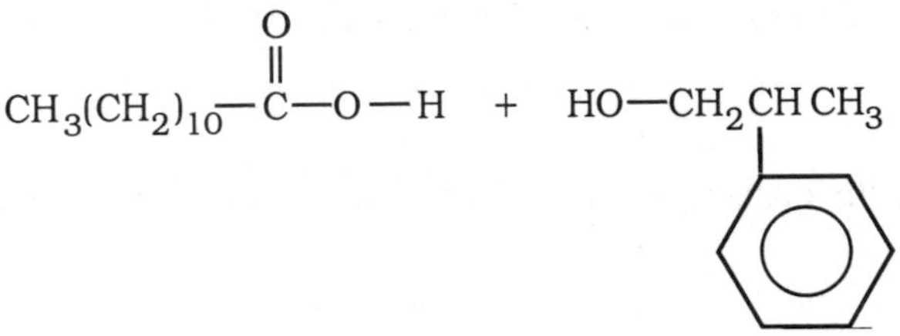

15.50 Draw the structure of repeating monomer unit in the polyester formed in the condensation of oxalic acid and 1,3-propanediol.

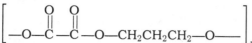

Solution:

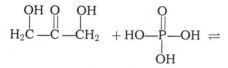

Esters of Inorganic Acids (Section 15.8)

15.52 Dihydroxyacetone reacts with phosphoric acid to form the important monoester called dihydroxy-acetone phosphate. Complete the reaction for its formation.

$$\underset{\underset{H_2C}{}}{OH}\ \underset{C}{\overset{O}{\|}}\ \underset{CH_2}{OH}\ +\ HO-\underset{\underset{OH}{|}}{\overset{O}{\overset{\|}{P}}}-OH\ \rightleftharpoons$$

Solution:

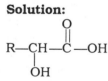

$$H_2C-C-CH_2-O-P-OH\ +\ H_2O$$

Chemistry Around Us, Key Chemicals, and How Reactions Occur

15.54 Draw the characteristic structure of an alpha hydroxy acid.

Solution:

$$R-\underset{\underset{OH}{|}}{CH}-\overset{O}{\overset{\|}{C}}-OH$$

15.56 What serious diseases may be partially prevented or treated with aspirin?

Solution:
Atherosclerosis, esophageal and colorectal cancers may be treated with aspirin according to recent studies.

15.58 In the saponification of esters, why is one of the products a carboxylate salt rather than a carboxylic acid?

Solution:
The NaOH used in saponificaiton reacts with the carboxylic acid that is formed to give the carboxylate salt.

PROGRAMMED REVIEW

Section 15.1 Nomenclature of Carboxylic Acids

The characteristic ending for carboxylic acid common names is (a) _____ _____. The characteristic ending for carboxylic acid IUPAC names is (b) _____ _____. Long-chain carboxylic acids derived from fats are referred to as (c) _____ _____. The simplest aromatic carboxylic acid is (d) _____ _____.

Section 15.2 Physical Properties of Carboxylic Acids

Carboxylic acid dimers contain (a) _____ molecules hydrogen bonded together. Carboxylic acids have (b) _____ boiling points than those of alcohols with comparable molecular weights. In comparing water solubility, carboxylic acids are (c) _____ soluble than aldehydes of comparable molecular weight. Low-molecular-weight carboxylic acids have sharp or (d) _____ odors.

Section 15.3 The Acidity of Carboxylic Acids

The negative ion formed as a carboxylic acid dissociates in water is the (a) _____ ion. In terms of acid strength, carboxylic acids are (b) _____ acids. At body pH carboxylic acids exist primarily as (c) _____ ions. Sodium hydroxide reacts with a carboxylic acid to produce a (d) _____ plus water.

Section 15.4 Salts of Carboxylic Acids

The characteristic ending for the names of salts of carboxylic acids is (a) _____. In terms of water solubility, carboxylate salts are (b) _____ soluble than the corresponding carboxylic acids. Sodium salts of long-chain carboxylic acids are useful as (c) _____. Certain carboxylic acid salts are used in foods as (d) _____ to prevent the growth of mold.

Section 15.5 Carboxylic Esters

Carboxylic esters are formed from carboxylic acids and (a) _____. The process of ester formation is called (b) _____. The single bond between a carbonyl carbon and an oxygen in an ester is called the (c) _____ linkage.

Section 15.6 Nomenclature of Esters

An ester is comprised of a carboxylic acid portion and an alcohol component. The first word in naming an ester is derived from the (a) _____ portion of the structure. The IUPAC ending for an ester name is (b) _____. An ester formed from acetic acid would be named such that the second word of the name is (c) _____. An ester derived from ethyl alcohol and benzoic acid would be called (d) _____.

Section 15.7 Reactions of Esters

Ester hydrolysis may be catalyzed by (a) _____. The products of ester hydrolysis are a carboxylic acid and an (b) _____. The basic cleavage of an ester is termed (c) _____. The products of basic cleavage of an ester are an alcohol and a (d) _____ _____.

Section 15.8 Esters of Inorganic Acids

Reaction of an alcohol with phosphoric acid can produce a (a) _____ ester. Phosphoric acid in which two hydrogens have been replaced by R groups may be referred to as a (b) _____. The letters ATP stand for adenosine (c) _____.

SELF-TEST QUESTIONS

Multiple Choice

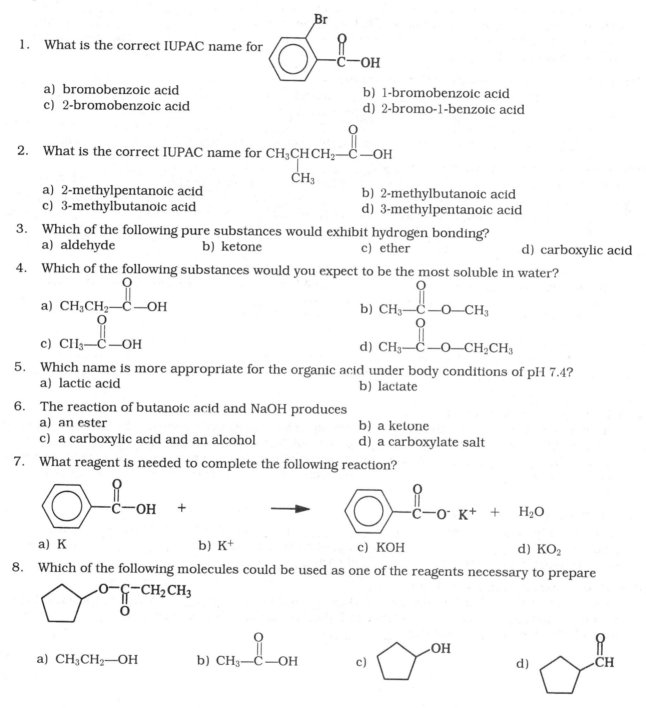

1. What is the correct IUPAC name for

    a) bromobenzoic acid
    c) 2-bromobenzoic acid

    b) 1-bromobenzoic acid
    d) 2-bromo-1-benzoic acid

2. What is the correct IUPAC name for CH_3CHCH_2—$\overset{O}{\overset{\|}{C}}$—OH

    $\overset{}{\underset{CH_3}{|}}$

    a) 2-methylpentanoic acid
    c) 3-methylbutanoic acid

    b) 2-methylbutanoic acid
    d) 3-methylpentanoic acid

3. Which of the following pure substances would exhibit hydrogen bonding?
    a) aldehyde   b) ketone   c) ether   d) carboxylic acid

4. Which of the following substances would you expect to be the most soluble in water?

    a) CH_3CH_2—$\overset{O}{\overset{\|}{C}}$—OH

    b) CH_3—$\overset{O}{\overset{\|}{C}}$—O—$CH_3$

    c) CH_3—$\overset{O}{\overset{\|}{C}}$—OH

    d) CH_3—$\overset{O}{\overset{\|}{C}}$—O—$CH_2CH_3$

5. Which name is more appropriate for the organic acid under body conditions of pH 7.4?
    a) lactic acid   b) lactate

6. The reaction of butanoic acid and NaOH produces
    a) an ester   b) a ketone
    c) a carboxylic acid and an alcohol   d) a carboxylate salt

7. What reagent is needed to complete the following reaction?

    a) K   b) K^+ c) KOH   d) KO_2

8. Which of the following molecules could be used as one of the reagents necessary to prepare

    a) CH_3CH_2—OH   b) CH_3—$\overset{O}{\overset{\|}{C}}$—OH   c)   d)

9. What is the organic product of the reaction

$$CH_3-\overset{\overset{\displaystyle O}{\|}}{C}-OH + CH_3-OH \xrightarrow{\ H^+\ }$$

a) $CH_3-\overset{\overset{\displaystyle O}{\|}}{C}-CH_2-OH$

b) $CH_3-\overset{\overset{\displaystyle O}{\|}}{C}-O-CH_2-OH$

c) $CH_3-\overset{\overset{\displaystyle O}{\|}}{C}-CH_3$

d) $CH_3-\overset{\overset{\displaystyle O}{\|}}{C}-O-CH_3$

10. The ester formed by reacting propanoic acid and isopropyl alcohol is
   a) propyl propanoate
   b) isopropyl propanoic acid
   c) isopropyl propanoate
   d) 2-propyl propanoate

11. The IUPAC name for the ester formed in the reaction of isopropyl alcohol and benzoic acid is
   a) benzyl isopropyl ester
   b) benzyl isopropanoate
   c) isopropyl benzoate
   d) isopropyl benzoic acid

12. Which of the following materials might be obtained as one of the products from the reaction

$$CH_3-\overset{\overset{\displaystyle O}{\|}}{C}-O-CH_2CH_2CH_3 + NaOH \longrightarrow$$

a) CH_3CH_2-OH

b) $CH_3CH_2-\overset{\overset{\displaystyle O}{\|}}{C}-O^-Na^+$

c) $CH_3-\overset{\overset{\displaystyle O}{\|}}{C}-O^-Na^+$

d) $CH_3-\overset{\overset{\displaystyle O}{\|}}{C}-OH$

Matching
Select the best match for each of the following:

13. $CH_3-\overset{\overset{\displaystyle O}{\|}}{C}-OH$

14.

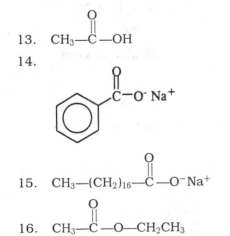

a) a preservative used in pop

b) a soap

c) present in vinegar

d) fingernail polish remover

15. $CH_3-(CH_2)_{16}-\overset{\overset{\displaystyle O}{\|}}{C}-O^-Na^+$

16. $CH_3-\overset{\overset{\displaystyle O}{\|}}{C}-O-CH_2CH_3$

For each reaction on the left, choose the correct description from the responses on the right.

17. ester $+ H_2O \xrightarrow{H^+}$ a) dissociation

18. ester $+ NaOH \longrightarrow$ b) esterification

19. carboxylic acid $+ H_2O \longrightarrow$ c) hydrolysis

20. carboxylic acid $+$ alcohol $\xrightarrow[\text{heat}]{H^+}$ d) saponification

True-False

21. The boiling points of carboxylic acids are lower than those of the corresponding alcohols.

22. Salts of carboxylic acids are not usually soluble in water.

23. Hydrogen bonding increases both the boiling point and the water solubility of carboxylic acids.

24. CH_3COOH has a higher boiling point than $CH_3-\overset{\overset{\displaystyle O}{\|}}{C}H$

25. Carboxylic acids are generally strong acids.

26. Both nitric acid and phosphoric acid can react with alcohols to form esters.

27. Certain phosphate esters are present in the body.

SOLUTIONS

A. Answers to Programmed Review

15.1	a) -ic acid	b) -oic acid	c) fatty acids	d) benzoic acid
15.2	a) two	b) higher	c) more	d) unpleasant
15.3	a) carboxylate	b) weak	c) carboxylate	d) salt
15.4	a) -ate	b) more	c) soaps	d) preservatives
15.5	a) alcohols	b) esterification	c) ester	
15.6	a) alcohol	b) -ate	c) acetate	d) ethyl benzoate
15.7	a) H^+	b) alcohol	c) saponification	d) carboxylate salt
15.8	a) phosphate	b) diester	c) triphosphate	

B. Answers to Self-Test Questions

1. c	10. c	19. a
2. c	11. c	20. b
3. d	12. c	21. F
4. c	13. c	22. F
5. b	14. a	23. T
6. d	15. b	24. T
7. c	16. d	25. F
8. c	17. c	26. T
9. d	18. d	27. T

Chapter 16

Amines and Amides

CHAPTER OUTLINE

16.1 Classification of Amines
16.2 Nomenclature of Amines
16.3 Physical Properties of Amines
16.4 Chemical Properties of Amines
16.5 Amines as Neurotransmitters

16.6 Other Biologically Important Amines
16.7 The Nomenclature of Amides
16.8 Physical Properties of Amides
16.9 Chemical Properties of Amides

LEARNING OBJECTIVES

When you have completed your study of this chapter, you should be able to:
1. Given structural formulas, classify amines as primary, secondary, or tertiary.
2. Assign common and IUPAC names to simple amines and amides.
3. Discuss how hydrogen bonding influences the physical properties of amines and amides.
4. Write equations for the reactions of amines with water, the formation of amine salts, the reactions of amine salts with strong bases, and the reactions of amines with acid chlorides and anhydrides.
5. Give the products of acidic and basic hydrolysis of amides.
6. List uses for important amides and amines.

ANSWERS AND SOLUTIONS TO EVEN-NUMBERED PROBLEMS

Classification of Amines (Section 16.1)

16.2   Classify each of the following as a primary, secondary, or tertiary amine:

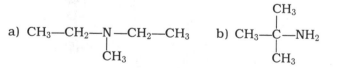

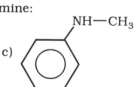

Solution:

a) tertiary   b) primary   c) secondary

16.4   Draw structural formulas for the four amines that have the molecular formula C_3H_9N. Label each one as primary, secondary, or tertiary.

Solution:

(1) CH_3—CH_2—CH_2—NH_2 primary

(2) CH_3—$\underset{\underset{NH_2}{|}}{CH}$—$CH_3$ primary

(3) CH_3—NH—CH_2—CH_3 secondary

(4) CH_3—$\underset{\underset{CH_3}{|}}{N}$—$CH_3$ tertiary

Nomenclature of Amines (Section 16.2)

16.6   Give each of the following amines a common name by adding the ending -amine to alkyl group names:

a) CH_3—$\underset{\underset{NH_2}{|}}{CH}$—$CH_3$

b) 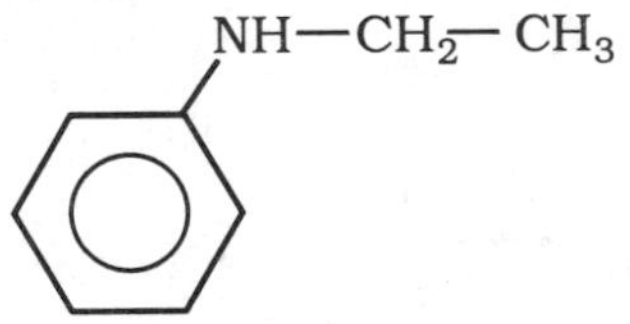 [benzene ring]—NH—CH_2—CH_3

c) CH_3—CH_2—CH_2—$\underset{\underset{CH_3—CH_3}{|}}{N}$—$CH_3$

Solution:

a) isopropylamine   b) ethylphenylamine   c) ethylmethylpropylamine

16.8   Give each of the following amines an IUPAC name by treating the amino group as a substituent:

a)  [cyclohexene ring with NH_3]

b) $\underset{\underset{NH_2}{|}}{CH_2}$—$CH_2$—$\underset{\underset{NH_2}{|}}{CH}$—$CH_3$

c) CH_3—$\overset{\overset{O}{||}}{C}$—$CH_2$—$CH_2$—$NH_2$

Solution:

a) 4-aminocyclohexene   b) 1,3-diaminobutane   c) 4-amino-2-butanone

16.10   Name the following aromatic amines as derivatives of aniline:

a) 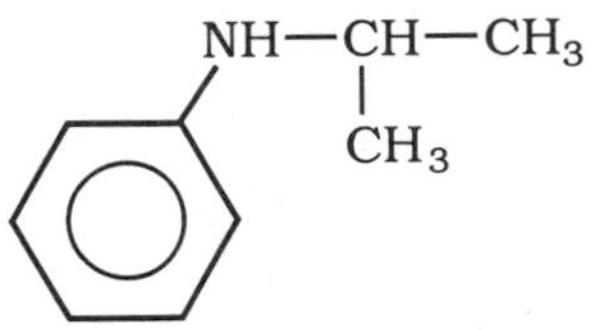 [benzene ring]—NH—$\underset{\underset{CH_3}{|}}{CH}$—$CH_3$

b) 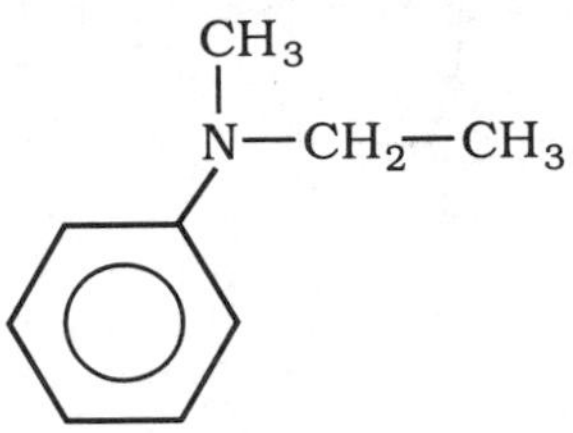 [benzene ring]—$\underset{\underset{N—CH_2—CH_3}{|}}{CH_3}$

Solution:

a) isopropylaniline

b) ethyl-*N*-methylaniline

16.12   Draw the structural formula for each of the following amines:

a)   3-amino-3-methyl-1-hexene   b) *p*-propylaniline   c) *N*,*N*-dimethylaniline

Solution:

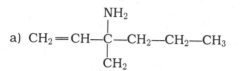

a) $CH_2{=}CH{-}\overset{\displaystyle NH_2}{\underset{\displaystyle CH_2}{C}}{-}CH_2{-}CH_2{-}CH_3$

b)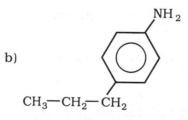

$CH_3{-}CH_2{-}CH_2$

c)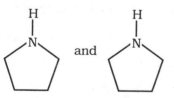

Physical Properties of Amines (Section 16.3)

16.14   Explain why all classes of low-molecular-weight amines are water-soluble.

Solution:

The nitrogen of the amine can form hydrogen bonds with the water, making the amines water soluble.

16.16   Why are the boiling points of tertiary amines lower than those of corresponding primary and secondary amines?

Solution:

Since there are no hydrogens on the nitrogen of the tertiary amines, there can be no hydrogen bonding between molecules.

16.18   Draw diagrams similar to Figure 16.1 to illustrate hydrogen bonding between the following compounds:

a) $CH_3{-}CH_2{-}NH{-}CH_3$ and H_2O

b)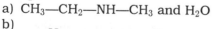

and

Solution:

a)

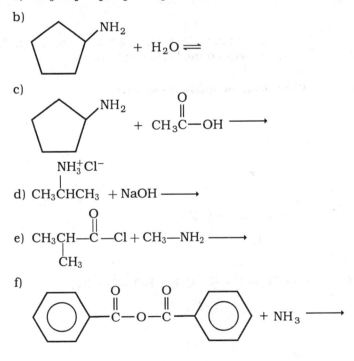

b)

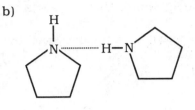

Chemical Properties of Amines (Section 16.4)

16.20 When diethylamine is dissolved in water, the solution becomes basic. Write an equation to account for this observation.

Solution:

$$CH_3-CH_2-\underset{\underset{CH_3}{\overset{|}{CH_2}}}{\overset{|}{N}}H + H_2O \rightleftharpoons CH_3-CH_2-\underset{\underset{CH_3}{\overset{|}{CH_2}}}{\overset{|}{N}}H_2 + OH^-$$

16.22 Complete the following equations. If no reaction occurs, write "no reaction."

a) $CH_3CH_2CH_2CH_2-NH_2 + HCl \longrightarrow$

b)

cyclopentyl$-NH_2$ $+ H_2O \rightleftharpoons$

c)

cyclopentyl$-NH_2$ $+ CH_3\overset{O}{\overset{||}{C}}-OH \longrightarrow$

d) $CH_3\underset{\overset{|}{NH_3^+Cl^-}}{CH}CH_3 + NaOH \longrightarrow$

e) $CH_3\underset{\overset{|}{CH_3}}{CH}-\overset{O}{\overset{||}{C}}-Cl + CH_3-NH_2 \longrightarrow$

f)

$C_6H_5-\overset{O}{\overset{||}{C}}-O-\overset{O}{\overset{||}{C}}-C_6H_5 + NH_3 \longrightarrow$

Solution:

a) $CH_3CH_2CH_2CH_2—NH_3^+Cl$

b)

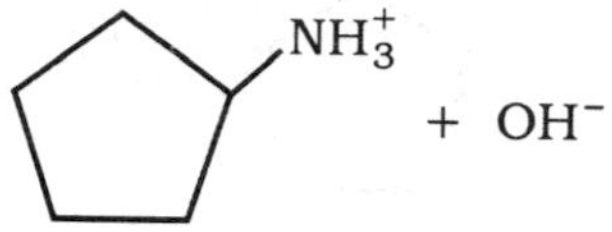

$+ \ OH^-$

c)

$+ \ CH_3\overset{\overset{\displaystyle O}{\|}}{C}—O^-$

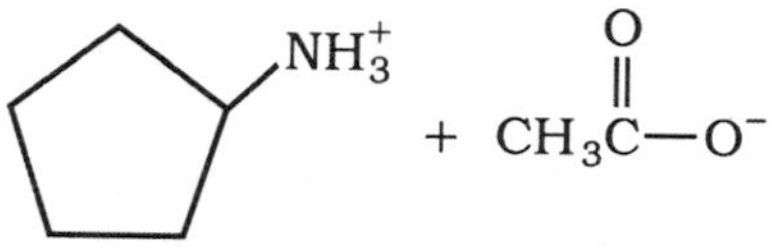

d) $CH_3\overset{\overset{\displaystyle NH_2}{|}}{C}HCH_3 + NaCl + H_2O$

e) $CH_3\overset{\underset{\displaystyle CH_3}{|}}{C}H—\overset{\overset{\displaystyle O}{\|}}{C}—NH—CH_3 + HCl$

f)

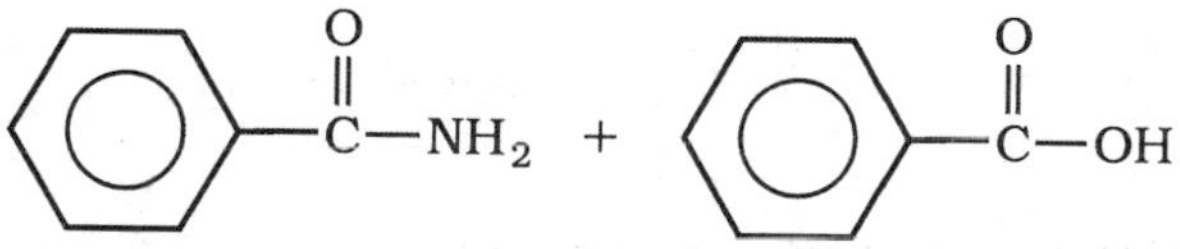

16.24 Why are amine drugs commonly administered in the form of their salts?

Solution:
The amine salts are much more soluble in water (or the aqueous fluids in the body); therefore, the active drug substance is administered in the form of the salt so the body can utilize it.

16.26 Write equations for two different methods of synthesizing the following amide:

$CH_3—\overset{\overset{\displaystyle O}{\|}}{C}—\overset{\underset{\displaystyle CH_3}{|}}{N}—CH_3$

Solution:

(1) $CH_3—\overset{\overset{\displaystyle O}{\|}}{C}—Cl + CH_3—NH—CH_3 \longrightarrow CH_3—\overset{\overset{\displaystyle O}{\|}}{C}—\overset{\underset{\displaystyle CH_3}{|}}{N}—CH_3 + HCl$

(2) $CH_3—\overset{\overset{\displaystyle O}{\|}}{C}—O—\overset{\overset{\displaystyle O}{\|}}{C}—CH_3 + CH_3—NH—CH_3 \longrightarrow CH_3—\overset{\overset{\displaystyle O}{\|}}{C}—\overset{\underset{\displaystyle CH_3}{|}}{N}—CH_3 + HO—\overset{\overset{\displaystyle O}{\|}}{C}—CH_3$

16.28   What reactants are needed to carry out the following conversions?

a)

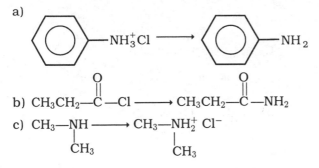

b) $CH_3CH_2{-}\overset{\overset{\displaystyle O}{\|}}{C}{-}Cl \longrightarrow CH_3CH_2{-}\overset{\overset{\displaystyle O}{\|}}{C}{-}NH_2$

c) $CH_3{-}\underset{\underset{\displaystyle CH_3}{|}}{NH} \longrightarrow CH_3{-}\underset{\underset{\displaystyle CH_3}{|}}{NH_2^+}\ Cl^-$

Solution:

a) NaOH   b) NH_3 c) HCl

Amines as Neurotransmitters (Section 16.5)

16.30   What term is used to describe the gap between one neuron and the next?

Solution:
Gaps between neurons are synapses.

16.32   Name the two amines often associated with the biochemical theory of mental illness.

Solution:
Norepinephrine and serotonin

Other Biologically Important Amines (Section 16.6)

16.34   List two neurotransmitters classified as catecholamines.

Solution:
Norepinephrine and dopamine

16.36   Describe one clinical use of epinephrine.

Solution:
Epinephrine is used clinically in local anesthetics to constrict blood vessels in the area of injection.

16.38   What is the source of alkaloids?

Solution:
Plants are the major source of alkaloids.

16.40   Give the name of an alkaloid for the following:
a) Found in cola drinks   b) Used to reduce saliva flow during surgery
c) Present in tobacco   d) A cough suppressant
e) Used to treat malaria   f) An effective pain killer

Solution:
a) caffeine   b) atropine   c) nicotine
d) codeine   e) quinine   f) morphine

The Nomenclature of Amides (Section 16.7)

16.42 Assign IUPAC names to the following amides:

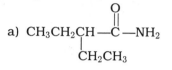

a) CH₃CH₂CH—C—NH₂
          |
          CH₂CH₃

b) 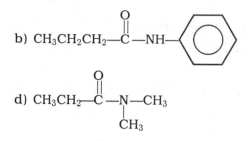 CH₃CH₂CH₂—C—NH—

c)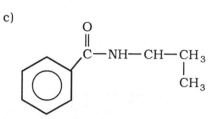

d) CH₃CH₂—C—N—CH₃
              |
              CH₃

Solution:
a) 2-ethylbutanamide
c) *N*-isopropylbenzamide

b) *N*-phenylbutanamide
d) *N,N*-dimethylpropanamide

16.44 Draw structural formulas for the following amides:

a) benzamide b) *N*-methylethanamide c) *N*-methyl-*N*-phenylbutanamide

Solution:

a)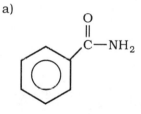

b) CH₃—C—NHCH₃

c) CH₃CHCH₂—C—NHCH₃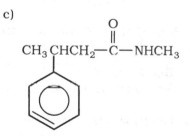

Physical Properties of Amides (Section 16.8)

16.46 Draw diagrams similar to those in Figures 16.4 and 16.5 to illustrate hydrogen bonding between the following molecules:

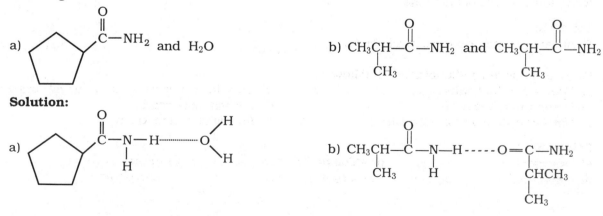

a) [cyclopentane]C—NH₂ and H₂O

b) CH₃CH—C—NH₂ and CH₃CH—C—NH₂
        | |
        CH₃   CH₃

Solution:

a) [cyclopentane]C—N—H·······O(H)(H)
                   |
                   H

b) CH₃CH—C—N—H-----O=C—NH₂
        | | |
        CH₃   H   CHCH₃
                          |
                          CH₃

Chemical Properties of Amides (Section 16.9)

16.48   Complete the following reactions:

a) $CH_3CH_2\overset{\overset{\displaystyle O}{\|}}{C}$—NH—$CH_3$ + NaOH ⟶

b) CH_3CH_2—$\overset{\overset{\displaystyle O}{\|}}{C}$—NH—$CH_3$ + H_2O + HCl ⟶

Solution:

a) CH_3CH_2—$\overset{\overset{\displaystyle O}{\|}}{C}$—$O^-$ Na^+ + CH_3—NH_2

b) CH_3CH_2—$\overset{\overset{\displaystyle O}{\|}}{C}$—OH + CH_3—$NH_3^+Cl^-$

16.50   One of the most successful mosquito repellants has the following structure and name. What are the products of the basic hydrolysis of N,N-diethyl-m-toluamide?

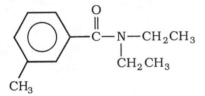

Solution:

—$\overset{\overset{\displaystyle O}{\|}}{C}$—$O_3^-$—$Na^+$ + CH_3CH_2—NH—CH_2CH

Chemistry Around Us and Over the Counter

16.52   What chemical substance is responsible for watery eyes and other symptoms of hay fever?

   Solution:
   Histamines, powerful amines released to the body in response to allergens

16.54   What are the three categories of products that are available to combat the symptoms of hay fever?

   Solution:
   Decongestants, antihistamines, and allergy preventatives

16.56   List the advantages and disadvantages of acetaminophen as an aspirin substitute.

   Solution:
   Acetaminophen does not irritate the intestinal tract and is not correlated with Reye's syndrome. Acetaminophen does not have any anti-inflammatory activity.

16.58   What compound can relieve the symptoms of Parkinson's disease in some patients?

Solution:
L-dopa has shown to relieve symptoms in some patients; however, side effects can include the onset of the symptoms of schizophrenia.

PROGRAMMED REVIEW

Section 16.1 Classification of Amines

An amine having one alkyl or aromatic group bonded to nitrogen is classified as a (a) _____ amine. A tertiary amine has (b) _____ alkyl or aromatic groups bonded to nitrogen. The amine CH_3NHCH_3 is classified as a (c) _____ amine.

Section 16.2 Nomenclature of Amines

In the IUPAC system, the $-NH_2$ group is called the (a) _____ group. The IUPAC name for ethylamine is (b) _____. The common name for $CH_3CH_2CH_2-NH-CH_3$ is (c) _____. Aromatic amines are named as derivatives of (d) _____.

Section 16.3 Physical Properties of Amines

(a) _____ amines do not form hydrogen bonds among themselves. A primary amine has a boiling point somewhat (b) _____ than the boiling point of an alcohol with similar molecular weight. Tertiary amines have boiling points similar to those of (c) _____ with similar molecular weights. In terms of water solubility, amines with fewer than six carbons are generally (d) _____.

Section 16.4 Chemical Properties of Amines

The most distinguishing feature of amines is their behavior as weak (a) _____. Methylamine reacts with water to form $CH_3-NH_3^+$ and (b) _____. Amines react with acids to form (c) _____ _____. Amine salts tend to be (d) _____ soluble in water than the parent amines. Amine salts in which (e) _____ alkyl groups are bonded to nitrogen are called quaternary ammonium salts. A primary amine can react with an acid chloride to form an (f) _____.

Section 16.5 Amines as Neurotransmitters

A (a) _____ acts as a chemical bridge in nerve impulse transmission between nerve cells. Another name for adrenalin, the "fight or flight" hormone, is (b) _____. A class of drugs structurally similar to adrenalin which are stimulants of the central nervous system is (c) _____. Because alkaloids contain (d) _____ they are usually weak bases. An alkaloid present in tobacco is (e) _____. The biological source of codeine is the (f) _____ plant. Heroin is a chemical derivative of (g) _____.

Section 16.6 Nomenclature of Amides

The characteristic IUPAC ending for amide names is (a) _____. The IUPAC name for

$$CH_3-CH_2-\overset{\overset{\displaystyle O}{\|}}{C}-NH_2$$

is (b) _____. In the name N-methylbenzamide, the letter N signifies the methyl group is attached to the (c) _____.

Section 16.7 Physical Properties of Amides

Low-molecular-weight amides are water soluble because of their ability to form (a) _____ bonds with water. Most amides exist as solids because of their ability to form a network of intermolecular (b) _____ bonds. (c) _____ amides often have lower boiling points than monosubstituted amides.

Section 16.8 Chemical Properties of Amides

Hydrolysis of amides under acidic conditions produces a (a) _____ and an (b) _____ salt.
Hydrolysis of amides under basic conditions produces an (c) _____ and a (d) _____ salt.

SELF-TEST QUESTIONS

Multiple Choice

1. The amine CH_3—$\overset{\overset{\displaystyle CH_3}{|}}{CH}$—$NH_2$ is classified as
   a) primary   b) secondary   c) tertiary   d) quaternary

2. Which of the following is a secondary amine?

   a) CH_3—CH_2—$\overset{\overset{\displaystyle NH_2}{|}}{CH}$—$CH_3$ b) CH_3—$\overset{\overset{\displaystyle CH_3}{|}}{N}$—$CH_2$—$CH_3$

   c) CH_3—CH_2—NH_2 d)

   (cyclopentane ring with NH)

3. A common name for CH_3CH_2NH—CH_2CH_3 is
   a) ethylaminoethane   b) diethylamine   c) aminoethylethane   d) diethylammonia

4. What is the IUPAC name for CH_3—$\overset{\overset{\displaystyle NH_2}{|}}{CH}$—$CH_2$—$\overset{\overset{\displaystyle CH_3}{|}}{CH}$—$CH_3$
   a) 4-amino-2-methylpentane   b) 4-amino-2-isopropylpropane
   c) 2-amino-4-methylpentane   d) 2-amino-1-isopropylpropane

5. The reaction of CH_3—$\overset{\overset{\displaystyle NH_2}{|}}{CH}$—$CH_3$ with HCl produces

   a) CH_3—$\overset{\overset{\displaystyle NH_3 + Cl^-}{|}}{CH}$—$CH_3$ b) CH_3—$\overset{\overset{\displaystyle NH}{||}}{C}$—$CH_3$ c) CH_3—$\overset{\overset{\displaystyle HN—Cl}{|}}{CH}$—$CH_3$ d) CH_3—$\overset{\overset{\displaystyle H_2N—Cl}{|}}{CH}$—$CH_3$

6. The reaction of a primary amine with a carboxylic acid chloride produces a/an
   a) secondary amine   b) tertiary amine   c) amide   d) carboxylate salt

7. What is the IUPAC name for CH_3—CH_2—$\overset{\overset{\displaystyle O}{||}}{C}$—$NH_2$
   a) 1-aminopropanamide   b) propanamide
   c) 1-aminobutanamide   d) butanamide

8. What is the IUPAC name for CH_3—CH_2—CH_2—$\overset{\overset{\displaystyle O}{||}}{C}$—$NH$—$CH_3$
   a) pentanamide   b) 1-methylpentanamide
   c) 1-methylbutanamide   d) N-methylbutanamide

9. One of the products produced when $CH_3CH_2CH_2-\overset{\overset{O}{\|}}{C}-NHCH_3$ is treated with NaOH is

   a) CH_3-NH_2

   b) $CH_3-NH_3^+Cl^-$

   c) $CH_3-CH_2-CH_2-NH_2$

   d) $CH_3-CH_2-\overset{\overset{O}{\|}}{C}-NH_2$

10. One of the products produced when $CH_3-\overset{\overset{O}{\|}}{C}-NH-CH_2-CH_3$ is treated with HCl and H_2O is

    a) $CH_3-\overset{\overset{O}{\|}}{C}-O^-$

    b) $CH_3-\overset{\overset{O}{\|}}{C}-NH_2$

    c) $CH_3-CH_2-NH_3^+Cl^-$

    d) $CH_3-CH_2-NH_2$

Matching

For each description on the left, select an amide from the list on the right.

11. _____ a tranquilizer

12. _____ an aspirin substitute

13. _____ an antibiotic

a) penicillin

b) benzamide

c) acetaminophen

d) valium

For each description on the left, select the correct alkaloid.

14. _____ present in coffee

15. _____ present in tobacco

16. _____ used as a cough suppressant

17. _____ used to dilate the pupil of the eye

a) atropine

b) nicotine

c) codeine

d) caffeine

True-False

18. Triethylamine is a tertiary amine.

19. The structure of 1-methylaniline is ⟨benzene ring⟩—NH-CH$_3$

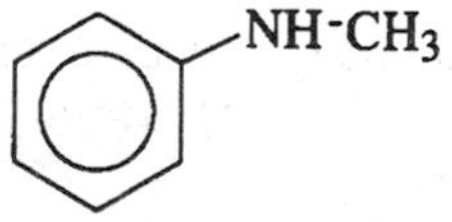

20. Tertiary amines have higher boiling points than primary and secondary amines.

21. Low-molecular-weight amines have a characteristic pleasant odor.

22. Both primary and tertiary amines can hydrogen bond with water molecules.

23. Amines under basic conditions exist as amine salts.

24. Disubstituted amides have lower boiling points than unsubstituted amides.

25. Amides with fewer than six carbons are water soluble.

26. Amide molecules are neither basic nor acidic.

27.   Serotonin is an important neurotransmitter.

28.   Amphetamine is a powerful nervous system stimulant.

SOLUTIONS

A. Answers to Programmed Review

16.1	a) primary	b) three	c) secondary
16.2	a) amino	b) aminoethane	c) methylpropylamine d) aniline
16.3	a) tertiary	b) lower	c) alkanes d) soluble
16.4	a) bases e) four	b) OH⁻ f) amide	c) amine salts d) more
16.5	a) neurotransmitter e) nicotine	b) epinephrine f) pappy	c) amphetamines g) morphine d) amines
16.6	a) -amide	b) propanamide	c) nitrogen
16.7	a) hydrogen	b) hydrogen	c) disubstituted
16.8	a) carboxylic acid	b) amine	c) amine d) carboxylate

16.4 b) OH^-

B. Answers to Self-Test Questions

1. a	11. d	21. F
2. d	12. c	22. T
3. b	13. a	23. F
4. c	14. d	24. T
5. a	15. b	25. T
6. c	16. c	26. T
7. b	17. a	27. T
8. d	18. T	28. T
9. a	19. F	
10. c	20. F	

Chapter 17

Carbohydrates

CHAPTER OUTLINE

17.1 Classes of Carbohydrates
17.2 Stereochemistry of Carbohydrates
17.3 Fischer Projections
17.4 Monosaccharides

17.5 Properties of Monosaccharides
17.6 Important Monosaccharides
17.7 Disaccharides
17.8 Polysaccharides

LEARNING OBJECTIVES

When you have completed your study of this chapter, you should be able to:
1. Describe the four major functions of carbohydrates in living organisms.
2. Classify carbohydrates as monosaccharides, disaccharides, or polysaccharides.
3. Identify molecules possessing chiral carbon atoms and use Fischer projections to represent D and L compounds.
4. Classify monosaccharides as aldoses or ketoses and classify them according to the number of carbon atoms they contain.
5. Write reactions for monosaccharide oxidation and glycoside formation.
6. Describe uses for important monosaccharides.
7. Draw the structures and list sources and uses for important disaccharides.
8. Write reactions for the hydrolysis of disaccharides.
9. Determine, on the basis of structure, whether a disaccharide is a reducing sugar.
10. Describe the structures and list sources and uses for important polysaccharides.

ANSWERS AND SOLUTIONS TO EVEN-NUMBERED PROBLEMS

Classes of Carbohydrates (Section 17.1)

17.2 Describe whether each of the following substances serves primarily as an energy source, a form of stored energy, or a structural material (some serve as more than one):
  a) cellulose   b) sucrose, table sugar   c) glycogen   d) starch

Solution:
a) a structural material in plants
b) energy storage in plants and energy source for animals
c) energy storage in animals
d) energy storage in plants

17.4   Define *carbohydrate* in terms of the functional groups present?

Solution:
A carbohydrate is polyhydroxy aldehyde or ketone, or a substance that yields such compounds upon hydrolysis.

Stereochemistry of Carbohydrates (Section 17.2)

17.6   Why are carbon atoms 1 and 3 of glyceraldehyde not considered chiral?

Solution:
They do not have 4 different groups attached. Carbon #1 has only three groups. Carbon #3 has four groups, but two of them are hydrogens.

17.8   Which of the following molecules can have enantiomers? Identify any chiral carbon atoms.

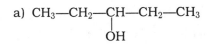

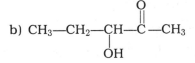

c)

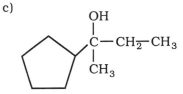

Solution:
a) no chiral carbons   b) yes, carbon #3 is chiral   c) yes, carbon #2 is chiral

Fischer Projections (Section 17.3)

17.10   Explain what the following Fischer projection denotes about the three-dimensional structure of the compound:

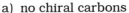

$$\begin{array}{c} CHO \\ H\!\!-\!\!\!-\!\!OH \\ CH_2CH_3 \end{array}$$

Solution:

The chiral carbon is at the intersection of the straight line. The H and OH project above the plane of the paper. The CHO and the $-CH_2-CH_3$ are below the plane of the paper.

17.12 Identify each of the following as a D or L form and draw the structural formula of the enantiomer:

a)

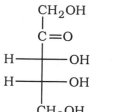

b)

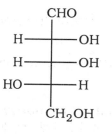

Solution:

a) D, the L enantiomer is

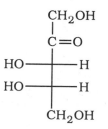

b) L, the D enantiomer is

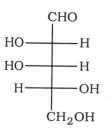

17.14 How many chiral carbon atoms are there in each of the following sugars? How many stereoisomers exist for each compound?

a)

$$\underset{|}{\overset{OH}{C}}H_2-\underset{|}{\overset{OH}{C}}H-\underset{|}{\overset{OH}{C}}H-\overset{O}{\overset{||}{C}}-H$$

b)

$$\underset{|}{\overset{OH}{C}}H_2-\overset{O}{\overset{||}{C}}-\underset{|}{\overset{OH}{C}}H-\underset{|}{\overset{OH}{C}}H-\underset{|}{\overset{OH}{C}}H-CH_3$$

Solution:

a) There are two chiral carbons and 4 stereoisomers (2^2).

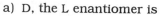

chiral carbons

b) There are three chiral carbons and 8 stereoisomers (2^3).

$$\underset{|}{\overset{OH}{C}}H_2-\overset{O}{\overset{||}{C}}-\underset{\underset{\uparrow}{|}}{\overset{OH}{C}}H-\underset{\underset{\uparrow}{|}}{\overset{OH}{C}}H-\underset{\underset{\uparrow}{|}}{\overset{OH}{C}}H-CH_3$$

chiral carbons

17.16 How many aldopentoses are possible? How many are the D form and how many the L form?

Solution:

An aldoheptose has 3 chiral carbons. There will be 8, (2^3), stereoisomers. Half of them (4) will be the D form and 4 will be the L form.

17.18 What physical property is characteristic of molecules which are optically active?

Solution:
Optically active molecules rotate the plane of polarized light.

Monosaccharides (Section 17.4)
17.20 Classify each of the following monosaccharides as an aldo- or keto-triose, tetrose, pentose, or hexose:

a)

b)

Solution:
a) an aldohexose

b) ketopentose

Properties of Monosaccharides (Section 17.5)
17.22 Explain why certain carbohydrates are called sugars.

Solution:
Those carbohydrates that have a sweet taste are called sugars.

17.24 Identify each of the following as an α or β form and draw the structural formula of the other anomer:

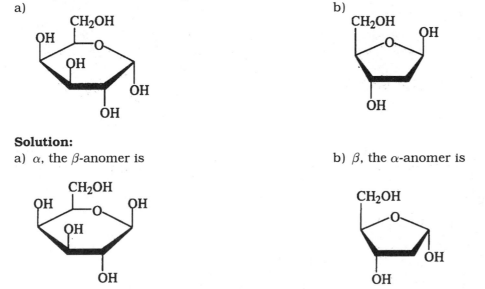

a)

b)

Solution:
a) α, the β-anomer is

b) β, the α-anomer is

17.26 The structure of talose differs from galactose only in the direction of the $-$OH group at position 2. Draw and label Haworth structures for α- and β-talose.

Solution:

α-talose   β-talose

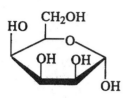

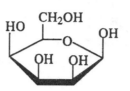

17.28 An unknown sugar failed to react with a solution of Cu^+. Classify the compound as a reducing or non-reducing sugar.

Solution:

It is a non-reducing sugar.

17.30 Complete the following reactions:

a)

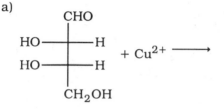

b)

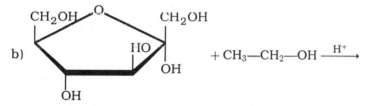

c)

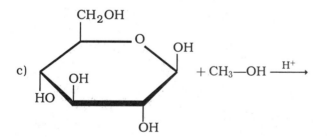

Solution:

a)

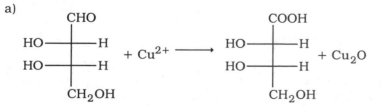

b)

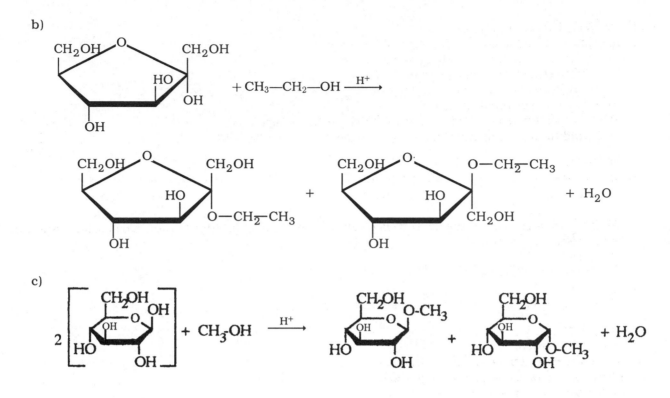

c)

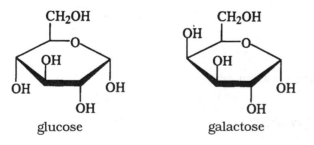

Important Monosaccharides (Section 17.6)

17.32 Explain why D-glucose can be injected directly into the bloodstream to serve as an energy source.

Solution:
D-glucose is the same compound as blood sugar.

17.34 How do the hexoses glucose and galactose differ structurally?

Solution:
The position of the OH and H on carbon #4 is reversed.

CH$_2$OH   CH$_2$OH

glucose   galactose

17.36 Explain why fructose can be used as a low-calorie sweetener.

Solution:
Fructose can be used as a low calorie sweetener because the caloric content per gram is about the same as for sucrose, but since fructose is so much sweeter, less is needed to provide the same sweetness.

Disaccharides (Section 17.7)

17.38 Identify a disaccharide that fits each of the following:
   a) The most common household sugar
   b) Converted to lactic acid when milk sours
   c) An ingredient of synthetic mother's milk
   d) Formed in germinating grain
   e) Hydrolyzes when cooked with acidic foods to give invert sugar
   f) Found in high concentrations in sugar cane

   Solution:
   a) sucrose   b) lactose   c) lactose
   d) maltose   e) sucrose   f) sucrose

17.40 Write word equations for the hydrolysis of the following disaccharides to monosaccharides. Name each of the resulting products.
   a) sucrose
   b) maltose
   c) lactose
   d) cellobiose, composed of two glucose units linked $\beta(1 \rightarrow 4)$

   Solution:
   In acidic solution,
   a) sucrose + water → glucose + fructose
   b) maltose + water → glucose + glucose
   c) lactose + water → galactose + glucose
   d) cellobiose + water → glucose + glucose

17.42 Sucrose and honey are commonly used sweeteners. Suppose you had a sweet-tasting water solution that contained either honey or sucrose. How would you chemically determine which sweetener was present?

   Solution:
   Honey contains a reducing sugar; sucrose does not. Adding Benedict's solution would give a positive test with honey.

17.44 Using structures, show why
   a) lactose is a reducing sugar
   b) sucrose is not a reducing sugar

   Solution:
   If the disaccharide has a hemiacetal, it is a reducing disaccharide.
   a) Lactose has a hemiacetal and an acetal.

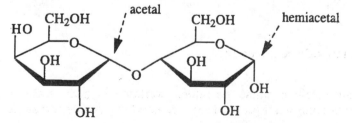

b) Sucrose has an acetal and ketal. It has no hemiacetal.

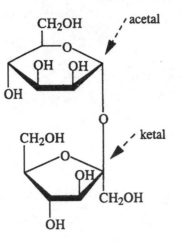

17.46 Which of the following disaccharides are reducing sugars? Explain.

a)

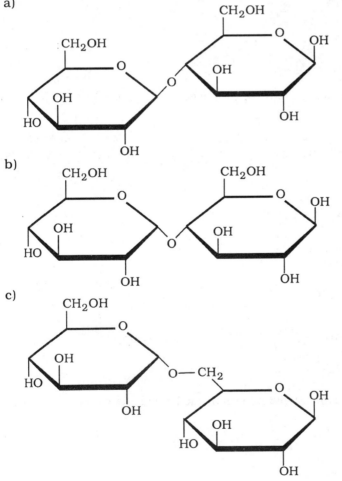

b)

c)

Solution:
All of the disaccharides are reducing sugars because in each structure, a hemi-acetal group is located in the right-hand ring.

17.48 Raffinose is a trisaccharide found in some plants.
a) What are the three monosaccharides formed upon hydrolysis of raffinose?

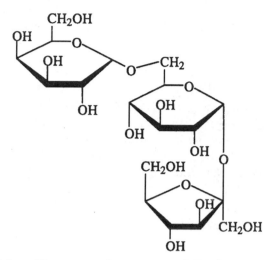

b) Is raffinose a reducing sugar? Explain.
c) What types of glycosidic linkage exist between the three monosaccharide units?

Solution:
a) galactose, glucose, and fructose
b) No, there are no hemiacetal nor hemiketal groups.
c) $\alpha(1\rightarrow6)$ and $\alpha1\rightarrow\beta2$

Polysaccharides (Section 17.8)
17.50 Name a polysaccharide that fits each of the following:
a) The unbranched polysaccharide in starch
b) A polysaccharide widely used as a textile fiber
c) The most abundant polysaccharide in starch
d) The primary constituent of paper
e) A storage form of carbohydrates in animals

Solution:

a) amylose	b) cellulose	c) amylopectin
d) cellulose	e) glycogen	

Chemistry Around Us and Key Chemicals
17.52 Name three chemical substances used as artificial sweeteners in the United States.

Solution:
saccharin, aspartame and acesulfame-K

17.54   Explain how eating an abundance of heavily sugared foods might possibly contribute to malnutrition.

Solution:
Heavily sugared foods would supply the needed calories but might lack other necessary nutrients.

17.56   Describe how dietary fiber might play a role in weight control.

Solution:
Fiber may cause people to feel full and therefore eat less.

PROGRAMMED REVIEW

Section 17.1   Classes of Carbohydrates
Carbohydrates may be defined as (a) _____ aldehydes or ketones or substances that yield such compounds upon hydrolysis. Carbohydrates formed by the combination of two monosaccharides are known as (b) _____. Carbohydrates formed by the combination of many monosaccharide units are known as (c) _____.

Section 17.2   Stereochemistry of Carbohydrates
Isomers which are mirror images of each other are called (a) _____. A chiral carbon atom has (b) _____ different groups attached. The number of stereoisomers possible in a structure containing n chiral carbon atoms is (c) _____.

Section 17.3   Fischer Projections
In a Fischer projection, a chiral carbon is represented by the (a) _____ of two lines. The two bonds coming toward the viewer in a Fischer projection are drawn (b) _____. A small capital D is used to indicate that an −OH group is on the (c) _____ in a Fischer projection. The (d) _____ enantiomer rotates a plane of polarized light to the left. The (e) _____ enantiomers of monosaccharides are preferred by the human body.

Section 17.4   Monosaccharides
Sugars with four carbon atoms are known as (a) _____. The presence of an aldehyde group in a monosaccharide may be designated by the use of the prefix (b) _____. Of the eight aldopentoses (c) _____ belong to the D series. A ketose is a carbohydrate containing a (d) _____ group.

Section 17.5   Physical Properties of Monosaccharides
Monosaccharides and disaccharides are also called (a) _____ because they taste sweet. Monosaccharides are very soluble in water because they contain several (b) _____ groups which hydrogen bond with water.

Section 17.6   Important Monosaccharides
The monosaccharide component of deoxyribonucleic acid is (a) _____. The common monosaccharide with an −OH group directed up at position 4 is (b) _____. The monosaccharide known as blood sugar is (c) _____ The sweetest of the common sugars is (d) _____

Section 17.7   Disaccharides
A disaccharide containing an $\alpha(1 \rightarrow 4)$ linkage is (a) _____. A disaccharide containing a fructose component is (b) _____. The disaccharide (c) _____ is sometimes referred to as milk sugar. The hydrolysis of sucrose produces a mixture referred to as (d) _____ sugar.

Section 17.8 Polysaccharides

Starch has a linear form called (a) _____ and a branched form called (b) _____. Glycogen has both $\alpha(1 \rightarrow 4)$ and (c) _____ linkages. Cellulose is a linear polymer of glucose units linked (d) _____.

SELF-TEST QUESTIONS

Multiple Choice

1. Which of the following is a monosaccharide?
    a) amylose   b) ribose   c) cellulose   d) lactose

2. Which of the following is a polysaccharide?
    a) amylose   b) lactose   c) maltose   d) galactose

3. How many chiral carbon atoms are in

$$CH_3-\overset{\overset{\displaystyle CH_3}{|}}{CH}-CH_2-\overset{\overset{\displaystyle Br}{|}}{CH}-CH_2-\overset{\overset{\displaystyle OH}{|}}{CH_2}$$

    a) 0   b) 1   c) 2   d) 3

4. How many chiral carbon atoms are in

$$\begin{array}{c} \overset{\displaystyle O}{\overset{\displaystyle \|}{C}}\!\cdot\!H \\ H-C-OH \\ HO-C-H \\ H_2C-OH \end{array}$$

    a) 0   b) 1   c) 2   d) 3

5. How many stereoisomers are possible for

$$CH_2-\overset{\overset{\displaystyle OH}{|}}{CH}-\overset{\overset{\displaystyle OH}{|}}{CH}-CHO$$ (with OH on first CH$_2$)

    a) 0   b) 2   c) 4   d) 8

6. Glucose is a/an
    a) ketopentose   b) ketohexose   c) aldopentose   d) aldohexose

7. Fructose is a/an
    a) ketopentose   b) ketohexose   c) aldopentose   d) aldohexose

8. A positive Benedict's test is indicated by the formation of
    a) Ag   b) CuOH   c) Cu_2O d) Cu_2^+

Matching

For each monosaccharide described on the left, select the best response from the right.

9. _____ given intravenously   a) fructose

10. _____ present with glucose in invert sugar   b) galactose

11. _____ combines with glucose to form lactose   c) glucose

12. _____ found in genetic material   d) ribose

For each disaccharide described on the left, select the best match from the responses on the right.

13. _____ used as household sugar   a) glycogen

14. _____ found in milk   b) sucrose

15. _____ formed in germinating grain   c) maltose

                                                    d) lactose

For each disaccharide on the left, select the correct hydrolysis products from the right.

16. _____ sucrose   a) glucose and galactose

17. _____ maltose   b) glucose and fructose

18. _____ lactose   c) only glucose

                                                    d) hydrolysis does not occur

Select the correct polysaccharide for each description on the left.

19. _____ a storage form of carbohydrates in animals   a) amylopectin

20. _____ most abundant polysaccharide in starch   b) amylose

21. _____ primary constituent of paper   c) glycogen

                                                         d) cellulose

True-False

22. A D enantiomer is the mirror image of an L enantiomer.

23. In a D carbohydrate, the hydroxyl group on the chiral carbon farthest from the carbonyl group points to the left.

24. The L carbohydrates are preferred by the human body.

25. Sugars that contain a hemiacetal group are reducing sugars.

26. In β-galactose, the hydroxyl group at carbon 1 points up.

27. Maltose contains a glycosidic linkage.

28. The glucose ring of lactose can exist in an open-chain form.

29. Sucrose contains a hemiacetal group.

30. Linen is prepared from cellulose.

SOLUTIONS

A. Answers to Programmed Review

17.1   a) polyhydroxy   b) disaccharides   c) polysaccharides

17.2   a) enantiomers   b) four   c) 2^n

17.3   a) intersection   b) horizontally   c) right   d) levorotatory
       e) D

17.4   a) tetroses   b) aldo   c) four   d) ketone

17.5   a) sugars   b) hydroxy

17.6   a) deoxyribose   b) galactose   c) glucose   d) fructose

17.7   a) maltose   b) sucrose   c) lactose   d) invert

17.8   a) amylose   b) amylopectin   c) $\alpha(1 \rightarrow 6)$ d) $\beta(1 \rightarrow 4)$

B. Answers to Self-Test Questions

1. b
2. a
3. b
4. c
5. c
6. d
7. b
8. c
9. c
10. a

11. b
12. d
13. b
14. d
15. c
16. b
17. c
18. a
19. c
20. a

21. d
22. T
23. F
24. F
25. T
26. T
27. T
28. T
29. F
30. T

Chapter 18

Lipids

CHAPTER OUTLINE

LEARNING OBJECTIVES

When you have completed your study of this chapter, you should be able to:
1. Classify lipids as saponifiable or non-saponifiable and list five major functions of lipids.
2. Describe four general characteristics of fatty acids.
3. Draw structural formulas of triglycerides given the formulas of the component parts.
4. Describe the structural similarities and differences of fats and oils.
5. Write key reactions for the
   a. Acid or enzyme catalyzed hydrolysis of a triglyceride.
   b. Saponification of a triglyceride.
   c. Hydrogenation of an oil.
6. Draw structural formulas and describe uses for phosphoglycerides and sphingolipids.
7. Identify the structural characteristic typical of steroids and list three important groups of steroids in the body.
8. Name the major categories of steroid hormones.
9. Describe the biological importance and therapeutic uses of the prostaglandins.

ANSWERS AND SOLUTIONS TO EVEN-NUMBERED PROBLEMS

Classification of Lipids (Section 18.1)

18.2   List two major functions of lipids in the human body.

Solution:

Energy storage and structural components are the two main functions, along with some hormones.

18.4   Classify the following as saponifiable or non-saponifiable lipids:
a) A steroid   b) A wax   c) A triglyceride
d) A phosphoglyceride   e) A glycolipid   f) A prostaglandin

Solution:
a) non-saponifiable   b) saponifiable   c) saponifiable
d) saponifiable   e) saponifiable   f) non-saponifiable

Fatty Acids (Section 18.2)

18.6   Describe the structure of a micelle formed by the association of fatty acid molecules in water. What forces hold the micelle together?

Solution:

A micelle of fatty acid molecules in water is a spherical cluster of fatty acids with the hydrophilic, polar ends on the surface of the cluster and the hydrophobic, nonpolar parts in the interior of the cluster. Dispersion forces hold the nonpolar parts of the micelle together.

18.8   Name two essential fatty acids and explain why they are called essential.

Solution:

Linoleic acid and linolenic acid are essential fatty acids. The are called essential because they cannot be synthesized by humans and are needed by the body.

The Structure of Fats and Oils (Section 18.3)

18.10   How are fats and oils structurally similar? How are they different?

Solution:

They both are triglycerides (esters) between glycerol and fatty acids. Fats are solid at room temperature and are derived from the more saturated fatty acids. Oils are liquids at room temperature and are derived from the more unsaturated fatty acids.

18.12   From Figure 18.7, arrange the following substances in order of increasing percentage of unsaturated fatty acids: eggs, canola oil, lard, olive oil, butter.

Solution:

butter, lard, egg, olive oil, canola oil

18.14   The percent fatty acid composition of two triglycerides is reported below. Predict which triglyceride has the lower melting point.

	Palmitic acid	Stearic acid	Oleic acid	Linoleic acid
Triglyceride A	20.4	28.8	38.6	10.2
Triglyceride B	9.6	7.2	27.5	55.7

Solution:
Triglyceride B. It has a lower percentage of saturated fatty acids (palmitic, stearic) and a higher percentage of unsaturated (oleic, linoleic). It will have the lower melting point.

Chemical Properties of Fats and Oils (Section 18.4)

18.16   In general terms name the products of the reactions below:
   a) Acid hydrolysis of a fat
   b) Acid hydrolysis of an oil
   c) Saponification of a fat
   d) Saponification of an oil

Solution:
   a) saturated fatty acids and glycerol
   c) saturated fatty acid salts and glycerol

   b) unsaturated fatty acids and glycerol
   d) unsaturated fatty acid salts and glycerol

18.18   What process is used to prepare a number of useful products such as margarines and cooking shortenings from vegetable oils?

Solution:
hydrogenation

18.20   Write reactions to show how each of the following products might be prepared from a typical triglyceride present in the source given. Use Table 18.1 as an aid:
   a) glycerol from beef fat
   c) a margarine from corn oil

   b) stearic acid from beef fat
   d) soaps from lard

Solution:
a)

$$H_2C-O-\overset{O}{\overset{\|}{C}}-(CH_2)_{16}-CH_3$$
$$HC-O-\overset{O}{\overset{\|}{C}}-(CH_2)_{16}-CH_3 + HCl + 3H_2O \longrightarrow \overset{OH}{H_2C}-\overset{OH}{CH}-\overset{OH}{CH_2} + 3\left[CH_3(CH_2)_{16}-\overset{O}{\overset{\|}{C}}-OH\right]$$
$$H_2C-O-\overset{O}{\overset{\|}{C}}-(CH_2)_{16}-CH_3$$

(glycerol)   (stearic acid)

b)

$$H_2C-O-\overset{O}{\overset{\|}{C}}-(CH_2)_{16}-CH_3$$
$$HC-O-\overset{O}{\overset{\|}{C}}-(CH_2)_{16}-CH_3 + HCl + 3H_2O \longrightarrow \overset{OH}{H_2C}-\overset{OH}{CH}-\overset{OH}{CH_2} + 3\left[CH_3(CH_2)_{16}-\overset{O}{\overset{\|}{C}}-OH\right]$$
$$H_2C-O-\overset{O}{\overset{\|}{C}}-(CH_2)_{16}-CH_3$$

c)

$$H_2C-O-\overset{O}{\overset{\|}{C}}-(CH_2)_6-(CH_2-CH=CH)_2-(CH_2)_4-CH_3$$
$$HC-O-\overset{O}{\overset{\|}{C}}-(CH_2)_6-(CH_2-CH=CH)_3-CH_2-CH_3 + 5H_2 \xrightarrow{Ni}$$
$$H_2C-O-\overset{O}{\overset{\|}{C}}-(CH_2)_2-(CH_2-CH=CH)_2-(CH_2)_4-CH_3$$

$$H_2C-O-\overset{O}{\overset{\|}{C}}-(CH_2)_{16}-CH_3$$
$$HC-O-\overset{O}{\overset{\|}{C}}-(CH_2)_{16}-CH_3$$
$$H_2C-O-\overset{O}{\overset{\|}{C}}-(CH_2)_2-(CH_2-CH=CH_2)_2-(CH_2)_4-CH_3$$

Note: A total hydrogenation is shown. It is common to do a partial hydrogenation and leave some double bonds.

d)

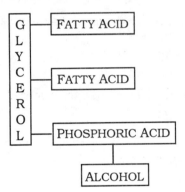

$H_2C-O-\overset{O}{\overset{\|}{C}}-(CH_2)_{16}-CH_3$
$H\overset{}{C}-O-\overset{O}{\overset{\|}{C}}-(CH_2)_{16}-CH_3$ $+$ 3 NaOH $\longrightarrow$ $H_2\overset{OH}{\overset{|}{C}}-\overset{OH}{\overset{|}{C}H}\cdot\overset{OH}{\overset{|}{C}H_2}$ $+$ 3 $\left[CH_3(CH_2)_{16}\overset{O}{\overset{\|}{C}}-O^-\ Na^+ \right]$
$H_2C-O-\overset{O}{\overset{\|}{C}}-(CH_2)_{16}-CH_3$

(glycerol)   (soap)

Waxes (Section 18.5)

18.22   Draw the structure of a wax formed from myristic acid and cetyl alcohol.

Solution:

$CH_3-(CH_2)_{12}-\overset{O}{\overset{\|}{C}}-O-CH_2-(CH_2)_{14}-CH_3$

Phosphoglycerides (Section 18.6)

18.24   Draw the general block diagram structure of a phosphoglyceride.

Solution:

```
G ┐  ┌──────────────┐
L ├──┤  FATTY ACID   │
Y │  └──────────────┘
C ┐  ┌──────────────┐
E ├──┤  FATTY ACID   │
R │  └──────────────┘
O ┐  ┌──────────────────┐
L ├──┤ PHOSPHORIC ACID   │
  │  └──────────────────┘
       │
       ┌──────────┐
       │ ALCOHOL   │
       └──────────┘
```

18.26   Describe two biological roles served by the lecithins.

Solution:
Lecithins are components of cell membranes and they aid in lipid transport.

18.28   What is the structural difference between a lecithin and a cephalin?

Solution:
They differ in the aminoalcohol ester on the phosphate. Cephalins are esters of serine or ethanolamine, and lecithins are choline esters.

Sphingolipids (Section 18.7)

18.30   Draw the general block diagram structure of a sphingolipid.

Solution:

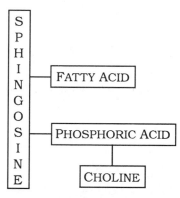

18.32   Give another name for cerebrosides. In what tissues are they found?

Solution:
Cerebrosides are also called glycolipids and are found in brain tissue.

Biological Membranes (Section 18.8)
18.34   What three classes of lipids are found in membranes?

Solution:
Phosphoglycerides, sphingomyelins, and cholesterol are the three classes of lipids in the membranes.

18.36   Describe the major features of the fluid-mosaic model of cell membrane structure.

Solution:
The membrane is a bi-layer having some flexibility as the lipid molecules may have some lateral movement. Protein molecules "float" in the bi-layer.

Steroids (Section 18.9)
18.38   Why is it suggested that some people restrict cholesterol in-take in their diet?

Solution:
According to many studies, lowering high blood cholesterol levels reduces the risk of heart disease and heart attack.

18.40   Explain how bile salts aid in the process of digestion.

Solution:
Bile salts emulsify lipid globules into smaller lipid micelles. This increases the surface area exposed to the digestive reactions and speeds up the hydrolysis of the fats.

18.42   What is the major component in gallstones?

Solution:
cholesterol (~80%)

Steroid Hormones (Section 18.10)

18.44 Name the two groups of adrenocorticoid bormones, give a specific example of each group, and explain the function of those compounds in the body.

Solution:

(1) Mineralocorticoids, such as aldosterone, regulate ion concentration in body fluids.

(2) Glucocorticoids, such as cortisol, cortisone or prednisolone, enhance carbohydrate metabolism.

18.46 Name the primary male sex hormone and the three principal female sex hormones.

Solution:

Testosterone is the primary male sex hormone. Estradiol, estrone and progesterone are the three principal female sex hormones.

18.48 What role do the estrogens and progesterone serve in preparation for pregnancy?

Solution:

Estrogens are involved in egg development; progesterone causes changes in the wall fo the uterus to prepare it to accept a fertilized egg and maintain the resulting pregnancy.

Prostaglandins (Section 18.11)

18.50 What compound serves as a starting material for prostaglandin synthesis?

Solution:

arachidonic acid

18.52 Name three therapeutic uses of prostaglandins.

Solution:

(1) induce labor   (2) treat asthma   (3) treat peptic ulcers

Chemistry Around Us and Key Chemicals

18.54 What hormones are formed from DHEA?

Solution:

testosterone and estrogen

18.56 Why is the synthetic hormone progestin usually taken along with estrogen therapy?

Solution:

It aids in reducing the risk of endometrial cancer by causing the shedding of the endometrium monthly.

18.58 What side effects might possibly result from consuming excessive amounts of Olestra-containing foods?

Solution:

Some gastrointestinal distress (diarrhea) has been reported. Also, many oil-soluble vitamins might be carried away from the body as the fat-substitute is excreted.

18.60 What factors result in an increased blood cholesterol level?

Solution:
cigarette smoking, obesity, and an inactive life-style

PROGRAMMED REVIEW

Section 18.1 Classification of Lipids

Saponifiable lipids all contain an (a) _____ functional group. Prostaglandins belong to the class of (b) _____ lipids. Simple lipids contain two components, a fatty acid and an (c) _____. Saponifiable lipids may be classified as either simple or (d) _____.

Section 18.2 Fatty Acids

In aqueous solution fatty acids form spherical clusters called (a) _____. Fatty acids usually have an (b) _____ number of carbon atoms. Long-chain (c) _____ fatty acids are usually liquids at room temperature. Fatty acids containing no carbon-carbon double bonds are referred to as (d) _____.

Section 18.3 The Structure of Fats and Oils

The alcohol component of a fat is (a) _____. Fats and oils may also be referred to as triacyl-glycerols or (b) _____. Fats are usually derived from (c) _____ sources. Oils consist of triesters containing (d) _____ fatty acids.

Section 18.4 Chemical Properties of Fats and Oils

A reaction of a fat with water under acidic conditions is called (a) _____. Saponification of an oil produces soaps and (b) _____. Semisolid cooking shortenings are produced from oils by a (c) _____ reaction. Oils containing two or more double bonds are referred to as (d) _____.

Section 18.5 Waxes

The (a) _____ functional group is present in waxes. Waxes consist of an alcohol component and a (b) _____ component. In terms of water solubility, waxes are (c) _____.

Section 18.6 Phosphoglycerides

Phosphoglycerides serve as major components of cell (a) _____. The amino alcohol portion of lecithin is (b) _____. Soybean lecithin is used in foods as an (c) _____ agent. Cephalins are particularly abundant in (d) _____ tissue.

Section 18.7 Sphingolipids

Sphingolipids contain an alcohol component called (a) _____. Sphingolipids are abundant in nerve and (b) _____ tissue. A type of sphingolipid which contains a carbohydrate component is called a (c) _____.

Section 18.8 Biological Membranes

Two major cell types are found in living organisms, prokaryotic and the more complex (a) _____. Membrane-enclosed bodies within cells are called (b) _____. A widely accepted structure for membranes is called the (c) _____-_____ model. Lipids in a membrane are organized in a (d) _____, a structure consisting of two sheets of lipid molecules arranged so that the hydrophobic portions are facing each other.

Section 18.9 Steroids

The number of rings contained in the steroid ring system is (a) _____. The steroid (b) _____ has been implicated in hardening of the arteries. Steroids emptied from the gallbladder to aid in digestion are called (c) _____. The major component of gallstones is (d) _____.

Section 18.10 Steroid Hormones

A class of adrenocorticoid hormones which regulates the concentration of ions in body fluids is the (a) _____. The hormone which regulates body levels of Na^+ is (b) _____. The major glucocorticoid, (c) _____, functions to increase glucose and glycogen concentrations in the body. Female sex hormones are produced in the (d) _____.

Section 18.11 Prostaglandins

Prostaglandins are synthesized from (a) _____ fatty acids. Prostaglandins are similar to (b) _____ in the sense that they are involved in a host of body processes.

SELF-TEST QUESTIONS

Multiple Choice

1. All simple lipids are
   a) salts of fatty acids
   b) esters of fatty acids with various alcohols
   c) esters of fatty acids with alcohol and other additional compounds
   d) esters of fatty acids with glycerol

2. The esters of fatty acids and a long chain alcohol are known as
   a) waxes   b) phospholipids   c) compound lipids   d) fats

3. Which of the following is a glycerol-containing lipid?
   a) sphingolipid   b) glycolipid   c) phospholipid   d) prostaglandin

4. Which of the following is a complex lipid?
   a) steroid   b) sphingomyelin   c) prostaglandin   d) triacylglycerol

5. Which fatty acid is most likely to be found in an oil?
   a) $CH_3(CH_2)_7CH{=}CH(CH_2)_7COOH$ b) $CH_3(CH_2)_{14}COOH$
   c) $CH_3(CH_2)_{16}COOH$ d) $CH_3(CH_2)_{18}COOH$

6. Which of the following food sources would most likely be highest in saturated fatty acids?
   a) cottonseed   b) corn   c) beef   d) sunflower seeds

7. Generally, the structural difference between a fat and an oil is the
   a) alcohol
   b) chain length of fatty acids
   c) degree of fatty-acid unsaturation
   d) degree of fatty-acid chain-branching

8. In triglycerides, fatty acids are joined to glycerol by
   a) ester linkages   b) ether linkages   c) phosphate linkages   d) hydrogen bonds

9. Which of the following is an essential fatty acid?
   a) stearic acid   b) myristic acid   c) linoleic acid   d) palmitic acid

Matching

Match the following formulas to the correct lipid classification given as a response.

10.

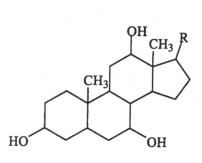

a) steroid

b) phosphoglyceride

c) fat or oil

d) wax

11.

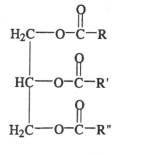

12.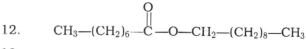

$CH_3-(CH_2)_6-\overset{O}{\overset{\|}{C}}-O-CH_2-(CH_2)_8-CH_3$

13.

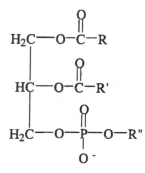

Materials can be obtained from lipids, and the lipids changed by chemical processes. Choose the correct process to accomplish each change described below.

14.   obtain a high molecular weight alcohol from a plant wax

15.   obtain glycerol from an oil

16.   obtain soaps from an oil

17.   raise the melting point of an oil

a) hydrogenation

b) acid-catalyzed hydrolysis

c) saponification

d) more than one listed process would work

In the following reaction, L represents a lipid, and A, B, C and D represent possible hydrolysis products.

18. A = fatty acids, B = glycerol,
    C = phosphoric acid, D = choline

19. A = fatty acids, B = glycerol,
    no other products

20. A = fatty acids, B = sphingosine,
    C = phosphoric acid, D = choline

21. A = fatty acids, B = sphingosine,
    C = carbohydrate, no other product forms

a) simple lipid

b) glycolipid

c) phospholipid

d) sphingolipid

True-False

22. Cell membranes contain about 60% lipid and 40% carbohydrate.

23. Cell membranes are thought to be relatively flexible.

24. A compound containing nine carbon atoms could not be a steroid.

25. In their physiological action, the prostaglandins resemble hormones.

26. All of the male and female sex hormones are steroids.

27. The hormone aldosterone exerts its influence at the pancreas.

SOLUTIONS

A. Answers to Programmed Review

18.1 a) ester	b) non-saponifiable	c) alcohol	d) complex
18.2 a) micelles	b) even	c) unsaturated	d) saturated
18.3 a) glycerol	b) triglycerides	c) animal	d) unsaturated
18.4 a) hydrolysis	b) glycerol	c) hydrogenation	d) polyunsaturated
18.5 a) ester	b) fatty acid	c) insoluble	
18.6 a) membranes	b) choline	c) emulsifying	d) brain
18.7 a) sphingosine	b) brain	c) glycolipid	
18.8 a) eukaryotic	b) organelles	c) fluid-mosaic	d) bilayer
18.9 a) four	b) cholesterol	c) bile salts	d) cholesterol
18.10 a) mineralocorticoids	b) aldosterone	c) cortisol	d) ovaries
18.11 a) unsaturated	b) hormones		

B. Answers to Self-Test Questions

1. b	10. a	19. a
2. a	11. c	20. d
3. c	12. d	21. b
4. b	13. b	22. F
5. a	14. d	23. T
6. c	15. c	24. T
7. c	16. c	25. T
8. a	17. a	26. T
9. c	18. c	27. F

Chapter 19

Proteins

CHAPTER OUTLINE
19.1 The Amino Acids
19.2 Zwitterions
19.3 Reactions of Amino Acids
19.4 Important Peptides
19.5 Characteristics of Proteins

19.6 The Primary Structure of Proteins
19.7 The Secondary Structure of Proteins
19.8 The Tertiary Structure of Proteins
19.9 The Quaternary Structure of Proteins
19.10 Protein Hydrolysis and Denaturation

LEARNING OBJECTIVES

When you have completed your study of this chapter, you should be able to:
1. Identify the characteristic parts of α-amino acids and classify the R groups according to polarity and acid-base character.
2. Draw structural formulas to illustrate the various ionic forms assumed by amino acids.
3. Write reactions to represent the oxidation of cysteine to give the disulfide cystine and the formation of peptides and proteins.
4. Correctly represent peptide and protein structures using three-letter abbreviations for amino acids.
5. Describe proteins in terms of the following characteristics: size, function, classification as fibrous or globular, and classification as simple or conjugated.
6. Explain what is meant by the primary, secondary, tertiary, and quaternary structure of proteins.
7. Describe the role of hydrogen bonding in the secondary structure of proteins and side-chain interactions in the tertiary structure of proteins.
8. Describe the conditions that can cause proteins to hydrolyze or become denatured.

ANSWERS AND SOLUTIONS TO EVEN-NUMBERED PROBLEMS

The Amino Acids (Section 19.1)
19.2 Draw the structure of butanoic acid. Label the alpha carbon.

Solution:

CH$_3$—CH$_2$—CH$_2$—C$\overset{\displaystyle O}{\overset{\|}{}}$—OH

\
alpha carbon

19.4   Draw structural formulas for the following amino acids, identify the chiral carbon atom in each one, and circle the four different groups attached to the chiral carbon.
a) threonine   b) aspartate   c) serine   d) phenylalanine

Solution:

a)

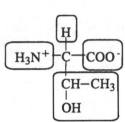

b)

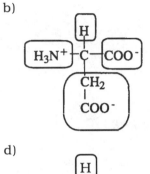

c)

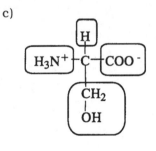

d)

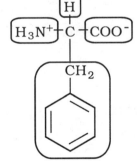

19.6   Draw Fischer projections representing the D and L forms of the following:
a) cysteine   b) glutamate

Solution:

a)

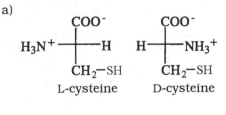

b)

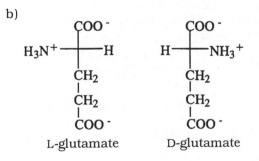

Zwitterions (Section 19.2)
19.8   What characteristics indicate that amino acids exist as zwitterions?

Solution:
They are very water soluble and are crystalline solids at room temperature with high melting points.

19.10 Write structural formulas to show the form the following amino acids would have in a solution with a pH higher than the amino acid isoelectric point:
    a) glutamine   b) proline

Solution:
The amino acid will have lost the proton from the $-NH_3^+$ at pH above the isoelectric point, giving an overall negatively charged ion.

a) $H_2N-CH-COO^-$
              CH_2
              CH_2
              $C=O$
              NH_2

b) $HN-CH-COO^-$
          $CH_2\ CH_2$
             CH_2

19.12 Write ionic equations to show how serine acts as a buffer against the following added ions:
    a) OH^- b) H_3O^+

Solution:

a)

$H_3N^+-CH-COO^- + OH^- \longrightarrow H_2N-CH-COO^- + H_2O$
           CH_2 CH_2
           HO HO

b)

$H_3N^+-CH-COO^- + H_3O^+ \longrightarrow H_3N^+-CH-COOH + H_2O$
           CH_2 CH_2
        $HC-CH_3$ $HC-CH_3$
          CH_3 CH_3

Reactions of Amino Acids (Section 19.3)

19.14 Write a complete structural formula and an abbreviated formula for the tripeptide formed from aspartate, cysteine, and valine in which the C-terminal residue is cysteine and the N-terminal residue is valine.

Solution:

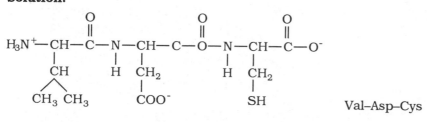

Val–Asp–Cys

19.16   How many tripeptide isomers are possible that contain one residue of each of valine, phenylalanine and lysine?

Solution:
There are six possible arrangements of three different amino acids.

Important Peptides (Section 19.4)

19.18   What special role does the amino acid cysteine have in the peptides vasopressin and oxytocin?

Solution:
Both vasopressin and oxytocin have 9 amino acid residues and both have a C-terminal amide. Both have cysteine as the #1 and the #6 amino acid residues. The two cysteine residues in the peptides form a disulfide linkage between them, forming a loop structure. Only two of the nine amino acid residues are different in the two peptides.

Characteristics of Proteins (Section 19.5)

19.20   Explain why the presence of certain proteins in body fluids such as urine or blood can indicate that cellular damage has occurred in the body.

Solution:
Protein molecules are formed inside of living cells, and are too large to pass through healthy cell membranes. Finding excessive protein in extra-cellular fluids such as blood or urine indicates cell membrane damage, which allowed the proteins to escape.

19.22   Explain why a protein is least soluble in an aqueous medium that has a pH equal to the isoelectric point of the protein.

Solution:
At the isoelectric point the protein has no net charge. At pH higher than the isoelectric point, the protein is negatively charged, and at lower pH, positively charged. The charged protein is more soluble in water than the non-charged protein found at the isoelectric point.

19.24   List the eight principal functions of proteins.

Solution:
Catalysis, structural, storage, protection, regulatory, nerve impulse transmission, motion, and transport.

19.26   For each of the following two proteins listed in Table 19.4, predict whether it is more likely to be a globular or a fibrous protein. Explain your reasoning.
a) collagen   b) lactate dehydrogenase

Solution:
Collagen is most likely to be fibrous because it is a structural protein. Lactate dehydrogenase is likely to be a globular protein as it has a catalytic function and must be soluble to be transported to needed locations.

19.28   Differentiate between simple and conjugated proteins.

Solution:
A conjugated protein contains the amino acid residues found in a simple protein plus other organic or inorganic components (prosthetic groups).

The Primary Structure of Proteins (Section 19.6)

19.30   Describe what is meant by the term *primary structure of proteins*.

Solution:
The linear sequence of the amino acid residues in the protein determines the primary structure.

19.32   Write the structure for a protein backbone. Make the backbone long enough to attach four R groups symbolizing amino acid side chains.

Solution:

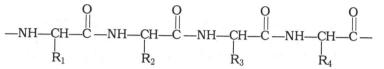

The Secondary Structure of Proteins (Section 19.7)

19.34   Describe what is meant by "supracoiling" in proteins.

Solution:
Two or more α-helices combine along their length to form a cable-like bundle.

The Tertiary Structure of Proteins (Section 19.8)

19.36   How do hydrogen bonds involved in tertiary protein structures differ from those involved in secondary structures?

Solution:
Secondary structure results from hydrogen bonding between groups along the "backbone" of the protein. Tertiary structure results from bonding between the R "side groups". Thus, if the hydrogen bonds are between "backbone" residues, secondary structures result. If the hydrogen bond (or other bonds) occur between "side groups" on the residues tertiary structures result.

19.38   Refer to Table 19.1 and list the type of side-chain interaction expected between the side chains of the following pairs of amino acid residues:
   a) tyrosine and glutamine   b) aspartate and lysine
   c) leucine and isoleucine   d) phenylalanine and valine

Solution:
   a) Hydrogen bonding   b) Salt bridges
   c) Hydrophobic interactions   d) Hydrophobic interactions

19.40   A globular protein in aqueous surroundings contains the following amino acid residues: phenylalanine, methionine, glutamate, lysine, and alanine. Which amino acid side chains would be directed toward the inside of the protein and which would be directed toward the aqueous surroundings?

Solution:
The nonpolar side groups will point inward in phenylalanine, methionine, and alanine. The polar side groups will point outward toward the water-based surrounding in glutamate and lysine.

The Quaternary Structure of Proteins (Section 18.9)

19.42   What types of forces give rise to quaternary structure?

Solution:

Ionic attractions, disulfide bridges, hydrogen bonds, and hydrophobic forces between sub-units give rise to quaternary structure.

19.44   What is meant by the term *subunit*?

Solution:

A subunit is a polypeptide chain that is part of a larger protein. Subunits have primary, secondary, and tertiary structural features. The arrangement of the subunits to form the larger protein is known as the quartenary structure.

Protein Hydrolysis and Denaturation (Section 19.10)

19.46   Suppose a sample of a protein is completely hydrolyzed and another sample of the same protein is denatured. Compare the final products of each process.

Solution:

Completely hydrolyzed protein gives the individual amino acids that made up the protein. In denatured protein, the secondary, tertiary, and quaternary structures are modified, leaving the original primary structure of the protein.

19.48   In what way is the protein in a raw egg the same as that in a cooked egg?

Solution:

The protein in a raw egg is the same as that in a cooked egg as to the composition and sequence of amino acids. The primary structure is the same. The 2°, 3°, and 4° structures are modified.

19.50   Once cooked, egg whites remain in a solid form. However, egg whites that are beaten to form meringue will partially change back to a jellylike form if allowed to stand for a while. Explain these behaviors using the concept of reversible protein denaturation.

Solution:

Whipping air into the egg white to make meringue is a reversible denaturation. As the air escapes, the egg white reverts to a structure similar to the original. Cooking is an irreversible denaturation.

19.52   Explain how egg white can serve as an emergency antidote for heavy metal poisoning.

Solution:

The heavy metal ions denature the protein in the egg white by interacting with the $-SH$ and COO^- groups. The heavy metal ions are tied up in the denatured egg white. Vomiting should then be induced to remove the denatured proteins, so that the denatured egg white proteins are not subsequently digested, releasing the metal ions again.

Chemistry Around Us and Key Chemicals

19.54   List three drugs that should never be taken by a nursing mother.

Solutions:

Any three of the following: bromocriptine, cancer chemotherapy drugs, ergotamine, lithium, methotrexate, nicotine, and the illegal street drugs.

19.56   What aspect of Alzheimer's disease can be treated with some success?

Solution:

Memory loss can be treated; however, this is purely symptomatic treatment and does not control the disease.

19.58   What is the molecular basis of sickle-cell disease?

Solution:

In individuals affected with sickle-cell disease, abnormal hemoglobin molecules have a different sequence of amino acid residues than the normal hemoglobin. The slight difference in the amino acid sequence causes the hemoglobin to take on a flat sickle shape after it gives up its oxygen. The sickled cells will tend to clump together and block capillaries, which causes pain in the affected individual.

PROGRAMMED REVIEW

Section 19.1 The Amino Acids

All twenty common amino acids contain two functional groups, an amino group and a (a) _____ group. Amino acids found in living systems usually exist in the (b) _____ enantiomeric form. Amino acids may be represented by (c) _____ -letter abbreviations. There are (d) _____ amino acids which contain an acidic R group.

Section 19.2 Zwitterions

The net charge on a zwitterion is (a) _____. In basic solutions an amino acid has a net (b) _____ charge. The pH at which the zwitterionic form of an amino acid predominates is called the (c) _____.

Section 19.3 Reactions of Amino Acids

The SH-containing amino acid which can be oxidized to a disulfide is (a) _____. The amide linkage between two amino acid components is also called a (b) _____ bond. The amino acid shown on the right side of a peptide is the (c) _____ residue. Polypeptide chains with more than (d) _____ amino acids are usually called proteins.

Section 19.4 Important Peptides

The antidiuretic hormone (ADH) is also known as (a) _____. The pituitary gland synthesizes the (b) _____ hormone which regulates the productivity of the adrenal gland.

Section 19.5 Characteristics of Proteins

Proteins which serve a catalytic function are called (a) _____. Proteins made up of long string-like molecules are classified as (b) _____ proteins. Proteins comprised solely of amino acids are called (c) _____ proteins. (d) _____ groups are the non-amino acid parts of conjugated proteins.

Section 19.6 The Primary Structure of Proteins

The primary structure of a protein is held together by (a) _____ bonds. The order in which amino acid residues are linked together is the (b) _____ structure of a protein.

Section 19.7 The Secondary Structure of Proteins

The two types of protein secondary structure are the (a) _____ and the (b) _____. Secondary structure of a protein is held intact by (c) _____ bonds.

Section 19.8 The Tertiary Structure of Proteins Secondary

The tertiary structure of proteins results from interactions between the (a) _____ of the amino acids. Interactions between two nonpolar groups are termed (b) _____ interactions. The attraction of oppositely charged side chains gives rise to (c) _____ bridges. Two alcohol-containing side chains may interact to form (d) _____ bonds.

Section 19.9 The Quaternary Structure of Proteins

The polypeptide chains comprising a quaternary structure are called (a) _____. The quaternary structure of hemoglobin consists of (b) _____ chains.

Section 19.10 Protein Hydrolysis and Deaaturation

Protein hydrolysis results in the formation of smaller peptides and (a) _____. The natural three-dimensional structure of a protein is called the (b) _____ state. (c) _____ is the process by which a protein loses its characteristic structure.

SELF-TEST QUESTIONS

Multiple Choice

1. The main distinguishing feature between various amino acids is
    a) the length of the carbon chain
    b) the number of amino groups
    c) the compostition of the side chain
    d) the number of acid groups

2. The amino acid valine is represented below. Which of the lettered carbon atoms is the alpha carbon atom?

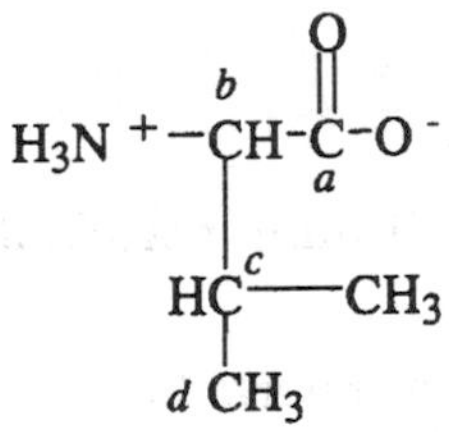

$$
\begin{array}{c}
\overset{\displaystyle O}{\underset{\displaystyle a}{\overset{b}{H_3N\ ^+-CH-}\overset{\|}{C}-O^-}} \\
\overset{c}{HC}\!-\!CH_3 \\
d\ CH_3
\end{array}
$$

    a) a b) b c) c d) d

3. The compound

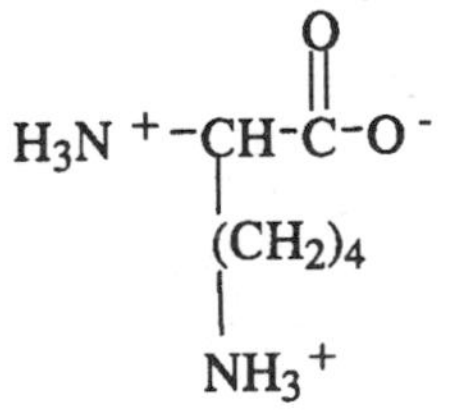

$$
\begin{array}{c}
H_3N\ ^+-\overset{\displaystyle O}{\overset{\|}{CH-C}}-O^- \\
(CH_2)_4 \\
NH_3{}^+
\end{array}
$$

    is a/an _____ amino acid.

    a) acidic
    b) basic
    c) neutral
    d) more than one response is correct

4. Which of the following would probably represent alanine at its isoelectric point?

a) N_3N^+—CH—C$\overset{O}{\overset{\|}{}}$—O$^-$
   |
   CH$_3$

b) N_2N—CH—C$\overset{O}{\overset{\|}{}}$—OH
   |
   CH$_3$

c) N_2N—CH—C$\overset{O}{\overset{\|}{}}$—O$^-$
   |
   CH$_3$

d) N_3N^+—CH—C$\overset{O}{\overset{\|}{}}$—OH
   |
   CH$_3$

5. A linkage present in all peptides is

a) H_2N—CH_2

b) H_2N—CH—
        |
        R

c) —C$\overset{O}{\overset{\|}{}}$—NH—

d) —C$\overset{O}{\overset{\|}{}}$—O—R

6. Prosthetic groups are found in
   a) all proteins   b) simple proteins   c) conjugated proteins   d) no proteins

7. A protein that is relatively spherical in shape and fairly soluble in water is a _____ protein.
   a) simple   b) conjugated   c) fibrous   d) globular

8. Which of the following characteristics of a protein would be classified as a primary structural feature?
   a) amino acid sequence
   b) pleated-sheet configuration
   c) α-helix configuration
   d) the shape of the protein molecule

9. Which protein serves as antibodies?
   a) myoglobin   b) hemoglobin   c) collagen   d) immunoglobulin

10. Which of the following side-group interactions involves nonpolar groups?
    a) salt bridges   b) hydrogen bonds   c) disulfide bonds   d) hydrophobic bonds

11. Which of the following does *not* contribute to the tertiary structure of proteins?
    a) salt bridges   b) hydrogen bonds   c) disulfide bonds   d) peptide bonds

12. The quaternary structure of hemoglobin involves _____ subunits.
    a) two   b) four   c) six   d) eight

13. Denaturation of a protein involves a breakdown of the
    a) primary structure
    b) secondary and tertiary structures
    c) primary and secondary structures
    d) primary, secondary and tertiary structures

14. Ions of heavy metals (Hg^{2+} or Pb^{2+}) denature proteins by combining with
    a) $-NH_2$ groups   b) $-SH$ groups

    c) $-OH$ groups   d) —C—NH— groups
                                                     $\overset{\|}{O}$

Matching
Match the following definitions to the correct words given as responses.

15. _____ A dipolar amino acid molecule
containing both a + and − charge.

16. _____ A substance composed of 25 amino
acids linked together.

17. _____ The pH at which amino acids exist in
the form that has no net charge.

a) polypeptide

b) ninhydrin

c) isoelectric point

d) zwitterion

Select the correct polypeptide for each description on the left.

18. _____ stimulates milk production

19. _____ controls carbohydrate metabolism

20. _____ decreases urine production

a) myosin

b) insulin

c) oxytocin

d) vasopressin

For each description on the left, select the correct protein class.

21. _____ hemoglobin belongs to this class

22. _____ proteins that function as enzymes

23. _____ insulin belongs to this class

24. _____ keratin belongs to this class

25. _____ collagen belongs to this class

a) regulatory proteins

b) transport proteins

c) structural proteins

d) catalytic proteins

For each of the bonds given on the left, choose a response that indicates the type of protein structure the bond is involved in forming.

26. _____ hydrogen bonds

27. _____ amide bonds

28. _____ hydrophobic bonds

29. _____ salt bonds

a) primary structure

b) secondary structure

c) tertiary structure

d) more than one response iscorrect

SOLUTIONS

19.1	a) carboxylate	b) L	c) three	d) two
19.2	a) zero	b) negative	c) isoelectric point	
19.3	a) cysteine	b) peptide	c) C-terminal	d) 50
19.4	a) vasopressin	b) adrenocorticotropic		
19.5	a) enzymes	b) fibrous	c) simple	d) prosthetic

19.6   a) peptide   b) primary

19.7   a) α-helix   b) β-pleated sheet

19.8   a) R groups   b) hydrophobic   c) salt   d) hydrogen

19.9   a) subunits   b) four

19.10 a) amino acids   b) native   c) denaturation

B. Answers to Self-Test Questions

1. c	11. d	21. b
2. b	12. b	22. d
3. b	13. b	23. a
4. a	14. b	24. c
5. c	15. d	25. c
6. c	16. a	26. d
7. d	17. c	27. a
8. a	18. c	28. c
9. d	19. b	29. c
10. d	20. d	

Chapter 20

Enzymes

CHAPTER OUTLINE

20.1 The General Characteristics of Enzymes
20.2 Enzyme Nomenclature and Classification
20.3 Enzyme Cofactors
20.4 Mechanism of Enzyme Action

20.5 Enzyme Activity
20.6 Factors Affecting Enzyme Activity
20.7 Enzyme Inhibition
20.8 Regulation of Enzyme Activity
20.9 Medical Application of Enzymes

LEARNING OBJECTIVES

When you have completed your study of this chapter, you should be able to:
1. Describe the general characteristics of enzymes and why enzymes are vital in body chemistry.
2. Determine the function and/or substrate of an enzyme on the basis of its name.
3. Identify the general function of cofactors.
4. Use the lock-and-key theory to explain specificity in enzyme action.
5. List two ways of describing enzyme activity.
6. Identify the factors that affect enzyme activity.
7. Compare the mechanisms of competitive and noncompetitive enzyme inhibition.
8. Describe the three methods of cellular control over enzyme activity.
9. Discuss the importance of measuring enzyme levels in the diagnosis of disease.

ANSWERS AND SOLUTIONS TO EVEN-NUMBERED PROBLEMS

The General Characteristics of Enzymes (Section 20.1)
20.2   List two ways that enzyme catalysis of a reaction is superior to normal laboratory conditions.

Solution:
Enzymatically catalyzed reactions do not require very acid conditions common in laboratory reactions. Lab conditions for most reactions must be hot as compared to enzyme run reactions. Further, the enzymatically catalyzed reactions normally are much faster than those run in the laboratory without an enzyme catalyst.

20.4   Define what is meant by the term *enzyme specificity*.

Solution:
Enzymes are specific for a general type of reaction and in some cases for a specific, single substrate.

Enzyme Nomenclature and Classification (Section 20.2)

20.6   What name is given to the systematic nomenclature system for enzymes?

Solution:
International Enzyme Commission (IEC) system

20.8   Match the following enzymes and substrates:

Enzymes	*Substrates*
a) urease	fumarate
b) fumarase	arginine
c) fructase	lactose
d) arginase	urea
e) lactase	fructose

Solution:
a) urease–urea   b) fumarase–fumarate   c) fructase–fructose
d) arginase–arginine   e) lactase–lactose

20.10   Because one substrate may undergo a number of reactions, it is often convenient to use an enzyme nomenclature system that includes both the substrate name (or general type) and the type of reaction catalyzed. Identify the substrate and type of reaction for the following enzyme names:
a) succinate dehydrogenase   b) L-amino acid reductase   c) cytochrome oxidase

Solution:
a) substrate: succinate acid; reaction: removal of hydrogen
b) substrate: L-amino acid; reaction: reduction
c) substrate: cytochrome; reaction oxidation

Enzyme Cofactors (Section 20.3)

20.12   What are the relationships between the terms *cofactor*, *active enzyme*, and *apoenzyme*?

Solution:
The *active enzyme* consists of a protein part, called an *apoenzyme*, and a nonprotein part called a *cofactor*. An organic cofactor is called a *coenzyme*.

20.14   List some typical inorganic ions that service as cofactors.

Solution:
Mg^{2+}, Zn^{2+}, Fe^{2+}, Ca^{2+}, and others that are mostly metal ions.

Mechanism of Enzyme Action (Section 20.4)

20.16 What is an enzyme-substrate complex?

Solution:

The binding of a substrate to an enzyme at the active site of the enzyme (ES complex).

20.18 Compare the lock-and-key theory with the induced-fit theory.

Solution:

In the lock and key model, the shape of the enzyme must fit the size and shape of the substrate in order for the substrate to bind to the active site. In the induced fit model, the enzyme can change its shape to fit the substrate that has attached to the active site.

Enzyme Activity (Section 20.5)

20.20 What observation may be used in experiments to determine enzyme activity?

Solution:

By monitoring some physical or chemical property of a substrate or a product involved, in the enzyme-catalyzed reaction, the rate of the reaction may be determined. The rate is a direct measure of the enzyme activity.

20.22 What is an enzyme international unit? Why is the international unit a useful method of expressing enzyme activity in medical diagnoses?

Solution:

An enzyme international unit (IU) is the quantity of enzyme that converts 1 micromol (pmol) of substrate per minute under a specified set of conditions. Many medical diagnoses need to know the actual enzyme activity present in a patient compared to a "standard" activity. The IU allows such a comparison.

Factors Affecting Enzyme Activity (Section 20.6)

20.24 Write a single sentence to summarize tbe information of each graph in Exercise 20.23.

Solution:

a) The rate reaction increases with increased substrate concentration until the enzyme is 100% utilized, at which point no further increase in rate is observed.
b) Increasing the enzyme concentration causes a linear increase in the rate of reaction.
c) There is an optimum pH, at which the rate is a maximum. The rate decreases as the pH increases or decreases from this optimum pH value.
d) Over a narrow temperature range, the rate increases at higher temperature, but when the temperature reaches a limit, the enzyme is deactivated and the rate goes to zero.

20.26 How might V_{max} for an enzyme be determined?

Solution:

Prepare a series of reaction containers, each containing the same amount of enzyme, with a differing substrate concentration. Measure the rate in each container. If the saturation concentration of the substrate is present, then any higher substrate concentrations will show no increase in rate, and that rate is V_{max}.

20.28   Using Table 20.4, list three enzymes and their optimum pH.

   Solution:
   Examples: pepsin at 1.5; sucrase at 6.2; catalase at 7.3

Enzyme Inhibition (Section 20.7)

20.30   Distinguish between irreversible and reversible enzyme inhibition.

   Solution:
   An irreversible inhibitor forms a strong covalent, usually permanent, bind with a specific functional group of the enzyme. The resulting molecule is an inactive enzyme. A reversible inhibitor does not attach permanently but binds reversibly to the enzyme resulting in an equilibrium state. Reversible inhibition can be removed by shifting the equilibrium.

20.32   List an antidote for each of the two poisons in Exercise 20.31 and describe how each functions.

   Solution:
   a) The antidote for cyanide is sodium thiosulfate. It must be administered before the cyanide can act as an enzyme inhibitor. It oxidizes the CN^- to CNS^- (thiocyanate), which does not bind as strongly to the iron.
   b) The antidote for heavy metal ions is injection of a chelating agent, such as EDTA, to tie up the metal ions, thereby preventing them from inhibiting enzyme activity.

Regulation of Enzyme Activity (Section 20.8)

20.34   List three mechanisms for the control of enzyme activity.

   Solution:
   (1) Activation of a zymogen or a proenzyme
   (2) Allosteric regulation
   (3) Genetic control over enzyme synthesis

20.36   What is another name for a zymogen?

   Solution:
   Proenzymes

20.38   Name and contrast the two types of modulators.

   Solution:
   Activators increase the activity of enzymes; inhibitors decrease the activity of enzymes.

20.40   What is meant by genetic control of enzyme activity and enzyme induction?

   Solution:
   Genetic control of enzyme activity is accomplished by the cell producing the enzyme in response to the existing conditions. If it is in response to a need for that enzyme, that synthesis is called enzyme induction.

Medical Application of Enzymes (Section 20.9)

20.42 How is each of the following enzyme assays useful in diagnostic medicine?
       a) CK   b) ALP   c) amylase

Solution:
Since certain enzymes are found almost exclusively inside tissue cells and are released only when these cells are damaged or destroyed, measuring the enzyme levels in different areas of the body can indicate if tissue damage or even cancer cells are evident.

a) CK assays are used in diagnosing heart attack.
b) ALP assays are useful in diagnosing liver or bone disease.
c) Amylase assays are useful in diagnosing diseases of the pancreas.

20.44 Why is an LDH assay a good initial diagnostic test?

Solution:
Since different body tissues contain differing amounts of LDH isoenzymes, a LDH assay of blood serum can pinpoint a likely malfunction of a particular organ.

Chemistry Around Us and Key Chemicals

20.46 What expression is sometimes used to refer to hereditary diseases caused by the absence of enzyme?

Solution:
Inborn errors of metabolism

20.48 What are extremozymes?

Solution:
Extremozymes are enzymes that can function under extremely high temperatures or pressures, or at abnormal pH values. Organisms that live around thermal vents on the ocean-bottom, or in hot, mineral springs or geysers have enzymes that must function at conditions not encountered by most living organisms.

20.50 How much topical antibiotic cream should be applied to a minor cut or insect bite?

Solution:
The amount necessary to cover an area the size of a finger tip.

20.52 How does ethanol serve as an antidote for ethylene glycol poisoning?

Solution:
Ethanol competes for the enzyme that oxidizes the alcohol groups of ethylene glycol. It is the oxidation product that makes ethylene glycol toxic.

PROGRAMMED REVIEW

Section 20.1 The General Characteristics of Enzymes

Enzymes speed up chemical reactions by (a) _____ activation energies. Enzyme (b) _____ is a characteristic of an enzyme that it catalyzes only certain reactions. A third important characteristic of enzymes is that their activity as catalysts can be (c) _____.

Section 20.2 Enzyme Nomenclature and Classification

The (a) _____ is the substance that undergoes a chemical change catalyzed by an enzyme. Systematic names for enzymes end in (b) _____. The IEC classification of enzymes groups them into (c) _____ categories. The common system for naming enzymes incorporates the (d) _____ or the type of reaction into the name.

Section 20.3 Enzyme Cofactors

A (a) _____ is a nonprotein molecule or ion required by an enzyme for catalytic activity. If a molecule required by an enzyme for catalytic activity is organic, it is called a (b) _____. The protein portion of an enzyme which requires an additional molecule or ion for catalytic activity is called the (c) _____.

Section 20.4 Mechanism of Enzyme Action

The (a) _____ _____ is the location on an enzyme where a substrate is bound and catalysis occurs. The combination formed when substrate and enzyme bond is called the enzyme-substrate (b) _____. The (c) _____-_____-_____ theory proposes that a substrate has a shape fitting that of the enzyme's active site. The (d) _____-_____ theory proposes that the conformation of an enzyme changes to accommodate an incoming substrate.

Section 20.5 Enzyme Activity

The number of molecules of substrate acted upon by one molecule of enzyme per minute is the (a) _____ number. Experiments that measure enzyme activity are called enzyme (b) _____. One standard (c) _____ _____ is the quantity of enzyme which catalyzes the conversion of 1 micromole of substrate per minute.

Section 20.6 Factors Affecting Enzyme Activity

Increasing the concentration of enzyme (a) _____ the rate of an enzyme-catalyzed reaction. As substrate concentration is increased, V_{max} is achieved when the enzyme is (b) _____ with substrate. The temperature at which enzyme activity is highest is the (c) _____ temperature. The (d) _____ pH is that at which enzyme activity is highest.

Section 20.7 Enzyme Inhibition

An (a) _____ is any substance that can decrease the rate of an enzyme-catalyzed reaction. Cyanide ion interferes with the operation of an iron-containing enzyme called (b) _____. An antidote for heavy-metal poisoning is (c) _____. Sulfa drugs are examples of (d) _____ enzyme inhibitors.

Section 20.8 Regulation of Enzyme Activity

A proenzyme or (a) _____ is the inactive precursor of an enzyme. An (b) _____ enzyme is one whose activity is changed by the binding of modulators. Enzyme regulation in which the enzyme that catalyzes the first step of a series of reactions is inhibited by the final product is called (c) _____ inhibition. Enzyme (d) _____ is the synthesis of enzymes in response to a temporary need of the cell.

Section 20.9 Medical Application of Enzymes

An enzyme useful in diagnosing prostate cancer is (a) _____. Multiple forms of the same enzyme are known as (b) _____. Examples of enzymes which occur in multiple forms are CK and (c) _____.

SELF-TEST QUESTIONS

Multiple Choice

1. Enzymes which act upon only one substance exhibit
   a) catalytic specificity
   b) binding specificity
   c) relative specificity
   d) absolute specificity

2. Which of the following enzyme properties is explained by the lock-and-key model?
   a) specificity
   b) high turnover rate
   c) high molecular weight
   d) susceptibility to denaturation

3. The induced-fit theory of enzyme action extends the lock-and-key theory in which of the following ways?
   a) assumes enzymes and substrates are rigid
   b) assumes the shape of substrates changes (conforms) to fit the enzyme
   c) assumes the enzyme shape changes to accommodate the substrate
   d) assumes enzymes have no active site

4. Enzyme turnover numbers are expressed in
   a) activity/mg
   b) units/mg
   c) units/minute
   d) molecules/minute

5. Under saturating conditions, an enzyme-catalyzed reaction has a velocity v. Which of the following would increase the rate of the reaction?
   a) a decrease in the substrate concentration
   b) an increase in the substrate concentration
   c) a decrease in the enzyme concentration
   d) an increase in the enzyme concentration

6. Which of the following has no effect on the rate of an enzyme-catalyzed reaction?
   a) the volume of the reaction mixture
   b) the temperature of the reaction mixture
   c) the pH of the reaction mixture
   d) the enzyme concentration in the reaction mixture

7. In noncompetitive inhibition,
   a) substrate and inhibitor bind at separate locations on the enzyme
   b) substrate and inhibitor bind at the same location on the enzyme
   c) the inhibitor forms a covalent bond with the enzyme
   d) the enzyme becomes permanently inactivated

8. The use of ethanol as a treatment for ethylene glycol poisoning is an excellent example of
   a) feedback inhibition
   b) competitive inhibition
   c) noncompetitive inhibition
   d) enzyme denaturation

9. An enzyme is sometimes generated initially in an inactive form called
   a) a coenzyme
   b) a cofactor
   c) an activator
   d) a zymogen

10. Enzymes which exist in more than one form are called
    a) proenzymes
    b) isoenzymes
    c) apoenzymes
    d) allosteric enzymes

11. Which of the following is an enzyme whose activity is changed by the binding of modulators?
    a) isoenzymes
    b) proenzymes
    c) allosteric enzymes
    d) apoenzymes

Matching

Select an enzyme that matches each description on the left.

12. _____ name is based on nature of reaction catalyzed

13. _____ name is based on enzyme substrate

14. _____ name is based on both the substrate and the nature of the reaction catalyzed

15. _____ name is an early nonsystematic type

a) pepsin

b) oxidase

c) succinate dehydrogenase

d) sucrase

Match the enzyme components below to the correct term given as a response.

16. _____ the protein portion of an enzyme

17. _____ an organic, but nonprotein portion of an enzyme

18. _____ a nonprotein molecule or ion required by an enzyme

a) cofactor

b) apoenzyme

c) coenzyme

d) proenzyme

Match each description on the left with a response from the right.

19. _____ reacts with cytochrome oxidase

20. _____ stops bacterial synthesis of folic acid

21. _____ inhibits succinate dehydrogenase

22. _____ inhibits bacterial cell-wall construction

a) penicillin

b) cyanide

c) sulfanilamide

d) malonate

Match the correct disease or condition given as a response to the enzyme useful in diagnosing the disease or condition.

23. _____ amylase

24. _____ lysozyme

25. _____ acid phosphatase

a) infectious hepatitis

b) pancreatic diseases

c) prostate cancer

d) monocytic leukemia

True-False

26. The presence of enzymes enables reactions to proceed at lower temperatures.

27. Enzymes are destroyed during chemical reactions in which they participate.

28. Some enzymes are carbohydrates.

29. All enzymes act only on a single substance.

30. Coenzymes are always organic compounds.

31. The optimum pH of most of the enzymes in the human body is near pH 7.

32. A plot of reaction rate (vertical axis) versus temperature of enzyme-catalyzed reactions most frequently yields a straight line.

33. Streptokinase is useful in dissolving blood clots.

34. Lactate dehydrogenase exists in five different isomeric forms.

SOLUTIONS

A. Answers to Programmed Review

20.1 a) lowering   b) specificity   c) regulated

20.2 a) substrate   b) -ase   c) six   d) substrate

20.3 a) cofactor   b) coenzyme   c) apoenzyme

20.4 a) active site   b) complex   c) lock-and-key   d) induced-fit

20.5 a) turnover   b) assays   c) international unit

20.6 a) increases   b) saturated   c) optimum   d) optimum

20.7 a) inhibitor   b) cytochrome oxidase c) EDTA   d) competitive

20.8 a) zymogen   b) allosteric   c) feedback   d) induction

20.9 a) acid phosphatase b) isoenzymes   c) LDH

B. Answers to Self-Test Questions

1. d	13. d	25. c
2. a	14. c	26. T
3. c	15. a	27. F
4. d	16. b	28. F
5. d	17. c	29. F
6. a	18. a	30. T
7. a	19. b	31. T
8. b	20. c	32. F
9. d	21. d	33. T
10. b	22. a	34. T
11. c	23. b	
12. b	24. d	

Chapter 21

Nucleic Acids and Protein Synthesis

CHAPTER OUTLINE

21.1 Components of Nucleic Acids
21.2 Structure of DNA
21.3 Replication of DNA
21.3 Ribonucleic Acid (RNA)
21.5 The Flow of Genetic Information

21.6 Transcription: RNA Synthesis
21.7 The Genetic Code
21.8 Translation and Protein Synthesis
21.9 Mutations
21.10 Recombinant DNA

LEARNING OBJECTIVES

When you have completed your study of this chapter, you should be able to:
1. Identify the components of nucleotides and correctly classify the sugars and bases.
2. Describe the structure of DNA.
3. Outline the process of DNA replication.
4. Contrast the structure of DNA and RNA and list the function of the three types of cellular RNA.
5. Describe the process by which RNA is synthesized in cells.
6. Explain how the genetic code functions in the flow of genetic information.
7. Outline the process by which proteins are synthesized in cells.
8. Describe how genetic mutations occur and how they influence organisms.
9. Describe the technology used to produce recombinant DNA.

ANSWERS AND SOLUTIONS TO EVEN-NUMBERED PROBLEMS

Components of Nucleic Acids (Section 21.1)

21.2 Which pentose sugar is present in DNA? In RNA?

Solution:
D-deoxyribose is present in DNA; D-ribose is found in RNA.

21.4   Indicate whether each of the following is a pyrimidine or a purine:
 a) guanine b) thymine c) uracil d) cytosine
 e) adenine

Solution:
 a) Purine b) Pyrimidine c) Pyrimidine d) Pyrimidine
 e) Purine

21.6   Write the structural formula for the nucleotide thymidine 5'-monophosphate. The base component is thymine.

Solution:

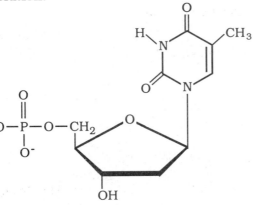

Structure of DNA (Section 21.2)

21.8   Identify the 3' and 5' ends of the DNA segment AGTCAT.

Solution:
 The 5' position is located at the left end: "A", and the 3' position at the right end: "T".

21.10   What data obtained from the chemical analysis of DNA supported the idea of complementary base pairing in DNA proposed by Watson and Crick?

Solution:
 The number of adenine and thymine bases present were consistently equal, and the numbers of cytosine and guanine bases present were also equal. This implied pairing between an A and a T, and pairing between a C and a G.

21.12   How many total hydrogen bonds would exist between the following strands of DNA and their complementary strands?
 a) CACGGT b) TTTAAA

Solution:

$$\begin{array}{cccccc} G & T & G & C & C & A \\ | & | & | & | & | & | \\ C & A & C & G & G & T \end{array}$$

 a) Sixteen. The DNA has the structure:
 There are four C-G pairs with three hydrogen bonds per pair, and two A-T pairs with two hydrogen bonds per pair.

b) Twelve. The DNA has the structure:

There are six A-T pairs with two hydrogen bonds per pair.

Replication of DNA (Section 21.3)

21.14 What is a chromosome? How many chromosomes are in a human cell?

Solution:

Chromosomes are structures within the nucleus of a cell that carry genetic information (DNA). Human cells contain 46 chromosomes.

21.16 Describe the function of the enzyme helicase in the replication of DNA.

Solution:

Helicase catalyzes the unwinding of the double helix into two strands.

21.18 In what direction is a new DNA strand formed?

Solution:

The new strand is formed from the 5' to the 3' of the new strand.

21.20 The base sequence CTTAAG represents a portion of a single strand of DNA.
a) Represent the complete double-stranded molecule for this portion of the strand.
b) Use the representation to illustrate the replication of the DNA strand. Be sure to clearly identify the nucleotide bases involved, new strands formed, and the daughter DNA molecules.

Solution:

a) The original double-stranded DNA:
$$
\begin{array}{c}
(1)\ \text{C T T A A G} \\
\text{| | | | | |} \\
(2)\ \text{G A A T T C}
\end{array}
$$

b) Replication:

old strand (1):	5' C T T A A G 3'	
	| | | | | |	daughter DNA (1)
new strand:	3' G A A T T C 5'	

new strand:	5' C T T A A G 3'	
	| | | | | |	daughter DNA (2)
old strand (2):	3' G A A T T C 5'	

Ribonucleic Acid (RNA) (Section 21.4)

21.22 List the three cellular regions where RNA is found.

Solution:

nucleus, ribosomes, mitochondria

21.24 Must the ratio of guanine to cytosine be 1:1 in RNA? Explain.

Solution:

No. RNA does not involve the complementary pairing to form the entire double helix. There does not need to be pairing between every C with a G and vice versa.

21.26 Which of the kinds of RNA delivers amino acids to the site of protein synthesis?

Solution:
transfer RNA (tRNA)

The Flow of Genetic Information (Section 21.5)

21.28 In the flow of genetic information, what is meant by the terms *transcription* and *translation*?

Solution:
Transcription is the transfer of genetic information from a DNA molecule to a mRNA. Translation is the conversion of the genetic code carried by mRNA into the amino acid sequence of a protein.

Transcription: RNA Synthesis (Section 21.6)

21.30 Write the base sequence for the mRNA that would be formed during transcription from the DNA strand with the base sequence GCCATATTG.

Solution:
The complementary ribonucleotides arrange as shown below:

DNA   (5') GCCATATTG (3')
mRNA   (3') CGGUAUAAC (5')

Rewriting the mRNA in 5' to 3' order gives: CAAUAUGGC

21.32 What is the relationship between exons, introns, and hnRNA?

Solution:
The hnRNA molecule is comprised of exons and introns. The exons carry the genetic code and are separated by intron units, which do not contain the code. The introns in the hnRNA are cut out and the exons are re-spliced to form mRNA.

The Genetic Code (Section 21.7)

21.34 Using the information in Table 21.2, determine the amino acid that is coded for be each of the following codons:

a) CAU   b) UCA   c) UCU

Solution:
a) histidine   b) serine   c) serine

21.36 Which of the following statements about the genetic code are true and which are false? Correct each false statement.
a) Each codon is composed of four bases.
b) Some amino acids are represented by more than one codon.
c) All codons represent an amino acid.
d) Each living species is thought to have its own unique genetic code.
e) The codon AUG at the beginning of a sequence is a signal for protein synthesis to begin at that codon.
f) It is not known whether or not the code contains stop signals for protein synthesis.

Solution:
a) False. Each codon is composed of 3 bases.
b) True.

c) False. Codons can represent an amino acid, or a "control word."
d) False. Each known living organism has the same set of codons.
e) True.
f) False. Three different codons are "stop" signals: UAA, UAG and UGA

Translation and Protein Synthesis (Section 21.5)

21.38 List the three major stages of protein synthesis.

Solution:
(1) Initiation of the polypeptide chain
(2) The elongation of the chain
(3) The termination of the completed polypeptide chain

21.40 What is meant by the terms A site and P site?

Solution:
The A site (aminoacyl site) is located on the mRNA-ribosome complex next to the P site. The P site (peptidyl site) the site on the ribosome occupied by the initiating codon, AUG.

Mutations (Section 21.9)

21.42 What is a genetic mutation?

Solution:
A genetic mutation is any change resulting in the wrong base sequence on the DNA.

21.44 What is the result of a genetic mutation that causes the mRNA sequence GCC to be replaced by CCC?

Solution:
The protein that is synthesized will have the amino acid, proline, at the site where an alanine should have been.

Recombinant DNA (Section 21.10)

21.46 Explain how recombinant DNA differs from normal DNA.

Solution:
Recombinant DNA is produced when foreign DNA is placed into a nucleus.

21.48 List three substances likely to be produced on a large scale by genetic engineering and give an important use for each.

Solution:
(1) Hormones such as human insulin, used for treatment of diabetes
(2) Vaccines for malaria or Hepatitis B
(3) Interferon to fight viral infections
(4) Human growth hormone to treat growth disorders

Chemistry Around Us and Key Chemicals

21.50 What two categories of drugs have been approved by the FDA to combat HIV?

Solution:
One group inhibits the enzyme that allows reverse transcription from the mRNA to the DNA. AZT is a reverse transcriptase inhibitor. The other group are protease inhibitors, which block the production of a key enzyme needed by the virus.

21.52   Why is the AIDS virus so lethal?

Solution:
(1) It is a retrovirus, being able to transcribe its genetic code onto the host DNA molecule.
(2) It attacks the helper T-cells that are important in the proper functioning of the immune system.

21.54   Describe the advantages of coupling PCR techniques with DNA fingerprinting.

Solution:
PCR techniques allows duplication of the DNA from a very small sample to make larger amounts of DNA for comparison and possible matching with a specific individual.

21.56   A map of the human genome is viewed as being of tremendous importance. Explain.

Solution:
Mapping the complete human genome would identify the gene that is responsible for synthesizing every protein in the human body, and allow correction of malfunctioning genetic codes.

21.58   What cellular source was used to provide the DNA that led to the cloned sheep Dolly?

Solution:
The DNA used to clone Dolly was from nuclei from adult mammary cells.

PROGRAMMED REVIEW

Section 21.1 Components of Nucleic Acids
The monomers of nucleic acids are (a) _____. The sugar present in RNA is (b) _____.
Adenine is one of two (c) _____ bases. Uracil is one of three (d) _____ bases.

Section 21.2 Structure of DNA
The nucleic acid "backbone" consists of a phosphate- (a) _____ chain. Phosphate groups in DNA link the 3' position of one sugar to the (b) _____ position of the next sugar. Two DNA strands with matched sequences are said to be (c) _____ to each other. The 3' end of the sequence AGC is located at the letter (d) _____.

Section 21.3 Replication of DNA
The hereditary material within a human cell consists of 46 packets of DNA called (a) _____. The process by which an exact copy of a DNA molecule is produced is called (b) _____. The point at which a DNA molecule unwinds in the copying process is called a (c) _____. DNA segments produced during the copying process are called (d) _____ fragments.

Section 21.4 Ribonucleic Acid (RNA)
The most abundant type of RNA in a cell is (a) _____. The region of a tRNA molecule that binds to mRNA during protein synthesis is called the (b) _____. The type of RNA which consists of two subunits is (c) _____. (d) _____ functions as a carrier of genetic information from the DNA of the cell nucleus to the site of protein synthesis.

Section 21.5 The Flow of Genetic Information

The flow of genetic information according to the (a) _____ dogma of molecular biology is from DNA to (b) _____ to protein. The process by which genetic information is passed from DNA to mRNA is called (c) _____. The process by which a specific protein results from information carried on mRNA is called (d) _____ of the code.

Section 21.6 Transcription: RNA Synthesis

The enzyme (a) _____ catalyzes the synthesis of RNA. Newly formed RNA is synthesized in the (b) _____ to _____ direction. Eukaryotic DNA segments that carry no amino acid code are called (c) _____ while coded segments are called (d) _____.

Section 21.7 The Genetic Code

Code words on mRNA are known as (a) _____. Code words consist of a sequence of (b) _____ nucleotide bases on mRNA. Of the 64 possible code words, 61 represent amino acids. The remaining three are signals for chain (c) _____. Most of the amino acids are represented by more than one codon, a condition known as (d) _____.

Section 21.8 Translation and Protein Synthesis

The first amino acid to be involved in protein synthesis in prokaryotic cells is (a) _____. The (b) _____ site is where an incoming tRNA carrying an amino acid attaches to the mRNA-ribosome complex. The movement of a ribosome along a mRNA is called (c) _____. Complexes of several ribosomes and mRNA are called polyribosomes or (d) _____.

Section 21.9 Mutations

A mutation is any change resulting in an incorrect sequence of bases on (a) _____. A mutation can lead to an incorrect sequence of (b) _____ for a protein. Chemicals that induce mutations are called (c) _____.

Section 21.10 Recombinant DNA

The application of recombinant DNA technology is sometimes referred to as (a) _____ engineering. Protective enzymes which can catalyze the cleaving of foreign DNA are called (b) _____ enzymes. A carrier of foreign DNA into a cell is called a (c) _____. Circular DNA often used as a carrier is called a (d) _____.

SELF-TEST QUESTIONS

Multiple Choice

1. A primary difference between DNA and RNA involves
   a) the phosphate
   b) the pentose sugar
   c) the type of bonding between sugar and base
   d) the type of bonding between sugar and phosphate

2. Which of the following bases is a purine?
   a) uracil   b) adenine   c) thymine   d) cytosine

3. A sample of nucleic acid is to be analyzed. For which of the following would an analysis be useful in deciding whether the sample was DNA or an RNA?
   a) guanine   b) cytosine   c) adenine   d) thymine

4. The structural backbone of all nucleic acids consists of alternating molecules of
    a) a sugar and a base   b) a base and phosphate
    c) a sugar and phosphate   d) purine and pyrimidine bases

5. If the nucleotide sequence in one strand of DNA were T-C-G, the complementary strand written in 5'-3'
    sequence would be
    a) A-C-G   b) T-G-C   c) C-G-A   d) A-G-C

6. A sample of double helical DNA contained 26% of the base thymine (T). The amount of adenine (A)
    should be
    a) 26%   b) 24%   c) 22%   d) 20%

7. The number of hydrogen bonds between A and T in double-stranded helical DNA is
    a) 1   b) 2   c) 3   d) 4

8. mRNA synthesis is called
    a) replication   b) translocation   c) translation   d) transcription

9. Histidine is represented by the codons CAC and CAU. This is an example of
    a) degeneracy of the genetic code   b) a nonsense codon
    c) punctuation in the genetic code   d) the universal nature of the genetic code

10. The genetic code consists of _____ three-letter codons.
    a) 16   b) 32   c) 64   d) 86

11. Segments of eukaryotic DNA which are coded for amino acids are called
    a) exons   b) introns   c) hnRNA   d) codons

12. Recombinant DNA is DNA
    a) formed by combining portions of DNA from two different sources
    b) found within plasmids
    c) with an unusual ability to recombine
    d) which contains two template strands

Matching
Characteristics are given below for various nucleic acids. Match each description to the correct type of nucleic
acid as a response.

13. _____ represents a large percent of cellular RNA   a) DNA

14. _____ serves as master template for the formation of   b) mRNA
             all nucleic acids in the body
                                                                  c) tRNA
15. _____ contains base sequences called codons
                                                                  d) rRNA
16. _____ has lowest molecular weight of all nucleic acids

17. _____ replicated during cell division

18. _____ delivers amino acids to site of protein synthesis

19. _____ contains anti-codon

20. _____ helps to serve as a site for protein synthesis

21. _____ carries the directions for protein synthesis to the
             site of protein synthesis

True-False

22. Purine bases contain both a five- and a six-member ring.

23. DNA and RNA contain identical nucleic acid backbones.

24. During DNA replication, only one DNA strand serves as a template for the formation of a complementary strand.

25. A complementary DNA strand formed during replication is identical to the strand serving as a template.

26. RNA is usually double-stranded.

27. Some amino acids are represented by more than one codon.

28. The genetic code has some code words that don't code for any amino acids.

29. The genetic code is thought to be different for humans and bacteria.

30. Incoming tRNA, carrying an amino acid, attach at the peptide site.

31. N-formylmethionine is always the firstamino acid in a growing peptide chain in prokaryotic cells.

32. Protein synthesis generally takes place within the cell nucleus.

33. Some genetic mutations might aid an organism rather than hinder it.

34. Viruses contain both nucleic acids and protein.

35. Restriction enzymes act at sites on DNA called palindromes.

SOLUTIONS

A. Answers to Programmed Review

21.1 a) nucleotides	b) ribose	c) purine	d) pyrimidine
21.2 a) sugar	b) 5'	c) complementary	d) C
21.3 a) chromosomes	b) replication	c) replication fork	d) Okazaki
21.4 a) rRNA	b) anticodon	c) rRNA	d) mRNA
21.5 a) central	b) mRNA	c) transcription	d) translation
21.6 a) RNA polymerase	b) 5', 3'	c) introns	d) exons
21.7 a) codons	b) three	c) termination	d) degeneracy
21.8 a) methionine	b) A	c) translocation	d) polysomes
21.9 a) DNA	b) amino acids	c) mutagens	
21.10 a) genetic	b) restriction	c) vector	d) plasmid

B. Answers to Self-Test Questions

1. b	4. c	7. b
2. b	5. c	8. d
3. d	6. a	9. a

10. c	19. c	28. T
11. a	20. d	29. F
12. a	21. b	30. F
13. d	22. T	31. T
14. a	23. F	32. F
15. b	24. F	33. T
16. c	25. F	34. T
17. a	26. F	35. T
18. c	27. T	

Chapter 22

Nutrition for Energy and Life

CHAPTER OUTLINE

22.1 Nutritional Requirements
22.2 The Macronutrients
22.3 Micronutrients I: Vitamins
22.4 Micronutrients II: Minerals
22.5 The Flow of Energy in the Biosphere

22.6 Metabolism and an Overview of Energy Production
22.7 ATP: The Primary Energy Carrier
22.8 Important Coenzymes in the Common Catabolic Pathway

LEARNING OBJECTIVES

When you have completed your study of this chapter, you should be able to:

1. Describe the differences between macronutrients and micronutrients in terms of amounts required and their functions in the body.
2. Show that you understand the concept of a Daily Value (DV).
3. Describe the primary functions in the body of each macronutrient.
4. Distinguish between and classify vitamins as water soluble or fat-soluble on the basis of name and behavior in the body.
5. List at least one good food source for each of the water soluble and fat-soluble vitamins.
6. List at least one food source and a primary function in the body for each major mineral.
7. Describe the major steps in the flow of energy in the biosphere.
8. Differentiate between metabolism, anabolism, and catabolism.
9. Outline the three stages in the extraction of energy from food.
10. Explain how ATP serves a central role in the production and use of cellular energy.
11. Explain the role of coenzymes in the common catabolic pathway.

ANSWERS AND SOLUTIONS TO EVEN-NUMBERED PROBLEMS

Nutritional Requirements (Section 22.1)

22.2   What is the principal component of dietary fiber?

Solution:
Cellulose. A complex carbohydrate that is not digested by the humans.

The Macronutrients (Section 22.2)

22.4   List two general functions in the body for each of the macronutrients.

Solution:
Carbohydrates are used for energy and for cell and tissue synthesis. Proteins provide amino acids for synthesis of proteins, and ingredients for hormones and enzymes. Lipids are used for energy, as energy storage and structural components.

22.6   List the types of macronutrients found in each of the following food items. List the nutrients in approximate decreasing order of abundance (most abundant first, etc.).
   a) potato chips   b) buttered toaster   c) plain toast with jam
   d) cheese sandwich   e) a lean steak   f) a fried egg

Solution:
Note: The abundance of macronutrients often depend on the methods of preparing or cooking.
   a) Carbohydrates, lipids, proteins   b) Carbohydrates, lipids, proteins
   c) Carbohydrates, proteins   d) Carbohydrates, lipids, proteins
   e) Proteins, lipids   f) Lipids, proteins, carbohydrates

22.8   Use Figure 22.2 and list the Daily Values (DV) for fat, saturated fat, total carbohydrate, and fiber based upon a 2000-Calorie diet.

Solution:
fat: 65 g; saturated fat: less than 20 g; carbohydrate: 300 g; fiber: 25 g

Micronutrients I: Vitamins (Section 22.3)

22.10   Why is there more concern about large doses of fat-soluble vitamins than of water-soluble vitamins?

Solution:
Fat-soluble vitamins are not excreted by the body into water-based fluids as easily as the water-soluble ones. Thus the body tends to accumulate the fat-soluble vitamins, giving rise to some undesirable effects.

22.12   What water-soluble vitamin deficiency is associated with the following?
   a) scurvy   b) beriberi   c) pernicious anemia   d) pellagra

Solution:
   a) Vitamin C   b) Vitamin B_1 c) Vitamin B_{12} d) Niacin

Micronutrients II: Minerals (Section 22.4)

22.14   What determines whether a mineral is classified as a major or trace mineral?

Solution:

The amount of mineral present in the body (greater or less than 5 g) distinguishes major minerals from trace minerals.

22.16   What are the general functions of trace minerals in the body?

Solution:

The trace minerals are components of vitamins (cobalt), enzymes (zinc, selenium), hormones (iodine), or specialized proteins (iron, copper).

The Flow of Energy in the Biosphere (Section 22.5)

22.18   Write a net equation for the photosynthetic process.

Solution:

$$6CO_2 + 6H_2O \rightarrow C_6H_{12}O_6 + 6O_2$$

22.20   Describe in general terms the steps in the flow of energy from the sun to molecules of ATP.

Solution:

During photosynthesis, energy in sunlight is stored in the sugars (carbohydrates). During respiration, the energy released from the sugars (carbohydrates) is carried in ATP molecules.

Metabolism and an Overview of Energy Production (Section 22.6)

22.22   What is a metabolic pathway?

Solution:

A metabolic pathway is an organized series of chemical reactions that coverts a starting material into a final product.

22.24   What stage of energy production is concerned primarily with the following:
a) Formation of ATP   b) Digestion of fuel molecules
c) Consumption of O_2 d) Generation of acetyl CoA

Solution:

a) Stage III   b) Stage I   c) Stage III   d) Stage II

ATP: The Primary Energy Carrier (Section 22.7)

22.26   In terms of energy production, what is the main purpose of the catabolic pathway?

Solution:

The main purpose of the catabolic pathway is to convert the chemical energy in foods to ATP.

22.28   Explain why ATP is referred to as the primary energy carrier.

Solution:

All types of work performed by the body require ATP as a source of energy.

22.30   What do the symbols P_i and PP_i represent?

Solution:

P_i is the symbol for the phosphate ion. PP_i represents the pyrophosphate ion.

22.32   Using Table 12.5, explain why phosphoenolpyruvate is called a high-energy compound but glycerol 3-phosphate is not.

Solution:
The hydrolysis of phosphoenolpyruvate liberates 14.8 kcal/mol, while the hydrolysis of glycerol 3-phosphate liberates only 2.2 kcal/mol.

22.34   Which portion of the ATP molecule is particularly responsible for its being described as a high-energy compound?

Solution:
The triphosphate portion

22.36   What is the role of mitochondria in the use of energy by living organisms?

Solution:
Mitochondria are the sites for most ATP synthesis in the cells.

22.38   Describe the structure of a mitochondrion and identify the location of enzymes important in energy production.

Solution:
A mitochondrion is a specialized structure within the living cell. The mitochondrion has an inner and outer membrane, the inner one having extensive folding. The enzymes for ADP → ATP conversion are located within these folds, as well as the enzymes for the citric acid cycle (oxidation of acetyl coenzyme A). The energy liberated by the oxidation can be used to make the ADP → ATP change.

Important Coenzymes in the Common Catabolic Pathway (Section 22.8)

22.40   What coenzymes are produced by the body from the following vitamins?
   a) riboflavin
   b) pantothenic acid
   c) nicotinamide

Solution:
   a) FAD   b) Coenzyme A   c) NAD^+

22.42   Which compound is oxidized the following reaction? Which compound is reduced?

$$^-OOC-CH_2-CH_2-COO^- + FAD \longrightarrow {}^-OOC-CH=CH-COO^- + FADH_2$$

Solution:
$^-OOC-CH_2CH_2-COO^-$ is oxidized. FAD is reduced.

22.44   Draw abbreviated structural formulas to show the reactive portion of FAD and the reduced compound $FADH_2$.

Solution:

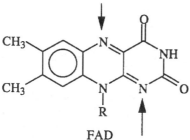

FAD

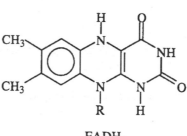

FADH₂

Chemistry Around Us and Over the Counter

22.46 What vitamin supplements are often recommended for vegetarians?

Solution:
Vitamin B-12, vitamin D, calcium, iron, zinc

22.48 What behavior is characteristic of individuals suffering from anorexia nervosa?

Solution:
People with anorexia have a psychological disorder causing them to fear becoming fat. They exhibit a preoccupation with food, a poor body image, low self-esteem and other psychological problems.

PROGRAMMED REVIEW

Section 22.1 Nutritional Requirements

Nutrients required in relatively large amounts by the body are called (a) _____. Those required in only small amounts are called (b) _____. Nutritional guidelines for the U. S. population are called (c) _____ _____ and abbreviated (d) _____.

Section 22.2 The Macronutrients

Nutrients classified as macronutrients include (a) _____, (b) _____, and (c) _____. Dietary carbohydrates are often classified as (d) _____ or (e) _____. The (f) _____ carbohydrates consist primarily of (g) _____, while the (h) _____ carbohydrates are collectively called (i) _____. Most of the lipids in food and in the body are (j) _____. Lipids called oils usually contain a high percentage of (k) _____ fatty acids. Linoleic acid is an example of an (l) _____ _____ _____. Proteins in food that contain all the (m) _____ amino acids in the proportions needed by the body are called (n) _____ proteins.

Section 22.3 Micronutrients I: Vitamins

(a) _____ are organic micronutrients that the body cannot produce in amounts needed for good health. The highly (b) _____ nature of the molecules of water-soluble vitamins renders them water soluble. All water-soluble vitamins except (c) _____ have been shown to function as (d) _____. Fat-soluble vitamins have very (e) _____ molecular structures that cause them to be insoluble in water, but soluble in the body lipids called (f) _____. The four fat-soluble vitamins are (g) _____.

Section 22.4 Micronutrients II: Minerals

(a) _____ are metals or nonmetals used in the body in the form of ions or compounds. (b) _____ minerals are found in the body in quantities greater than 5 g, while (c) _____ minerals are found in quantities (d) _____ than 5 g. The inorganic structural components and principal ions of the body utilize (e) _____ minerals, while (f) _____ minerals are important in some enzymes, vitamins and hormones.

Section 22.5 The Flow of Energy in the Biosphere

The (a) _____ is the ultimate source of energy used in all biological processes. The process of photosynthesis requires energy, CO_2 and (b) _____. The products of photosynthesis are carbohydrates and (c) _____. As plants and animals combine carbohydrates with oxygen, the products are (d) _____ and water.

Section 22.6 Metabolism and an Overview of Energy Production

The sum total of all the chemical reactions involved in maintaining a living cell is called (a) _____. (b) _____ consists of all reactions that lead to the breakdown of biomolecules. (c) _____ includes all reactions that lead up to the synthesis of biomolecules. Stage III in the extraction of energy from food is referred to as the common (d) _____ pathway. The whole purpose in the extraction of energy from food is to convert the chemical energy in foods to molecules of (e) _____.

Section 22.7 ATP: The Primary Energy Carrier

The base in ATP is (a) _____. Phosphate (P_i) is sometimes referred to as (b) _____ phosphate. If a hydrolysis reaction is energy-releasing, ΔG has a (c) _____ value. Compounds that liberate much free energy upon hydrolysis are called (d) _____ compounds.

Section 22.8 Important Coenzymes in the Common Catabolic Pathway

Stage II of the oxidation of foods produces (a) _____. Coenzyme A is derived from the B vitamin (b) _____. NAD^+ is derived from the vitamin (c) _____. The other major electron carrier (in addition to NAD^+) in the oxidation of fuel molecules is (d) _____.

SELF-TEST QUESTIONS

Multiple Choice

1. Nutrients required by the body in relatively large amounts are
   a) fiber   b) macronutrients   c) micronutrients   d) vitamins

2. The "D" in DV stands for
   a) daily   b) determined   c) dietary   d) deficiency

3. Which of the following is an example of a complex carbohydrate nutrient?
   a) starch   b) cellulose   c) glucose   d) lactose

4. Which of the following has an established DV?
   a) carbohydrates   b) lipids   c) proteins   d) fiber

5. Which of the following is a water-soluble vitamin?
   a) vitamin E   b) vitamin B_2 c) vitamin D   d) vitamin A

6. Which of the following is a trace mineral?
   a) phosphorus   b) sulfur   c) potassium   d) copper

7. It is recommended that no more than _____ percent of our calories be obtained from fats.

   a) 10   b) 30   c) 40   d) 50

8. Mitochondria are called "power stations of the cell" because

   a) most of the cellular energy is consumed there

   b) mitochondria are rich in fat molecules

   c) mitochondria exist at higher temperatures than other cell components

   d) mitochondria are the major sites of ATP synthesis

9. What product of the second stage is passed on to the third stage of the energy production processes?

   a) CO_2 b) H atoms   c) acetyl CoA   d) H_2O

10. FAD often accompanies the enzyme-catalyzed

   a) oxidation of $-CH_2CH_2-$; to $-CH=CH-$ b) oxidation of alcohol groups

   c) oxidation of aldehydes   d) transfer of acetyl groups

Matching

Select the stage in the extraction of energy from food where the following molecules are produced.

11. _____ majority of ATP   a) stage I

12. _____ amino acids   b) stage II

13. _____ acetyl CoA   c) stage III

14. _____ carbon dioxide   d) more than one response is correct

Select the mineral that best matches each description on the left.

15. _____ a major mineral found in the body in   a) cobalt

            largest amounts   b) iodine

16. _____ a vitamin component   c) calcium

17. _____ a hormone component   d) zinc

Select the coenzyme that best matches each description on the left.

18. _____ serves as an electron acceptor in the   a) coenzyme Q

            oxidation of an alcohol   b) coenzyme A

19. _____ a derivative of the vitamin riboflavin   c) FAD

20. _____ contains a sulfhydryl group $(-SH)$ d) NAD^+

Select the process that best matches each description on the left.

21. _____ energy-releasing solar process   a) cellular respiration

22. _____ conversion of CO_2 and H_2O to carbohydrates   b) photosynthesis

23. _____ oxidation of glucose to CO_2 and H_2O c) nuclear fusion

                                                        d) cellular anabolism

True-False

24. Carbon dioxide is a source of carbon for the earth's organic compounds.

25. ΔG for the hydrolysis of ATP is a positive value.

26. Ten amino acids are listed as being essential.

27. The base present in ATP is deoxyribose.

28. An ATP molecule at pH 7.4 (body pH) has a -3 charge.

29. On a mass basis, water is the most abundant compound found in the human body.

30. Sucrose, a disaccharide, is correctly classified as a complex carbohydrate.

31. Most nutritional studies indicate that typical diets in the U. S. contain too high a percentage of complex carbohydrates.

32. Lipids are digested faster than carbohydrates or proteins.

33. Concern about vitamin overdoses focuses primarily on water-soluble vitamins.

34. Most fat-soluble vitamins are known to function in the body as coenzymes.

35. Vitamin K is important in the process of blood clotting.

SOLUTIONS

A. Answers to Programmed Review

22.1	a) macronutrients	b) micronutrients	c) Daily Values	d) DV

22.2	a) carbohydrates	b) lipids	c) proteins	d) simple
	e) complex	f) simple	g) sugars	h) complex
	i) starch	j) triglycerides	k) unsaturated	l) essential fatty acid
	m) essential	n) complete		

22.3	a) vitamins	b) polar	c) vitamin C	d) coenzymes
	e) nonpolar	f) fats	g) A, D, E, and K	

22.4	a) minerals	b) major	c) trace	d) less
	e) major	f) trace		

22.5	a) sun	b) H_2O	c) O_2	d) CO_2

22.6	a) metabolism	b) catabolism	c) anabolism	d) catabolic
	e) ATP			

22.7	a) adenosine	b) inorganic	c) negative	d) high-energy

22.8	a) acetyl coenzyme A	b) pantothenic acid	c) nicotinamide	d) FAD

B. Answers to Self-Test Questions

1. b	5. b	9. c
2. c	6. d	10. a
3. a	7. b	11. c
4. c	8. d	12. a

13. b
14. c
15. c
16. a
17. b
18. d
19. c
20. b

21. c
22. b
23. a
24. T
25. F
26. T
27. F
28. F

29. T
30. F
31. F
32. F
33. F
34. F
35. T

Chapter 23

Carbohydrate Metabolism

CHAPTER OUTLINE

23.1 Digestion of Carbohydrates
23.2 Blood Glucose
23.3 Glycolysis
23.4 Fates of Pyruvate
23.5 The Citric Acid Cycle
23.6 The Electron Transport Chain

23.7 Oxidative Phosphorylation
23.8 Complete Oxidation of Glucose
23.9 Glycogen Metabolism
23.10 Gluconeogenesis
23.11 Hormonal Control of Carbohydrate
        Metabolism

LEARNING OBJECTIVES

When you have completed your study of this chapter, you should be able to:
 1. Identify the products of carbohydrate digestion.
 2. Explain the importance to the body of maintaining proper blood sugar levels.
 3. List the net reactants and products of the glycolysis pathway.
 4. Describe how the glycolysis pathway is regulated in response to cellular needs.
 5. Name the three fates of pyruvate.
 6. Identify the two major functions of the citric acid cycle.
 7. Describe how the citric acid cycle is regulated in response to cellular energy needs.
 8. Explain the function of the electron transport chain and describe how electrons move down the chain.
 9. List the major features of the chemiosmotic hypothesis.
10. Calculate the amount of ATP produced by the complete oxidation of a mole of glucose.
11. Explain the importance of the processes of glycogenesis and glycogenolysis.
12. Describe gluconeogenesis and the operation of the Cori cycle.
13. Describe how hormones regulate carbohydrate metabolism.

ANSWERS AND SOLUTIONS TO EVEN-NUMBERED PROBLEMS

Digestion of Carbohydrates (Section 23.1)

23.2   What are the products produced by the digestion of

a) starch   b) lactose   c) sucrose

Solution:

a) glucose   b) glucose and galactose   c) glucose and fructose

Blood Glucose (Section 23.2)

23.4   Describe what is meant by the term *blood sugar level* and *normal fasting level*.

Solution:

Blood sugar level is the concentration of glucose in the blood. The concentration is expressed in mg/100mL. Normal fasting level is the blood sugar level measured after fasting (no food) for 8 to 12 hours. The "normal" range is 70 to 110 mg/mL.

23.6   How do each of the following terms relate to blood sugar level?

a) hypoglycemia   b) hyperglycemia   c) renal threshold   d) glucosuria

Solution:

a) The blood sugar level is lower than normal.
b) The blood sugar level is higher than normal.
c) The blood sugar level, at which the glucose begins to be excreted into the urine.
d) The blood sugar level is higher than the renal threshold so that glucose is excreted in the urine.

23.8   Explain why severe hypoglycemia can be very serious.

Solution:

If the blood sugar is too low, severe hypoglycemia can cause convulsions and shock.

Glycolysis (Section 23.3)

23.10   What is the starting material for glycolysis? The product?

Solution:

Glycolysis begins with glucose and the final product is pyruvate.

23.12   How many steps in glycolysis require ATP? How many steps produce ATP?

Solution:

Of the 10 steps in glycolysis, steps 1 and 3 require ATP, and steps 7 and 10 produce ATP. Overall, more ATP is produced in the steps 7 and 10 than is consumed in steps 1 and 3.

23.14   Number the carbon atoms of glucose 1 through 6 and show the location of each carbon in the two molecules of pyruvate produced by glycolysis.

Solution:

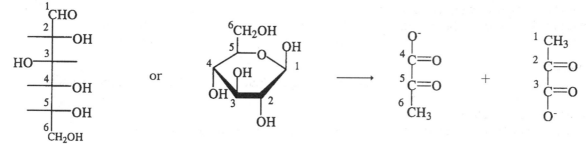

23.16 What effect does a high concentration of glucose 6-phosphate have on glycolysis?

Solution:
A high concentration of glucose 6-phosphate inhibits the action of hexokinase, the enzyme that catalyzes the formation of glucose 6-phosphate.

Fates of Pyruvate (Section 23.4)

23.18 Distinguish between the terms *aerobic* and *anaerobic*.

Solution:
Aerobic processes are done in the presence of oxygen. Anaerobic processes are done in the absence of oxygen (or a very limited amount of oxygen).

23.20 Explain how lactate formation allows glycolysis to continue under anaerobic conditions.

Solution:
When there is not enough oxygen to regenerate the NAD^+ needed to convert pyruvate to acetyl CoA, the pyruvate can be converted to lactate and produce the NAD^+ instead. This conversion does not produce as much energy as the aerobic conversion but it does produce 2 ATP per glucose, along with the NAD^+ needed for glycosis to proceed.

23.22 Explain how alcoholic fermentation allows yeast to survive with limited oxygen supplies.

Solution:
The enzymes in yeast can allow the pyruvate formed in glycolysis to be converted to alcohol and form NAD^+. The overall process of converting glucose into alcohol produces 2 ATP, along with the NAD^+. These are sufficient to maintain life for the yeast.

The Citric Acid Cycle (Section 23.5)

23.24 What is the primary function of the citric acid cycle in ATP production? What other vital role is served by the citric acid cycle?

Solution:
The citric acid cycle is the principal process for generating NADH and $FADH_2$, which are necessary for ATP synthesis. The citric acid cycle is also an important source of intermediates in the biosynthesis of amino acids.

23.26 Describe the citric acid cycle by identifying the following:
a) The fuel needed by the cycle
b) The form in which carbon atoms leave the cycle
c) The form in which hydrogen atoms and electrons leave the cycle

Solution:
a) acetyl CoA   b) CO_2 c) NADH and $FADH_2$

23.28 How many molecules of each of the following are produced by one run through the citric acid cycle?
a) NADH   b) $FADH_2$ c) GTP   d) CO_2

Solution:
a) 3   b) 1   c) 1   d) 2

23.30 Write a reaction for the step in the citric acid cycle that involves oxidation by FAD.

Solution:
Step 6: $^-OOC{-}CH_2{-}CH_2{-}COO^- + FAD \longrightarrow {}^-OOC{-}CH{=}CH{-}COO^- + FADH_2$
                   succinate   fumarate

23.32 The reactions in the citric acid cycle involve enzymes.
a) Identify the enzymes at the three control points.
b) Each control-point enzyme is inhibited by NADH and ATP. Explain how this enables the cell to be responsive to energy needs.
c) Two of the control-point enzymes are activated by ADP. Explain the benefit of this regulation to the cell.

Solution:
a) (step 1) Citrate synthetase, (step 3) isocitrate dehydrogenase, and (step 4) α-ketoglutarate dehydrogenase.
b) As the cellular need for ATP decreases, the resulting increase in ATP inhibits the production of ATP. When the cell needs more ATP, the reduced ATP concentration allowsthe cycle to proceed, making the needed ATP.
c) When ADP levels are high (ATP levels are low) the cell needs more ATP. The high ADP activates the enzyme for step 1 of the cycle allowing the cycle to proceed and reduces the ADP levels.

The Electron Transport Chain (Section 23.6)
23.34 Which coenzymes bring electrons to the electron transport chain?

Solution:
NADH and $FADH_2$

23.36 What is the role of the cytochromes in the electron transport chain?

Solution:
Electrons are passed along the electron transport chain, and an oxygen atom accepts the electrons. The combination with hydrogen produces water.

Oxidative Phosphorylation (Section 23.7)
23.38 The third stage in the oxidation of foods to provide energy involves oxidative phosphorylation. What is oxidized? What is phosphorylated?

Solution:
The coenzymes NADH and $FADH_2$ are oxidized to NAD^+ and FAD. ADP is phosphorylated to make ATP.

23.40   How many mol ATP are synthesized when 1 mol $FADH_2$ passes electrons to molecular oxygen?

Solution:
1.5 mol ATP are formed.

23.42   How many ATP molecules or equivalent can be formed by the oxidation of six acetyl CoA molecules in the common catabolic pathway?

Solution:
Each acetyl CoA molecule produces 3 NADH, 1 FADH and 1 GTP, which is equivalent to 10 ATP per acetyl CoA. Thus, 6 acetyl CoA molecules can make the equivalent of 60 ATP.

23.44   The change in free energy for the complete oxidation of glucose to CO_2 and H_2O is -686 kcal per mole of glucose. If all of this free energy could be used to drive the synthesis of ATP, how many moles of ATP could be formed from 1 mol of glucose?

Solution:
$$\frac{686 \text{ kcal/mol glucose}}{7.3 \text{ kcal/mol ATP}} = \frac{94 \text{ mol ATP}}{\text{mol glucose}}$$

23.46   What enzyme plays a key role in ATP synthesis?

Solution:
F_1-ATPase

Complete Oxidation of Glucose (Section 23.8)

23.48   A total of 32 mol of ATP can be produced by the complete oxidation of 1 mol of glucose in the liver. Indicate the number of these moles of ATP produced by
a) glycolysis
b) the citric acid cycle
c) the electron transport chain and oxidative phosphorylation

Solution:
a) 2 (net)   b) 2   c) 28

Glycogen Metabolism (Section 23.9)

23.50   Where is glycogen stored in the body?

Solution:
muscle tissue and liver

23.52   What high-energy compound is involved in the conversion of glucose to glycogen?

Solution:
Uridine triphosphate (UTP)

23.54 Why can liver glycogen, but not muscle glycogen, be used to raise blood sugar levels?

> **Solution:**
> Liver cells have the essential enzyme glucose 6-phosphatase, which is not found in muscle tissue.

Gluconeogenesis (Section 23.10)

23.56 What organ serves as the principal site for gluconeogenesis?

> **Solution:**
> liver

23.58 What are the sources of compounds that undergo gluconeogenesis?

> **Solution:**
> Lactate, the glycerol derived from fat hydrolysis, and certain amino acids

Hormonal Control of Carbohydrate Metabolism (Section 23.11)

23.60 Describe the influence of glucagon and insulin on the following:
  a) Blood sugar level   b) Glycogen formation

> **Solution:**
> a) Insulin promotes the utilization of blood glucose, causing the level to decrease. Glucagon activates the breakdown of glycogen, thereby raising blood glucose levels.
> b) Insulin promotes glycogen formation, while glucagon stimulates the breakdown.

Chemistry Around Us and Over the Counter

23.62 Describe the cause and symptoms of lactose intolerance.

> **Solution:**
> Lactase deficiency causes lactose intolerance, which may be characterized by a bloated feeling, cramps, and diarrhea when lactose-containing foods are eaten. Ingesting the necessary enzymes that are sold over the counter in drug stores easily alleviates these symptoms.

23.64 Explain why you breathe heavily when you engage in strenuous exercise.

> **Solution:**
> The strenuous exercise depletes the oxygen supply in the muscle cells. The shift from aerobic to anaerobic conversion of pyruvate results in a decrease in the pH of the blood. The decrease in blood pH triggers the rapid, deep breathing.

23.66 What is carbohydrate loading?

> **Solution:**
> A period 3–4 days of heavy exercise on a low carbohydrate diet is followed by a period of 1–2 days of light exercise while on a high carbohydrate diet. This causes a rebound effect which results in the production if higher levels of growth hormone and insulin which increases the glycogen stores up to double the normal levels.

23.68 Contrast the characteristics of Type I and Type II diabetes.

Solution:

In Type I diabetes, which usually appears in children before the age of 10, practically no insulin is produced. Type II diabetes results from a gradual decrease in insulin production as the individual gets older.

23.70 What are the consequences of unchecked diabetes?

Solution:

Hyperglycemia, glucosuria, and lowered blood pH may develop; if left unchecked, diabetes can eventually lead to blindness, coma and death.

PROGRAMMED REVIEW

Section 23.1 Digestion of Carbohydrates

The major function of dietary carbohydrate is to serve as an (a) _____ source. Digestion of polysaccharides produces (b) _____. The focal point of carbohydrate metabolism is the monosaccharide (c) _____.

Section 23.2 Blood Glucose

The amount of glucose in the blood is referred to as the blood sugar (a) _____. If a blood sugar concentration is below normal, a condition called (b) _____ exists. When the blood glucose concentration is above normal, the condition is referred to as (c) _____. The blood glucose concentration at which glucose is excreted in the urine is called the (d) _____.

Section 23.3 Glycolysis

Glycolysis is the conversion of glucose to (a) _____. Glycolysis occurs within the (b) _____ of the cell. The glycolysis pathway is regulated by three (c) _____ . The phosphorylation of glucose is controlled by (d) _____ inhibition.

Section 23.4 Fates of Pyruvate

Under aerobic conditions, pyruvate is converted to (a) _____. Under anaerobic conditions within the body, pyruvate is converted to (b) _____. Alcoholic fermentation involves the conversion of glucose to (c) _____. Each fate of pyruvate involves the regeneration of (d) _____ so that glycolysis can continue.

Section 23.5 The Citric Acid Cycle

The fuel of the citric acid cycle is (a) _____. Carbon atoms leave the cycle as molecules of (b) _____. Four oxidation-reduction reactions in the cycle produce (c) _____ molecule(s) of NADH and (d) _____ molecule(s) of $FADH_2$.

Section 23.6 The Electron Transport Chain

The enzymes for the electron transport chain are located within the inner membrane of the (a) _____. Molecular oxygen is reduced in the electron transport chain and (b) _____ is formed. A group of iron-containing enzymes called (c) _____ are located in the electron transport chain. Electrons are brought to the electron transport chain by $FADH_2$ and (d) _____.

Section 23.7 Oxidation Phosphorylation

Oxidative phosphorylation occurs at (a) _____ different locations of the electron transport chain. During oxidative phosphorylation ADP is converted to (b) _____. The theory which proposes a mechanism for oxidative phosphorylation is called the (c) _____ hypothesis. This hypothesis proposes that a (d) _____ flow occurs across the inner mitochondrial membrane.

Section 23.8 Complete Oxidation of Glucose

One molecule of cytoplasmic NADH in the muscles generates (a) _____ molecules of ATP, while (b) _____ molecules of ATP come from each molecule of mitochondrial NADH. The complete aerobic catabolism of 1 mol of glucose in the liver produces (c) _____ molecules of ATP. The majority of ATP molecules are formed as a result of oxidative (d) _____.

Section 23.9 Glycogen Metabolism

The synthesis of glycogen from glucose is called (a) _____. Energy for the synthesis of glycogen is provided by the high-energy compound (b) _____. The breakdown of glycogen to glucose is called (c) _____. The enzyme glucose 6-phosphatase which is necessary for the conversion of glycogen to glucose is found primarily in the (d) _____.

Section 23.10 Gluconeogenesis

Gluconeogenesis is the synthesis of (a) _____ from non-carbohydrate molecules. The majority of gluconeogenesis takes place in the (b) _____. A key intermediate in the conversion of lactate to glucose is (c) _____. The cyclic process in which glucose is converted to lactate and lactate is reconverted to glucose is called the (d) _____ cycle.

Section 23.11 Hormonal Control of Carbohydrate Metabolism

Insulin enhances the absorption of (a) _____ from blood into cells. The hormone (b) _____ works in opposition to insulin by raising blood glucose levels. Epinephrine stimulates (c) _____ breakdown in the muscles.

SELF-TEST QUESTIONS

Multiple Choice

1. The digestion of carbohydrates produces glucose, fructose and
   a) starch   b) amylose   c) lactose   d) galactose

2. The central compound in carbohydrate metabolism is
   a) glycogen   b) glucose   c) fructose   d) galactose

3. The glycolysis pathway is located within the _____ of cells.
   a) cytoplasm   b) nucleus   c) mitochondria   d) ribosomes

4. The net production of ATP in glycolysis from one molecule of glucose is _____ molecules.
   a) 0   b) 1   c) 2   d) 4

5. The net production of NADH in glycolysis from one molecule of glucose is _____ molecules.
   a) 0   b) 1   c) 2   d) 4

6. Enzymes regulate the glycolysis pathway at _____ control points.
   a) 2   b) 3   c) 4   d) 5

7. The muscle pain which follows prolonged and vigorous contraction of skeletal muscles is the result of the accumulation of
   a) lactate   b) citrate   c) NADH   d) pyruvate

8. How many molecules of pyruvate are produced from the glycolysis of one molecule of glucose?
   a) 0   b) 1   c) 2   d) 3

9. How many net molecules of ATP can be formed in the electron transport chain by oxidative phosphorylation from each molecule of cytoplasmic NADH produced during glycolysis?
   a) 1.0   b) 1.5   c) 2.0   d) 2.5

10. The complete oxidation of glucose in the liver results in the formation of _____ molecules of ATP.
    a) 18   b) 32   c) 34   d) 38

11. Which of the following exerts an effect on blood sugar levels opposite to that of insulin?
    a) cholesterol   b) glucagon   c) vasopressin   d) aldosterone

12. What disease is commonly associated with glucosuria?
    a) diabetes mellitus   b) hepatitis   c) hypoglycemia   d) hyperinsulinism

13. The first step of the citric acid cycle involves the reaction of oxaloacetate with _____ to form citrate.
    a) coenzyme A   b) acetyl CoA   c) CO_2 d) malate

14. The high-energy compound formed in the citric acid cycle is
    a) CTP   b) ATP   c) UTP   d) GTP

15. One turn of the citric acid cycle produces _____ molecules of NADH.
    a) 1   b) 2   c) 3   d) 4

16. The citric acid cycle is inhibited by
    a) ATP   b) CO_2 c) NAD^+ d) FAD

17. One product of the electron transport chain is
    a) O_2 b) H_2O c) CO_2 d) NADH

18. How many sites in the electron transport chain can support the synthesis of ATP?
    a) 1   b) 2   c) 3   d) 4

Matching

Match each description on the left with the condition or disease on the right.

19. _____ a high glucose level in the blood   a) glucosuria

20. _____ a low glucose level in the blood   b) hypoglycemia

21. _____ a high glucose level in the urine   c) hyperglycemia

                                                      d) galactosemia

Match each description on the left with a product on the right.

22. _____ a product of fermentation   a) acetaldehyde

23. _____ produced by carbohydrate oxidation   b) lactate
           under aerobic conditions
                                                      c) acetyl CoA

24. _____ produced by glycolysis under anaerobic   d) ethanol
           conditions

Match each characteristic on the left with the cycle or process on the right.

25. _____ glucose is converted to glycogen

26. _____ synthesis of glucose from non-carbo-
    hydrate sources

27. _____ breakdown of glycogen to glucose

a) glycogenesis

b) gluconeogenesis

c) glycogenolysis

d) glycolysis

True-False

28. Fructose and galactose are not metabolized by humans.

29. The blood sugar level in a hypoglycemic individual is higher than the normal fasting level.

30. Some ATP is formed during glycolysis.

31. The glycolysis pathway is inhibited by ATP.

32. Part of the Cori cycle involves the conversion of pyruvate to glucose.

33. About 90% of gluconeogenesis takes place in the liver.

34. Oxidative phosphorylation is a process coupled with the electron transport chain.

35. The chemiosmotic hypothesis pertains to the flow of cytoplasmic NADH across the mitochondrial membrane.

36. The net production of ATP from one mole of glucose in brain cells is greater than that in liver cells.

SOLUTIONS

A. Answers to Programmed Review

23.1 a) energy	b) monosaccharides	c) glucose	
23.2 a) level	b) hypoglycemia	c) hyperglycemia	d) renal threshold
23.3 a) pyruvate	b) cytoplasm	c) enzymes	d) feedback
23.4 a) acetyl CoA	b) lactate	c) ethanol	d) NAD^+
23.5 a) acetyl CoA	b) CO_2	c) three	d) one
23.6 a) mitochondrion	b) H_2O	c) cytochromes	d) NADH
23.7 a) three	b) ATP	c) chemiosmotic	d) proton
23.8 a) two	b) three	c) 32	d) phosphorylation
23.9 a) glycogenesis	b) UTP	c) glycogenolysis	d) liver
23.10 a) glucose	b) liver	c) pyruvate	d) Cori
23.11 a) glucose	b) glucagon	c) glycogen	

B. Answers to Self-Test Questions

1.	d	13.	b	25.	a
2.	b	14.	d	26.	b
3.	a	15.	c	27.	c
4.	c	16.	a	28.	F
5.	c	17.	b	29.	F
6.	b	18.	c	30.	T
7.	a	19.	c	31.	T
8.	c	20.	b	32.	T
9.	d	21.	a	33.	T
10.	b	22.	b	34.	T
11.	b	23.	c	35.	F
12.	a	24.	b	36.	F

Chapter 24

Lipid and Amino Acid Metabolism

CHAPTER OUTLINE

LEARNING OBJECTIVES

When you have completed your study of this chapter, you should be able to:
1. Describe the digestion, absorption, and distribution of lipids in the body.
2. Explain what happens during fat mobilization.
3. Identify the metabolic pathway by which glycerol is catabolized.
4. Outline the steps of β-oxidation for fatty acids.
5. Determine the amount of ATP produced by the complete catabolism of a fatty acid.
6. Name the three ketone bodies and list the conditions that cause their overproduction.
7. Describe the pathway for fatty acid synthesis.
8. Describe the source and function of the body's amino acid pool.
9. Write equations for transamination and deamination reactions.
10. Explain the overall results of the urea cycle.
11. Describe how amino acids can be used for energy production, the synthesis of triglycerides, and gluconeogenesis.
12. State the relationship between intermediates of carbohydrate metabolism and the synthesis of nonessential amino acids.

ANSWERS AND SOLUTIONS TO EVEN-NUMBERED PROBLEMS

Introduction and Blood Lipids (Section 24.1)

24.2   What are the products of triglyceride digestion?

Solution:
glycerol (glycerin), fatty acids, and monoglycerides

24.4   What are chylomicrons?

Solution:
They are the complex composed of insoluble lipids and proteins, lipoproteins.

24.6   List the major types of lipoproteins.

Solution:
Lipoproteins are classified into 4 categories: chylomicrons, very low density lipoproteins, low density lipoproteins, and high density lipoproteins.

Fat Mobilization (Section 24.2)

24.8   Describe the relationship between epinephrine and fat mobilization.

Solution:
Epinephrine stimulates triglyceride hydrolysis to fatty acids and glycerol in the adipose tissue.

24.10   Which cells utilize fatty acids over glucose to satisfy their energy needs?

Solution:
Resting muscle cells and the liver cells

Glycerol Metabolism (Section 24.3)

24.12   Describe the two fates of glycerol after it has been converted to an intermediate of glycolysis.

Solution:
It can go through glycolysis to give pyruvate. The pyruvate can be used either for cellular energy production directly or it can be used in gluconeogenesis to produce glucose.

Oxidation of Fatty Acids (Section 24.4)

24.14   How is a fatty acid prepared for catabolism? Where in the cell does fatty acid activation take place?

Solution:
The fatty acid is converted to a fatty acyl CoA in the cytoplasm of the cell.

24.16   Describe the role of FAD and NAD^+ in the oxidation of fatty acids.

Solution:
FAD and NAD^+ are reduced during the catabolism of lipids to $FADH_2$ and NADH. The greater the amounts of $FADH_2$ and NADH, the greater is energy availability for ATP synthesis.

24.18   Write an equation for the activation of palmitic acid (16 carbons).

Solution:

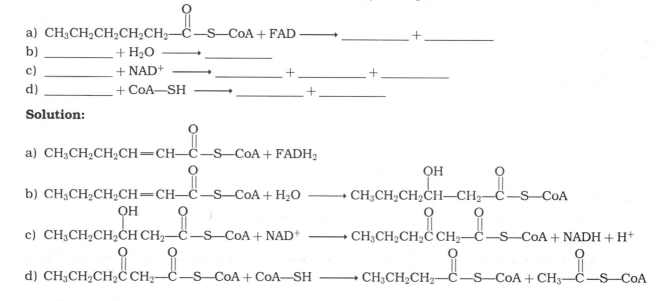

24.20 Why is the oxidation of fatty acids referred to as the fatty acid spiral rather than the fatty acid cycle?

Solution:
Each β-oxidation series results in a shorter fatty acid, two carbons shorter than the starting fatty acid. The product is similar to the starting material, but is not the same. As the process repeats, the size of the fatty acid is "spiraling" down. The process is a "spiral" not a "cycle", because "cycle" would imply that starting and ending product are the same.

24.22 Complete the following equations for one turn of the fatty acid spiral:

a) $CH_3CH_2CH_2CH_2CH_2$—$\overset{\overset{O}{\|}}{C}$—S—CoA + FAD $\longrightarrow$ _____ + _____

b) _____ + H_2O $\longrightarrow$ _____

c) _____ + NAD^+ $\longrightarrow$ _____ + _____ + _____

d) _____ + CoA—SH $\longrightarrow$ _____ + _____

Solution:

a) $CH_3CH_2CH_2CH=CH$—$\overset{\overset{O}{\|}}{C}$—S—CoA + $FADH_2$

b) $CH_3CH_2CH_2CH=CH$—$\overset{\overset{O}{\|}}{C}$—S—CoA + H_2O $\longrightarrow$ $CH_3CH_2CH_2\overset{\overset{OH}{|}}{C}H$—$CH_2$—$\overset{\overset{O}{\|}}{C}$—S—CoA

c) $CH_3CH_2CH_2\overset{\overset{OH}{|}}{C}H\, CH_2$—$\overset{\overset{O}{\|}}{C}$—S—CoA + NAD^+ $\longrightarrow$ $CH_3CH_2CH_2\overset{\overset{O}{\|}}{C}\, CH_2$—$\overset{\overset{O}{\|}}{C}$—S—CoA + NADH + H^+

d) $CH_3CH_2CH_2\overset{\overset{O}{\|}}{C}\, CH_2$—$\overset{\overset{O}{\|}}{C}$—S—CoA + CoA—SH $\longrightarrow$ $CH_3CH_2CH_2$—$\overset{\overset{O}{\|}}{C}$—S—CoA + CH_3—$\overset{\overset{O}{\|}}{C}$—S—CoA

The Energy from Fatty Acids (Section 24.5)

24.24 Determine the ATP yield per molecule of acetyl CoA that enters the citric acid cycle?

Solution:
10 ATP (Section 23.7)

24.26 Calculate the number of ATP molecules that can be produced from the complete oxidation of a ten-carbon fatty acid to carbon dioxide and water.

Solution:

Activation of fatty acid	=	−2 ATP
5 acetyl CoA x 10 ATP/acetyl CoA	=	50 ATP
4 NADH × 2.5 ATP/NADH	=	10 ATP
4 $FADH_2$ × 1.5 ATP/$FADH_2$	=	6 ATP
		64 ATP

Ketone Bodies (Section 24.6)

24.28   List the three substances known as ketone bodies.

Solution:
Acetoacetate, β-hydroxybutyrate, acetone

24.30   Where are ketone bodies formed? What tissues utilize ketone bodies to meet energy needs?

Solution:
Ketone bodies are formed in the liver. The brain, heart, and skeletal muscles can utilize ketone bodies for energy.

24.32   Why is ketosis frequently accompanied by acidosis?

Solution:
Two of the three ketone bodies are carboxylic acids. Their presence in the blood lowers the pH of the blood.

Fatty Acid Synthesis (Section 24.7)

24.34   What substance supplies the carbons for fatty acid synthesis?

Solution:
Acetyl CoA

24.36   Why can the liver convert glucose to fatty acids but cannot convert fatty acids to glucose?

Solution:
Human cells, including the liver, lack the enzyme to convert acetyl CoA to pyruvate, the starting material for gluconeogenesis.

Amino Acid Metabolism (Section 24.8)

24.38   What name is given to the cellular supply of amino acids

Solution:
Amino acid pool

24.40   What term is used to denote the breakdown and resynthesis of body proteins?

Solution:
Protein turnover

24.42   Name four important biomolecules other than proteins that are synthesized from amino acids.

Solution:
Purine and pyrimidine bases of nucleic acids, heme structures, choline and ethanolamine, and neurotransmitters such as acetylcholine and dopamine.

Amino Acid Catabolism: The Fate of the Nitrogen Atoms (Section 24.9)

24.44   Match the terms *transamination* and *deamination* to the following:
a) An amino group is removed from an amino acid and donated to an α-keto acid.
b) Ammonium ion is produced.

c) New amino acids are synthesized from other amino acids.

d) A keto acid is produced from an amino acid.

Solution:

a) transamination   b) deamination

c) transamination   d) transamination and deamination

24.46 Write an equation to illustrate transamination between leucine and pyruvate.

Solution:

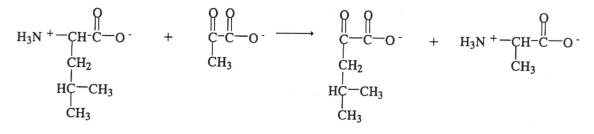

24.48 Write an equation to illustrate the oxidative deamination of glycine.

Solution:

$$NH_3^+—CH_2—COO^- + NAD^+ \longrightarrow HC\overset{\displaystyle O}{\overset{\|}{}}—COO^- + NH_4^+ + NADH$$

24.50 In what organ of the body is urea synthesized?

Solution:

the liver

24.52 Name the compound that enters the urea cycle by combining with ornithine.

Solution:

carbamoyl phosphate

24.54 Write a summary equation for the formation of urea.

Solution:

$NH_4^+ + CO_2 + aspartate + 2H_2O + 3ATP \rightarrow urea + fumarate + 2H^+ + 2ADP + AMP + 2P_i + PP_i$

Amino Acid Catabolism: The Fate of the Carbon Skeleton (Section 24.10)

24.56 Is any amino acid both glucogenic and ketogenic?

Solution:

Yes. As shown in Figure 24.10 in the text, 4 of the 20 amino acids are both glucogenic and ketogenic: isoleucine, tryptophan, phenylalanine, and tyrosine

24.58 Based on the following conversions, classify each amino acid as glucongenic or ketogenic.

a) aspartate oxaloacetate   b) leucine acetyl CoA   c) tyrosine acetoacetyl CoA

Solution:
a) Glucogenic   b) Ketogenic   c) Ketogenic

Amino Acid Biosynthesis (Section 24.11)

24.60 Differentiate between essential and nonessential amino acids.

Solution:
A nonessential amino acid can be synthesized in the body in adequate amounts, and could be absent from the diet. An essential amino acid cannot be synthesized by the body and must be present in the diet.

24.62 List two general sources of intermediates for the biosynthesis of amino acids.

Solution:
The glycolysis pathway and the intermediates in the citric acid cycle

Chemistry Around Us and Over the Counter

24.64 List three essential roles served by cholesterol.

Solution:
Cholesterol is essential for the formation of cell membranes, the insulation of nerves, the synthesis of several hormones, and/or the digestion of food.

24.66 Why may vegetarians have the problem of obtaining enough calories to maintain a healthy body?

Solution:
The caloric content of vegetables is much lower than that of meats.

24.68 What is the best source of a balanced amino acid supply for the body?

Solution:
Food protein in contrast to protein from supplements

24.70 Describe the treatment used for PKU disease.

Solution:
Control the diet, avoiding foods containing phenylalanine or which may be converted to phenylalanine, and add tyrosine to the diet.

PROGRAMMED REVIEW

Section 24.1 Blood Lipids

Lipoprotein aggregates found in the lymph and blood are called (a) _____. Lipoproteins with the greatest density are called (b) _____ lipoproteins. The protein component is greatest in the (c) _____ lipoproteins. The cholesterol content of lipoproteins is useful in assessing the risk of (d) _____.

Section 24.2 Fat Mobilization

Triglycerides are stored in (a) _____ tissue. Fat mobilization involves the (b) _____ of triglycerides followed by the entry of fatty acids and (c) _____ into the bloodstream.

Lipid and Amino Acid Metabolism   317

Section 24.3 Glycerol Metabolism

Glycerol is converted to (a) _____ phosphate, one of the intermediates of the (b) _____ pathway. Glycerol can be converted into pyruvate for energy production or to (c) _____ through gluconeogenesis.

Section 24.4 Oxidation of Fatty Acids

A fatty acid is activated by a reaction with (a) _____ in the presence of ATP. β-oxidation of a fatty acid produces molecules of (b) _____ . Every run through the β-oxidation process produces the reduced coenzymes (c) _____ and (d) _____ .

Section 24.5 The Energy from Fatty Acids

Activation of a fatty acid requires the equivalent of hydrolysis of (a) _____ molecules of ATP to ADP. A ten- carbon fatty acid requires (b) _____ trips through the fatty acid spiral and produces (c) _____molecules of acetyl CoA and (d) _____ molecules each of $FADH_2$ and NADH.

Section 24.6 Ketone Bodies

Ketone bodies are synthesized from (a) _____. The ketone body which can sometimes be detected on the breath of a diabetic is (b) _____. An abnormally low blood pH due to the presence of ketone bodies is called (c) _____. Ketonuria is the presence of ketone bodies in the (d) _____.

Section 24.7 Fatty Acid Synthesis

Biosynthesis of fatty acids occurs within the (a) _____ of the cell. Intermediates during fatty acid synthesis are attached to an acyl carrier (b) _____. The (c) _____ is the most important organ involved in fatty acid synthesis.

Section 24.8 Amino Acid Metabolism

In terms of amount used, the most important function of amino acids is to provide building blocks for the synthesis of (a) _____ in the body. The dynamic process in which body proteins are continuously hydrolyzed and re-synthesized is called protein (b) _____. Unlike carbohydrates and fatty acids, amino acids in excess of immediate body requirements cannot be (c) _____ for later use. The turnover rate for proteins is usually expressed as a (d) _____-life.

Section 24.9 Amino Acid Catabolism: The Fate of the Nitrogen Atoms

Enzymes which catalyze the transfer of amino groups are called (a) _____. The term oxidative (b) _____ is applied to an oxidation process resulting in the removal of an amino group. The organ where the urea cycle is located is the (c) _____. The fuel for the urea cycle is (d) _____ phosphate.

Section 24.10 Amino Acid Catabolism: The Fate of the Carbon Skeleton

(a) _____ amino acids are those whose carbon skeletons can be converted to intermediates used in the synthesis of glucose. (b) _____ amino acids are those whose carbon skeletons can be converted to acetyl CoA or acetoacetyl CoA. All 20 amino acids can be degraded into pyruvate, acetyl CoA, acetoacetyl CoA or intermediates of the (c) _____ cycle.

Section 24.11 Amino Acid Biosynthesis

Amino acids that can be synthesized in the amounts needed by the body are called (a) _____ amino acids. Amino acids which cannot be made in large enough amounts to meet bodily needs must be included in our diet and are called (b) _____ amino acids. The key starting materials for the synthesis of nine amino acids are intermediates of the (c) _____ pathway and the citric acid cycle. Three amino acids are synthesized from α-keto acids via reactions catalyzed by (d) _____.

SELF-TEST QUESTIONS

Multiple Choice

1. What reaction occurs during the digestion of triglycerides?
    a) hydrolysis   b) hydrogenation   c) hydration   d) oxidation

2. The products of triglyceride digestion are fatty acids, some monoglycerides, and
    a) triglycerides   b) diglycerides   c) glucose   d) glycerol

3. Lipoproteins which contain the greatest amount of protein are the
    a) chylomicrons   b) very-low density lipoproteins
    c) low-density lipoproteins   d) high-density lipoproteins

4. What reaction occurs during fat mobilization?
    a) reduction   b) hydrolysis   c) oxidation   d) hydration

5. One of the hormones involved in fat mobilization is
    a) aldosterone   b) vasopressin   c) insulin   d) epinephrine

6. The entry point for glycerol into the glycolysis pathway is
    a) dihydroxyacetone phosphate   b) acetyl CoA
    c) pyruvate   d) fructose 1,6-diphosphate

7. In order for fatty acids to enter the mitochondria for degradation, they must first be converted to
    a) acetyl CoA   b) fatty acyl CoA   c) pyruvate   d) malonate

8. Which of the following processes takes place during the fatty acid spiral?
    a) addition of water to a double bond
    b) oxidation of an OH group to a ketone
    c) addition of H atoms to a double bond
    d) more than one response is correct

9. During one run through the fatty acid spiral, which bond of the following fatty acid would be broken?

    $$CH_3-(CH_2)_4-CH_2\overset{d}{-}CH_2\overset{c}{-}CH_2\overset{b}{-}\overset{\overset{O}{\underset{a}{\parallel}}}{C}-OH$$

    a) bond a   b) bond b
    c) bond c   d) bond d

10. Which of the following is a product of the fatty acid spiral?
    a) pyruvate   b) acetyl CoA
    c) $CO_2 + H_2O$ d) more than one response is correct

11. How many runs through the fatty acid spiral would be required to completely break down one molecule of a 12- carbon fatty acid?
    a) a   b) 5   c) 6   d) 12

12. How many $FADH_2$ molecules are produced from one turn of the fatty acid spiral?
    a) 1   b) 2   c) 3   d) 4

13. How many ATP molecules ultimately result from one turn of the fatty acid spiral?
    a) 8   b) 12   c) 14   d) 17

14. The concentration of ketone bodies builds up when increased amounts of _____ are oxidized.
    a) amino acids   b) fatty acids   c) glucose   d) glycerol

15. Which of the following types of proteins tends to have the shortest half-life in the body?
    a) enzymes   b) plasma proteins
    c) connective tissue proteins   d) muscle proteins

16. The primary function of the urea cycle in the body is to
    a) produce ATP from ADP   b) convert amino acids into keto acids
    c) convert keto acids into amino acids   d) convert ammonium ions into urea

17. Which of the following substances is a product of a transamination reaction?
    a) amino acid   b) ammonia
    c) keto acid   d) more than one response is correct

18. The nitrogen of amino acids appears in the urine of mammals primarily as
    a) uric acid   b) urea   c) ammonia   d) N_2

19. Which of the following substances enters into the urea cycle?
    a) malonyl CoA   b) acetyl CoA
    c) phosphocreatine   d) carbamoyl phosphate

20. The carbon atom of urea is derived from
    a) CO_2 b) acetyl CoA   c) pyruvate   d) aspartate

21. Deamination of an amino acid produces
    a) ammonium ions   b) an α-keto acid
    c) CO_2 d) more than one response is correct

22. The carbon skeletons of amino acids are ultimately catabolized through
    a) glycolysis   b) glycogenolysis
    c) the citric acid cycle   d) the urea cycle

23. A number of amino acids can be synthesized in the body from intermediates of
    a) the citric acid cycle   b) the urea cycle
    c) oxidative phosphorylation   d) the fatty acid spiral

True-False

24. Fat has a caloric value more than twice that of glycogen and starch.

25. Upon complete oxidation to CO_2 and H_2O, fatty acids produce more net energy than a carbohydrate containing the same number of carbon atoms.

26. The glycerol resulting from triglyceride hydrolysis can be converted to pyruvate.

27. The glycerol resulting from triglyceride hydrolysis can be converted to glucose.

28. Starvation can lead to ketosis.

29. Ketosis results when too little acetyl coenzyme A is produced for the needs of the body.

30. Amino acids in excess of immediate body requirements are stored for later use.

31. Amino acids can be catabolized for energy production.

32. Urea formation is an energy-yielding process.

33. The human body excretes small amounts of nitrogen as ammonium ions.

34. The carbon atoms of some amino acids can be used in fatty acid synthesis.

35. The carbon atoms of some amino acids can be used in glucose synthesis.

36. Biosynthesis of fatty acids occurs in the mitochondria of the cell.

37. The liver is the most important organ in fatty acid synthesis.

SOLUTIONS

A. Answers to Programmed Review

24.1 a) chylomicrons	b) high-density	c) high-density	d) heart attack
24.2 a) adipose	b) hydrolysis	c) glycerol	
24.3 a) dihydroxyacetone	b) glycolysis	c) glucose	
24.4 a) coenzyme A	b) acetyl CoA	c) $FADH_2$	d) NADH
24.5 a) two	b) four	c) five	d) four
24.6 a) acetyl CoA	b) acetone	c) ketoacidosis	d) urine
24.7 a) cytoplasm	b) protein	c) liver	
24.8 a) protein	b) turnover	c) stored	d) half
24.9 a) transaminases	b) deamination	c) liver	d) carbamoyl
24.10 a) glucogenic	b) ketogenic	c) citric acid	
24.11 a) nonessential	b) essential	c) glycolysis	d) transaminases

B. Answers to Self-Test Questions

1. a	14. b	27. T
2. d	15. a	28. T
3. d	16. d	29. F
4. b	17. d	30. F
5. d	18. b	31. T
6. a	19. d	32. F
7. b	20. a	33. T
8. d	21. d	34. T
9. c	22. c	35. T
10. b	23. a	36. F
11. b	24. T	37. T
12. a	25. T	
13. c	26. T	

Chapter 25

Body Fluids

CHAPTER OUTLINE

25.1   Comparison of Body Fluids
25.2   Oxygen and Carbon Dioxide
       Transport
25.3   Chemical Transport to the Cells
25.4   Constituents of Urine
25.5   Fluid and Electrolyte Balance

25.6   Acid-Base Balance
25.7   Buffer Control of Blood pH
25.8   Respiratory Control of Blood pH
25.9   Urinary Control of Blood pH
25.10 Acidosis and Alkalosis

LEARNING OBJECTIVES

When you have completed your study of this chapter, you should be able to:
1. Compare the chemical composition of plasma, interstitial fluid, and intracellular fluid.
2. Explain how oxygen and carbon dioxide are transported within the bloodstream
3. Explain how materials move from the blood into the body cells and from the body cells into the blood.
4. List the normal and abnormal constituents of urine.
5. Discuss how proper fluid and electrolyte balance is maintained in the body.
6. Explain how acid-base balance is maintained in the body.
7. List the causes of acidosis and alkalosis.

ANSWERS AND SOLUTIONS TO EVEN-NUMBERED PROBLEMS

Comparison of Body Fluids (Section 25.1)

25.2   Which two of the following have nearly the same chemical composition: plasma, intracellular fluid, interstitial fluid?

Solution:
Plasma and interstitial fluid have nearly the same composition.

25.4   Which of the fluids of Exercise 25.2 contains the highest concentration of the following?
    a) protein   b) K^+ c) Na^+ d) HPO_4^{2-}
    e) HCO_3^- f) Mg^{2+} g) Cl^-

Solution:
a) intracellular fluids   b) intracellular fluids
c) plasma and interstitial fluids   d) intracellular fluids
e) plasma and interstitial fluids   f) intracellular fluids
g) plasma and interstitial fluids

25.6   What is the principal anion in the following?
    a) blood plasma   b) interstitial fluid   c) intracellular fluid

Solution:
a) Cl^- b) Cl^- c) HPO_4^{2-}

Oxygen and Carbon Dioxide Transport (Section 25.2)

25.8   What term is given to the reversible flow of chloride ion across the red blood cell membrane during the oxygen and carbon dioxide transport process?

Solution:
chloride shift

25.10   Discuss the significance of red blood cells in the O_2 and CO_2 transport reactions that occur at (a) the lungs and (b) the tissue cells.

Solution:
The red blood cell is the site for the CO_2/O_2 exchange with hemoglobin both in the lungs and at the tissue cells.

25.12   Write the equilibrium equation representing the reaction between oxygen and hemoglobin. In which direction will the equilibrium shift when the following occurs?
    a) The pH is increased.
    b) The concentration (pressure) of O_2 is increased.
    c) The concentration of CO_2 is increased (see Figure 25.5).
    d) The blood is in a capillary near actively metabolizing tissue cells.

Solution:
$HHb + O_2 \rightleftarrows HbO_2^- + H^+$
a) right (The H^+ decreases as pH increases.)
b) right
c) left (As CO_2 is increased the H^+ increases.)
d) left (In the metabolizing tissue, the O_2 is decreased and CO_2 is increased.)

Chemical Transport to the Cells (Section 25.3)

25.14   Which factor, blood pressure or osmotic pressure, has the greater influence on the direction of fluid movement through capillary walls at (a) the venous end of a capillary and (b) the arterial end of a capillary?

Solution:
a) Osmotic pressure   b) Blood pressure

Constituents of Urine (Section 25.4)

25.16 What organic material is normally excreted in the largest amounts in urine?

Solution:
Urea

25.18 Under what circumstances would the presence of a normal constituent of urine be considered abnormal?

Solution:
If the concentration of a normal constituent is outside the normal range, this would be an abnormal constituent.

Fluid and Electrolyte Balance (Section 25.5)

25.20 List four routes by which water leaves the body.

Solution:
Kidneys (urine), lungs (water vapor), skin (perspiration), and intestines (feces)

25.22 What is the primary way fluid balance is maintained in the body?

Solution:
A variation in the output of urine from the kidneys

25.24 Trace the events of the regulation of fluid balance by the thirst mechanism.

Solution:
As the fluid balance in the body decreases, the saliva glands decrease the output of saliva, resulting in a dryness in the mouth. This dryness is interpreted as "thirst" and water is drunk to get more water into the body.

Acid-Base Balance (Section 25.6)

25.26 List three systems that cooperate to maintain blood pH in an appropriate narrow range?

Solution:
(1) the buffers in the blood   (2) the urinary system   (3) the respiratory system

Buffer Central of Blood pH (Section 25.7)

25.28 Write equations to illustrate how the $H_2PO_4^-/HPO_4^{2-}$ buffer would react to added H^+ and added OH^-.

Solution:
The added H^+ converts HPO_4^{2-} to $H_2PO_4^-$ $HPO_4^{2-} + H^+ \rightarrow H_2PO_4^-$
The added OH^- converts $H_2PO_4^-$ to HPO_4^{2-} $H_2PO_4^- + OH^- \rightarrow HPO_4^{2-} + H_2O$

Respiratory Control of Blood pH (Section 25.8)

25.30 Describe the respiratory mechanism for controlling blood pH.

Solution:
Exhaled H_2O and CO_2 are derived from H_2CO_3. The more CO_2 and H_2O that are exhaled, the more carbonic acid is removed from the blood, thus elevating the blood pH.

25.32 Hypoventilation is a symptom of what type of acid-base imbalance in the body?

Solution:
Respiratory alkalosis

Urinary Control of pH (Section 25.9)

25.34 What happens to CO_2 in the blood when the kidneys function to control blood pH?

Solution:
CO_2 diffuses into the distal tubule cells to form carbonic acid, which then ionizes to H^+ and HCO_3^-. Thus, CO_2 concentration is lowered in the blood.

25.36 What ionic shift maintains electron charge balance when the kidneys function to increase the pH of blood?

Solution:
As the H^+ leaves the kidney tubule into the urine, a Na^+ ion is re-absorbed into the blood from the urine to replace the H^+.

Acidosis and Alkalosis (Section 25.10)

25.38 What type of acid-base imbalance in the blood is caused by hypoventilation? By hyperventilation?

Solution:
Hypoventilation causes respiratory acidosis (CO_2 not removed)
Hyperventilation causes respiratory alkalosis (too much CO_2 removed)

25.40 Explain how prolonged vomiting can lead to alkalosis.

Solution:
After the acidic contents of the stomach are expelled, the replenishing of the stomach acid makes the blood alkaline.

Chemistry Around Us and Over the Counter

25.42 What cellular materials does the drug EPO stimulate the body to produce?

Solution:
red blood cells

25.44 What dangerous side effects may accompany the use of the drug EPO?

Solution:
The risks of EPO use are stroke or heart attack due to possible blood clots.

25.46 How much blood is typically collected during blood donation?

Solution:
1 pint

25.48 How is heat stroke prevented?

Solution:
Drink plenty of water before and during the strenuous activity. Take rests (in the shade preferably) as needed, wear lightweight clothing that allows perspiration to evaporate.

PROGRAMMED REVIEW

Section 25.1 Comparison of Body Fluids
The majority of body fluids are located inside the cells and are called (a) _____ fluid. All body fluids not located inside the cells are collectively known as (b) _____ fluids. Chemically, the two extracellular fluids, plasma and (c) _____ fluid, are nearly identical. The principal cation found in plasma is (d) _____.

Section 25.2 Oxygen and Carbon Dioxide Transport
The oxygenated form of hemoglobin is called (a) _____. Hemoglobin combined with CO_2 is known as (b) _____. The majority of CO_2 is carried from body tissues to the lunds in the form of (c) _____ ions. The movement of chloride ions to maintain electrical neutrality within red blood cells is called the chloride (d) _____.

Section 25.3 Chemical Transport to the Cells
Fluid flow through capillary walls is governed by blood pressure and by (a) _____ pressure. Of these two factors, (b) _____ pressure is greater at the arterial end of a capillary and there is a tendency for a net flow of fluid to occur (c) _____ the capillary.

Section 25.4 Coastituents of Urine
The presence of large amounts of bile pigments in urine may be indicative of (a) _____. Ketonuria is a term used to describe the presence of (b) _____ within urine. The cation present in greatest amounts in urine is (c) _____. The anion most prevalent in urine is (d) _____.

Section 25.5 Fluid and Electrolyte Balance
Water intake is regulated by the (a) _____ mechanism. Water leaves the body through the intestines, kidneys, skin and (b) _____. (c) _____ is known as the antidiuretic hormone. The hormone (d) _____ stimulates the reabsorption of Na^+ ions.

Section 25.6 Acid-Base Balance
Blood pH is normally within the range of 7.35 to (a) _____. An increase in blood pH is called (b) _____. An abnormally low blood pH is called (c) _____. A constant blood pH is maintained by the interactive operation of three systems: buffer, (d) _____, and urinary.

Section 25.7 Buffer Control of Blood pH
Three major buffer systems of the blood are the bicarbonate buffer, the (a) _____ buffer, and the (b) _____ proteins. The most important of these is the bicarbonate buffer system, consisting of a mixture of HCO_3^- and (c) _____.

Section 25.8 Respiratory Control of Blood pH
The water and carbon dioxide which are exhaled are formed from (a) _____ acid. An increased rate of breathing called (b) _____ is caused by a (c) _____ blood pH. Slow, shallow breathing is called (d) _____.

Section 25.9 Urinary Control of Blood pH
The excretion of H^+ within the urine is accompanied by the conversion of CO_2 to (a) _____ within the blood. For every H^+ ion entering the urine, a (b) _____ ion passes into the tubule cells. The enzyme which catalyzes the combining of CO_2 and H_2O is (c) _____.

Section 25.10 Acidosis and Alkalosis

When blood pH is normal and balanced, it contains 20 parts of bicarbonate ions to 1 part of (a) acid. A condition of (b) _____ alkalosis is caused by hyperventilation. (c) _____ acidosis is a condition of acidosis resulting from causes other than hypoventilation. Excessive intake of baking soda may give rise to metabolic (d) _____.

SELF-TEST QUESTIONS

Multiple Choice

1. Which of the following fluids contain similar concentrations of Na^+ and Cl^- ions?
    a) blood plasma and interstitial fluid
    b) blood plasma and intracellular fluid
    c) interstitial fluid and intracellular fluid
    d) all three have about the same concentrations

2. A body fluid is analyzed and found to contain a low concentration of Na^+ and a high concentration of K^+. The fluid is most likely
    a) blood plasma   b) intracellular fluid
    c) interstitial fluid   d) all three have about the same concentrations

3. Which of the following reactions takes place in red blood cells at the lungs?
    a) $H_2CO_3 \rightarrow H_2O + CO_2$ b) $CO_2 + H_2O \rightarrow H_2CO_3$
    c) $H^+ + HbO_2^- \rightarrow HHb + O_2$ d) $H_2CO_3 \rightarrow H^+ + HCO_3^-$

4. Most oxygen is carried to various parts of the body, via the bloodstream, in the form of
    a) a dissolved gas   b) oxyhemoglobin
    c) bicarbonate ion   d) carbon dioxide

5. During respiration reactions at the lungs and cells, the function of the chloride shift is to
    a) eliminate toxic chlorine from the body
    b) maintain pH balance in red blood cells
    c) maintain charge balance in red blood cells
    d) activate the carbonic anhydrase enzyme in red blood cells

6. Which of the following would be considered to be an abnormal constituent in urine?
    a) ammonium ion   b) creatinine
    c) protein   d) bicarbonate salts

7. The vasopressin mechanism and the aldosterone mechanism both tend to regulate the
    a) fluid and electrolyte levels in the body   b) rate of glucose oxidation in the body
    c) rate of hemoglobin production in the body   d) CO_2 levels in the cells

8. Which of the following maintains a constant pH for the blood?
    a) respiration reactions associated with breathing
    b) kidney activity
    c) formation and excretion of perspiration
    d) more than one response is correct

9. Which buffer system is regulated in part by the kidneys and by the respiratory system?
    a) bicarbonate   b) phosphate
    c) protein   d) ammonium

10. The three major buffer systems of the blood are the plasma proteins, the bicarbonate buffer, and the
    a) succinate buffer   b) ammonium buffer
    c) lactate buffer   d) phosphate buffer

11. Exhaling CO_2 and H_2O
    a) raises blood pH   b) lowers blood pH
    c) increases the blood concentration of H_2CO_3 d) has no effect on H_2CO_3 concentration

12. Blood pH normally remains in the range
    a) 7.15–7.25   b) 7.25–7.35   c) 7.35–7.45   d) 7.45–7.55

Matching

For each cause of blood acid-base imbalance listed on the left, select the resulting condition from the responses.

13. _____ prolonged diarrhea   a) respiratory acidosis

14. _____ excessive intake of baking soda   b) respiratory alkalosis

15. _____ hypoventilation   c) metabolic acidosis

16. _____ hyperventilation   d) metabolic alkalosis

For each condition listed on the left, identify the resulting abnormal urine constituent on the right.

17. _____ hepatitis   a) ketone bodies

18. _____ starvation   b) glucose (in large amounts)

19. _____ diabetes mellitus   c) bile pigments

                                             d) more than one constituent results

Select a correct name for each formula on the left.

20. _____ HbO_2^- a) carbaminohemoglobin

21. _____ HHb   b) oxyhemoglobin

22. _____ $HHbCO_2$ c) deoxyhemoglobin

                                             d) carboxyhemoglobin

True-False

23. Plasma is classified as an intracellular fluid.

24. Osmotic pressure differences between plasma and interstitial fluid always tend to move fluid into the blood.

25. Osmotic pressure exceeds heart pressure at the venous end of the circulatory system.

26. An increased rate of breathing tends to lower blood pH.

27. The excretion of H^+ ions decreases urine pH.

28. The pH of blood is increased by the excretion of H^+ in urine.

29. Vomiting may give rise to metabolic acidosis.

30. The concentration of protein in plasma is much higher than the protein concentration in the interstitial fluid.

31. The body is 82-90% water.

32. Aldosterone is also called the antidiuretic hormone (ADH).

33. Death can result if blood pH fallsbelow 6.8.

34. H_2CO_3 is a moderately strong acid.

35. Urine is buffered by the phosphate buffer.

SOLUTIONS

A. Answers to Programmed Review

25.1 a) intracellular	b) extracellular	c) interstitial	d) Na^+
25.2 a) oxyhemoglobin	b) carbaminohemoglobin	c) HCO_3^-	d) shift
25.3 a) osmotic	b) blood	c) from	
25.4 a) jaundice	b) ketone bodies	c) Na^+	d) Cl^-
25.5 a) thirst	b) lungs	c) vasopressin	d) aldosterone
25.6 a) 7.45	b) alkalosis	c) acidosis	d) respiratory
25.7 a) phosphate	b) plasma	c) H_2CO_3	
25.8 a) carbonic	b) hyperventilation	c) low	d) hypoventilation
25.9 a) HCO_3^-	b) Na^+	c) carbonic anhydrase	
25.10 a) carbonic	b) respiratory	c) metabolic	d) alkalosis

B. Answers to Self-Test Questions

1. a	13. c	25. T
2. b	14. d	26. F
3. a	15. a	27. T
4. b	16. b	28. T
5. c	17. c	29. F
6. c	18. a	30. T
7. a	19. d	31. F
8. d	20. b	32. F
9. a	21. c	33. T
10. d	22. a	34. F
11. a	23. F	35. T
12. c	24. T	